AF361017

Structure, Properties and Applications of Polymeric Foams

Structure, Properties and Applications of Polymeric Foams

Editor

Aleksander Hejna

MDPI • Basel • Beijing • Wuhan • Barcelona • Belgrade • Manchester • Tokyo • Cluj • Tianjin

Editor
Aleksander Hejna
Department of Polymer
Technology
Gdańsk University of
Technology
Gdańsk
Poland

Editorial Office
MDPI
St. Alban-Anlage 66
4052 Basel, Switzerland

This is a reprint of articles from the Special Issue published online in the open access journal *Materials* (ISSN 1996-1944) (available at: www.mdpi.com/journal/materials/special_issues/Struct_Prop_Appl_Polym_Foam).

For citation purposes, cite each article independently as indicated on the article page online and as indicated below:

LastName, A.A.; LastName, B.B.; LastName, C.C. Article Title. *Journal Name* **Year**, *Volume Number*, Page Range.

ISBN 978-3-0365-2833-5 (Hbk)
ISBN 978-3-0365-2832-8 (PDF)

Contents

About the Editor

Aleksander Hejna

Dr. Aleksander Hejna graduated from the Faculty of Chemistry of Gdańsk University of Technology in 2013 with a MSc degree in Materials Engineering. Then, from 2013 to 2017, Dr. Hejna pursued his PhD thesis at Department of Polymer Technology. For his scientific activity during his PhD, he was awarded the Scholarship of the Ministry of Science and Higher Education for Doctoral Students for Outstanding Achievements.

Since 2019, Dr. Aleksander Hejna has been a researcher at the Faculty of Chemistry, Gdansk University of Technology, where he is Principal Investigator in two research projects. The first one is the project *Structure and properties of lignocellulosic fillers modified in situ during reactive extrusion*, which is funded by the National Science Center, under the Sonatina program. The second project, entitled *Development of technology for the manufacturing of foamed polyurethane-rubber composites for the use as damping materials*, is financed by the National Center for Research and Development within the Lider program.

During his research work, Mr. Aleksander has cooperated with many domestic and foreign research centers, among which are Universitat Politècnica de Catalunya (Barcelona, Spain), Institute for Color Science and Technology (ICST) (Tehran, Iran), Latvian State Institute of Wood Chemistry (Riga, Latvia), Universidad de Pais Vasco (San Sebastian, Spain) or Zhejiang University of Technology (Hangzhou, China).

His scientific output consists of almost 70 scientific publications, three chapters in monographs and 21 conference papers presented at international and national conferences. His research papers have been cited more than 950 times (according to the Scopus database), so his Hirsch index is 17. In addition, four solutions developed by Mr. Hejna and his team are the subject of patent applications, and two are under patent protection.

 *materials*

Editorial

Special Issue: Structure, Properties and Applications of Polymeric Foams

Aleksander Hejna

Department of Polymer Technology, Faculty of Chemistry, Gdańsk University of Technology, Gabriela Narutowicza 11/12, 80-233 Gdańsk, Poland; aleksander.hejna@pg.edu.pl

Citation: Hejna, A. Special Issue: Structure, Properties and Applications of Polymeric Foams. *Materials* **2021**, *14*, 1474. https://doi.org/10.3390/ma14061474

Received: 8 March 2021
Accepted: 16 March 2021
Published: 17 March 2021

Publisher's Note: MDPI stays neutral with regard to jurisdictional claims in published maps and institutional affiliations.

The Special Issue "Structure, Properties and Applications of Polymeric Foams" aimed to gather the numerous reports associated with the different aspects of polymeric foams. First of all, The Guest Editor would like to express gratitude to the Editors-in-Chief of *Materials* for the opportunity to create and run the Special Issue, as well as to all the authors from the whole world who contributed with their high quality papers, and reviewers, who helped to enhance the level of manuscripts. Finally, the Section Managing Editor, Ms. Yulia Zhao, should be acknowledged for the excellent management of the editorial process.

Focusing on the topic of the Special Issue, foaming of polymeric materials enables weight reduction, very important from the economic point of view, but most of all it should be considered as an excellent means to provide new properties to various polymeric materials [1,2]. Permanent developments in foaming technologies allow manufacturing of foams with micro- and even nano-sized pores, expanding the already very wide range of their applications. Moreover, recent advances allow the preparation of functionally graded foams, which are characterized by the engineered porosity distribution, imitating the cellular materials created by nature, e.g., plant stems or bones. Recent progress in processing of such materials was comprehensively summarized by Suethao et al. [3]. Nevertheless, it is important to remember about the most conventional applications of polymeric foams as insulation materials, which still require substantial research, e.g. considering the heat transfer mechanism, investigated by Blazejczyk et al. [4].

Moreover, having in mind the ongoing trends and law regulations, it is important to remember the environmental impact of polymeric foams. Recent technological developments often involve biodegradability of foams, application of environmentally friendly raw materials, and innovative recycling methods [5]. Such an approach was strongly emphasized by Abolins et al. [6], who described the innovative transformation of tall oil fatty acids into bio-polyols. Their application in manufacturing of rigid polyurethane foams could enable substitution of petroleum-based materials and reduction in foams' environmental impact without the deterioration of thermal insulation performance. Another interesting paper dealing with the biopolyols from renewable resources was published by Kosmela et al. [7]. They used biopolyol (obtained by crude glycerol liquefaction of *Enteromorpha* macroalgae and *Zostera marina* seagrass originated from Baltic Sea) to prepare rigid polyurethane foams. Except for the conclusions related to the benefits of the environmentally-friendly approach, it is worth mentioning the excellent application of micro-computed tomography for the evaluation of foams' cellular structure. When combined with the finite element analysis, it can be used to investigate the relationship between the macroscopic mechanical properties and microscopic foam structure, as presented in the excellent paper of Iizuka et al. [8]. Deeper analysis of this relationship can enable the adjustment of the manufacturing process to obtain the desired microstructure of foams, which directly affects their performance. As a result, tuning the microstructure of foams could be used to acquire desired global materials properties [9].

Alongside for the application of renewable raw materials, the use of recycled ones should be emphasized considering the current pro-ecological trends, as in the work of

Hejna et al. [10] dealing with structure and performance of flexible foamed polyurethane/ground tire rubber composites.

Summarizing, the Guest Editor believes that the collection of the articles presented as a Special Issue "Structure, Properties and Applications of Polymeric Foams" perfectly shows the multitude of foams' applications and potential research directions associated with these materials.

Funding: This research received no external funding.

Conflicts of Interest: The author declares no conflict of interest.

References

1. Llewelyn, G.; Rees, A.; Griffiths, C.; Jacobi, M. A Design of Experiment Approach for Surface Roughness Comparisons of Foam Injection-Moulding Methods. *Materials* **2020**, *13*, 2358. [CrossRef] [PubMed]
2. Standau, T.; Long, H.; Murillo Castellón, S.; Brütting, C.; Bonten, C.; Altstädt, V. Evaluation of the Zero Shear Viscosity, the D-Content and Processing Conditions as Foam Relevant Parameters for Autoclave Foaming of Standard Polylactide (PLA). *Materials* **2020**, *13*, 1371. [CrossRef] [PubMed]
3. Suethao, S.; Shah, D.U.; Smitthipong, W. Recent Progress in Processing Functionally Graded Polymer Foams. *Materials* **2020**, *13*, 4060. [CrossRef] [PubMed]
4. Blazejczyk, A.; Jastrzebski, C.; Wierzbicki, M. Change in Conductive–Radiative Heat Transfer Mechanism Forced by Graphite Microfiller in Expanded Polystyrene Thermal Insulation—Experimental and Simulated Investigations. *Materials* **2020**, *13*, 2626. [CrossRef] [PubMed]
5. Zhang, N.; Cao, H. Enhancement of the Antibacterial Activity of Natural Rubber Latex Foam by Blending It with Chitin. *Materials* **2020**, *13*, 1039. [CrossRef] [PubMed]
6. Abolins, A.; Pomilovskis, R.; Vanags, E.; Mierina, I.; Michalowski, S.; Fridrihsone, A.; Kirpluks, M. Impact of Different Epoxidation Approaches of Tall Oil Fatty Acids on Rigid Polyurethane Foam Thermal Insulation. *Materials* **2021**, *14*, 894. [CrossRef] [PubMed]
7. Kosmela, P.; Hejna, A.; Suchorzewski, J.; Piszczyk, Ł.; Haponiuk, J.T. Study on the Structure-Property Dependences of Rigid PUR-PIR Foams Obtained from Marine Biomass-Based Biopolyol. *Materials* **2020**, *13*, 1257. [CrossRef] [PubMed]
8. Iizuka, M.; Goto, R.; Siegkas, P.; Simpson, B.; Mansfield, N. Large Deformation Finite Element Analyses for 3D X-ray CT Scanned Microscopic Structures of Polyurethane Foams. *Materials* **2021**, *14*, 949. [CrossRef] [PubMed]
9. Rumianek, P.; Dobosz, T.; Nowak, R.; Dziewit, P.; Aromiński, A. Static Mechanical Properties of Expanded Polypropylene Crushable Foam. *Materials* **2021**, *14*, 249. [CrossRef] [PubMed]
10. Hejna, A.; Olszewski, A.; Zedler, Ł.; Kosmela, P.; Formela, K. The Impact of Ground Tire Rubber Oxidation with H_2O_2 and $KMnO_4$ on the Structure and Performance of Flexible Polyurethane/Ground Tire Rubber Composite Foams. *Materials* **2021**, *14*, 499. [CrossRef] [PubMed]

 *materials*

Review

Recent Progress in Processing Functionally Graded Polymer Foams

Supitta Suethao [1], Darshil U. Shah [2] and Wirasak Smitthipong [1,3,4,*]

[1] Specialized Center of Rubber and Polymer Materials in Agriculture and Industry (RPM), Department of Materials Science, Faculty of Science, Kasetsart University, Chatuchak, Bangkok 10900, Thailand; supitta.sue@gmail.com

[2] Centre for Natural Material Innovation, Department of Architecture, University of Cambridge, Cambridge CB2 1PX, UK; dus20@cam.ac.uk

[3] Office of Natural Rubber Research Program, Thailand Science Research and Innovation (TSRI), Chatuchak, Bangkok 10900, Thailand

[4] Office of Research Integration on Target–Based Natural Rubber, National Research Council of Thailand (NRCT), Chatuchak, Bangkok 10900, Thailand

* Correspondence: fsciwssm@ku.ac.th

Received: 11 August 2020; Accepted: 8 September 2020; Published: 13 September 2020

Abstract: Polymer foams are an important class of engineering material that are finding diverse applications, including as structural parts in automotive industry, insulation in construction, core materials for sandwich composites, and cushioning in mattresses. The vast majority of these manufactured foams are homogeneous with respect to porosity and structural properties. In contrast, while cellular materials are also ubiquitous in nature, nature mostly fabricates heterogeneous foams, e.g., cellulosic plant stems like bamboo, or a human femur bone. Foams with such engineered porosity distribution (graded density structure) have useful property gradients and are referred to as functionally graded foams. Functionally graded polymer foams are one of the key emerging innovations in polymer foam technology. They allow enhancement in properties such as energy absorption, more efficient use of material, and better design for specific applications, such as helmets and tissue restorative scaffolds. Here, following an overview of key processing parameters for polymer foams, we explore recent developments in processing functionally graded polymer foams and their emerging structures and properties. Processes can be as simple as utilizing different surface materials from which the foam forms, to as complex as using microfluidics. We also highlight principal challenges that need addressing in future research, the key one being development of viable generic processes that allow (complete) control and tailoring of porosity distribution on an application-by-application basis.

Keywords: porous polymers; cellular materials; microstructure; property gradient; functionally graded structure

---

1. Introduction

Polymer foams find a wide range of applications, including in pillows and mattresses, physical insulation, furniture, engineering materials, housing decoration, and electronic devices, etc. In comparison to metallic and inorganic (e.g., ceramic and glass) porous materials, polymeric porous materials are of interest as they are substantially lighter (because of their lower density), have lower cost, offer a wider range of compressive strengths (from elastic to flexible to semirigid to rigid), and are producible at considerably lower temperatures using a range of methods, including spray foaming [1–10]. This review will focus on foams of a polymeric nature only.

The polymer foams with supercritical fluids are attracting interest, especially for producing microporous foams. These are cellular polymer foams with approximate 10 μm in pore dimeter and 10^9 pores per cm^3 in pore density. These compact materials present high toughness, high impact strength, and high stiffness-to-weight ratio. In addition, polymer foams with supercritical CO_2 do not usually require the use of harmful organic solvents. Such an advantage provides the method suitable for processing porous structures from biocompatible polymers as scaffolds for biomedical applications [11–13].

From a market perspective, recently reported global market values of exported polymer foam are presented in the Table 1. The exported values of polymer foam have increased from 2017 to 2018 for all types of polymer foam (based on data from the International Trade Centre (ITC) [14]).

Table 1. Values of the polymer foam exported to the world during 2017–2018 [14].

Types of Polymer Foam	Exported Value in Million USD (% of Total)			
	2017		2018	
Polystyrene foam	1.276	(10.3%)	1.339	(9.9%)
Polyvinyl chloride foam	1.799	(14.5%)	2.001	(14.9%)
Polyurethanes foam	3.860	(31.1%)	4.167	(30.9%)
Other plastic foams	4.426	(35.7%)	4.852	(36.0%)
Rubber foams	1.053	(8.5%)	1.110	(8.2%)

Generally, polymer foams are porous materials that have two or more phases. In a two-phase polymer foam, the polymer matrix forms a continuous phase and the gaseous-porosity phase is composed of gas bubbles. The porous structure is produced by either a chemical or a physical blowing agent for gas bubble production in a polymer matrix. In the case of chemical blowing agents, a chemical reaction produces gas bubbles, usually through the decomposition of a chemical. By contrast, physical blowing agents are inert gases or supercritical fluids (mostly CO_2 and N_2), which can be dissolved into the polymer matrix during a saturation process [15–20].

Chemical blowing agents can be used for both liquid and solid polymers. Concerning liquid polymers, in particular natural latex, chemical blowing agents such as potassium oleate are used in the "Dunlop process" to manufacture rubber foams for pillows and mattresses, etc. [21–25]. In contrast, production with a solid polymer is performed by gas diffusion processes (induced by a chemical blowing agent) between the foam and molten polymer matrix. This type of process can be controlled by the formulation and process of polymer to be foamed [26].

2. Polymer Foaming Process

Normally, the process of polymer foam production with a physical blowing agent is divided into two principal steps. First, the polymer matrix is saturated with a physical blowing agent (gas or supercritical fluid) at constant conditions. Next, the supersaturated state is brought about by phase separation, induced either by rapidly increasing temperature or reducing pressure, for generating gas bubbles, and therefore cells, inside the polymer matrix [27–30]. The cells grow to reach the point that the viscosity of the polymer matrix is increased corresponding to the force opposing the expansion of the foam until it becomes sufficiently high [31]. The foam density depends on the gas loading or the gas fraction in the polymer matrix, the cell size and distribution count on the cell nucleation process, and the expansion process [32].

In the case of plastic foam (or non-rubber foams), there are three basic processing steps: (1) polymer/blowing agent solution mixing, (2) microcellular nucleation, and (3) cell growth and density stabilization. The first step of single-phase polymer/blowing agent solution mixing is formed by saturating the polymer with the blowing agent under certain conditions. The saturation point is determined by the solubility limit of the blowing agent in the polymer, while the time required for the solution formation is determined by the rate of diffusion of the blowing agent into the polymer matrix. Microcellular nucleation (Figure 1) is controlled by inducing a thermodynamic instability in the single-phase solution. This is usually succeeded by drastically reducing the solubility of the gas in solution by operating the pressure and/or temperature of the mixture [33–40]. Since the separation of the polymer and gas phases is thermodynamically more favorable, the resulting supersaturated mixture becomes the driving condition for the nucleation of numerous microcells. Continuous microcellular processing typically utilizes a rapid pressure drop to nucleate bubbles. This stage is very crucial to the overall process, because it dictates the cell morphology of the material and its resulting properties. Therefore, solubility as a function of pressure is important for the development of the process. The final stage in the production of microcellular plastics is cell growth. After cell nucleation has occurred, any available gas diffuses into the cell and increases its size, thereby reducing the density of the polymer matrix. Generally, the growth of cell depends on the time allowed for the cells to expand, the system temperature, the amount of gas available, the processing pressure, and the viscoelastic properties of the polymer/gas solution [41–45].

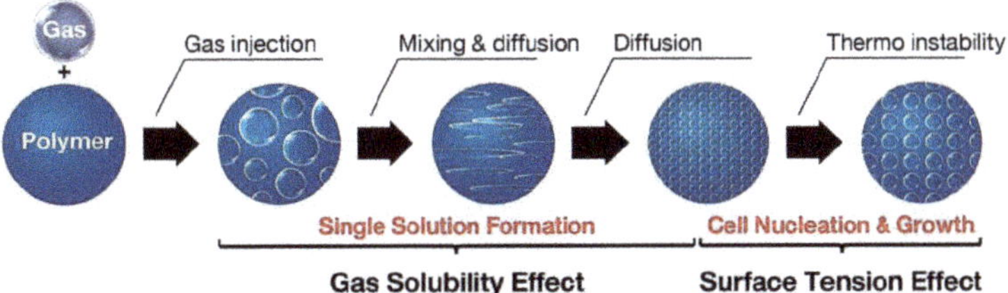

Figure 1. Overview of the microcellular foaming process.

Typical foaming processes can be classified into batch foaming, and extrusion and injection molding. The batch foaming process has lower process temperatures than those needed in other processes; this causes an increasing of the CO_2 solubility in amorphous polymers, resulting in higher cell densities and smaller cell sizes [46]. Such foam characteristics, i.e., cell size and cell density, affect directly the mechanical properties of polymer foam [47]. Generally, the batch foaming process is utilized to amorphous polymers, which starts from the rubbery state at the saturation condition. On the other hand, the semicrystalline polymers possess the non-uniform cell structure because of the inability of the polymer/blowing agent formation in the crystalline structure. This type of crystalline structure of the polymer needs to be destroyed before foaming due to the melting point reduction of the polymer using a co-solvent [47].

Concerning the process of polymer foaming with a physical blowing agent, the porous structure of the polymer/gas (fluid) system depends on the important parameters below [48,49]:

(i)   the degree of crystallinity of the polymer matrix,

(ii)   the amount of the dissolved gas (fluid),

(iii)   the degree of saturation of gas in the polymer,

(iv)   the interfacial energy of polymer/gas (fluid), and

(v)   the plasticization profile of the polymer/gas system (i.e., the melting point and the glass transition temperature, T_g, of polymer matrix).

Interestingly, among conventional polymer foams, the different cell sizes (graded density structure) attract far more attention compared with a uniform cell structure, because this type of functionally graded structure exhibits better mechanical properties compared with conventional foams [50,51]. However, fabrication of such functionally graded foams is complicated. Generally, the functionally graded structure of polymer foam can be obtained by foaming process, nanofiller, blowing agent, or polymer composition, etc. In recent years, various graded cellular materials have attracted interest [52,53]. The polyethylene foam with density gradient improved the mechanical properties due to the change of deformation mode [54]. The polyurethane honeycombs with four density gradients were modified from the uniform density equivalent using different parameters: fused filament fabrication 3D printing, density grading, energy absorbing, and damping profiles [55]. Several types of graded foam are produced for functional applications, such as impact strength, acoustic capabilities, energy absorption, etc. These new graded foams (polyurethane foam, acrylonitrile butadiene styrene foam, polyethylene foam, polypropylene foam, polylactic acid foam, and polymethyl methacrylate foam) are investigated in the structure–property relationship [56–61].

The main aims of this review paper focus on the importance of structure-property-processing relations in polymer foams. In particular, recent polymer foams with cell size gradients or functionally graded foams are of interest. Functionally graded foams are foams that incorporate various cell sizes in the same material and therefore possess a structure with a "cell size gradient". This structure could be mimicked from structural materials found in nature, such as bones and bamboo. These types of foams could be useful for tailor-made material products with functional properties ranging from thermal insulation, to high stiffness or strength at low weight, to buoyancy, and impact resistance [62–65]. Consequently, cellular structure with cell size gradients (different cell sizes) has received interest from both academic and industrial sectors.

3. Thermodynamic Aspects and Computer Modeling of Polymer Foam Processing

The mechanism of gas bubble nucleation inside the polymer matrix for the relevant foaming method is very complex, governed by multiple phenomena, including interfacial energy of polymer/gas system. The creation of gas nuclei can be related to either homogeneous or heterogeneous nucleation (Figure 2). Homogeneous nucleation possesses the spontaneous generating of gas molecules in the polymer matrix; on the other hand, heterogeneous nucleation exhibits the gas nuclei on the boundaries of two phases (polymer and another material like filler) [66–70]. In the case of plastic foaming without use of a chemical agent, bubble nucleation is often assumed to be homogeneous. However, in the process of rubber foaming with a chemical agent and filler, both homogeneous and heterogeneous nucleation occur.

Polymer/Gas Foam

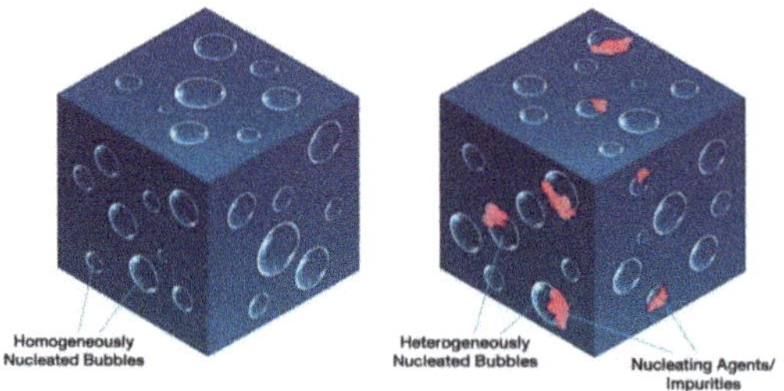

Figure 2. Schematic of homogeneous and heterogeneous nucleation in a polymer/gas system.

Nucleation refers to the initial stage of gas bubble formation from the initial polymer matrix. In this step, a gas bubble has to conquer the Gibb's free energy before the bubble can grow to an optimum scale. This step is explained by classical nucleation theory: the difference of Gibb's free energy of the polymer/gas system can be expressed as the sum of gain in the Gibb's free energy related to the formation of interface. Generally, in an isothermal system at chemical equilibrium, the difference of Gibb's free energy (ΔG) corresponds to the formation of polymer/gas system, which is expressed by the equation [31]:

$$\Delta G = \left(-4\pi r^3/3\right)\Delta P + 4\pi r^2 \gamma, \tag{1}$$

where r is the radius of spherical cluster, γ is the interfacial energy between gas and polymer, and ΔP is the difference of pressure. The next equation is obtained and relates to homogeneous nucleation, as ΔG is plotted against cluster size with a maximum at a critical radius, r_c, is obtained:

$$d\Delta G/dr = 0 \text{ thus } r_c = 2\gamma/\Delta P, \tag{2}$$

The maximum value of ΔG^* for homogeneous nucleation is derived by substituting Equation (2) into Equation (1) as [71]:

$$\Delta G^* = 16\pi\gamma^3/3\Delta P^2, \tag{3}$$

Decreasing the interfacial energy, or increasing the difference of pressure, results in increasing the nucleation rate. However, the interfacial energy of the polymer/gas system is complicated to measure. Normally, such interfacial energy is calculated corresponding to the surface energy of each material at equilibrium [31,72]; the limitation of this theory relates rather to the determination of this parameter.

The presence of fillers, suspension of chemical agents, impurities, or another material in the system is the cause of heterogeneous nucleation. Generally, the presence of particles or impurities decreases the Gibb's free energy ΔG and involves a reduction factor f as:

$$\Delta G_{het} = \Delta G_{hom}\left(f_{(m,w)}/2\right),$$

$$m = \cos\theta = (\gamma_{13} - \gamma_{23})/\gamma_{12} \text{ and } w = R/r_{cr} \text{ (relative curvature)}, \tag{4}$$

where R is the radius of particles; γ_{13}, γ_{23}, and γ_{12} are the interfacial energies of polymer/particle, gas/particle, and polymer/gas, respectively; and θ is the contact angle between the cell, polymer, and particles.

The type of nucleating agents affects the nucleation process of polymer foam, which can be explained by continuum conservation models. Concerning the polymer melt, thermodynamic fluctuations allow nucleus growth due to the surface and viscous forces. When the pressure inside the cell decreases, the gas concentration at the cell surface also decreases [15,66]. In the batch foaming process, the cell growth certainly depends on the temperature process. Based on classical nucleation theory, when the foaming nucleation temperature (T_{nuc}) decreases, the formation of smaller cells can occur. If T_{nuc} is below the glass transition temperature (T_g) of high viscosity polymer matrix, nanocell structures can appear. On the other hand, microcell structures appear when the T_{nuc} closes to the T_g of the polymer matrix (Figure 3).

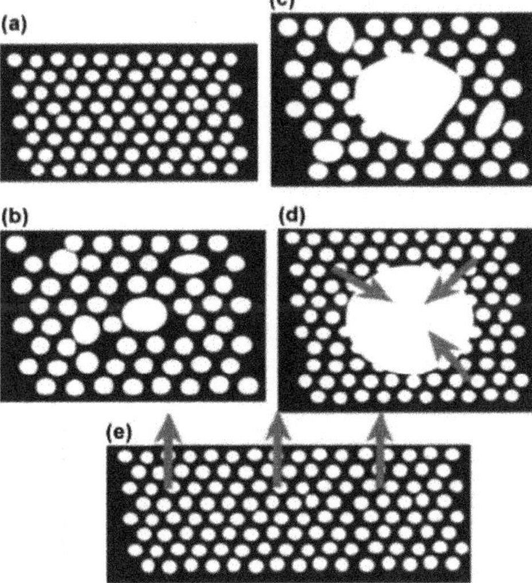

Figure 3. Illustration of cell formation at (a) $T_{nuc} \ll T_g$; (b) $T_{nuc} < T_g$; (c) $T_{nuc} = T_g$; (d) $T_{nuc} > T_g$ in polymer; and (e) $T_{nuc} > T_g$ near surfaces [73]. (T_{nuc}—foaming nucleation temperature, T_g—glass transition temperature.) Adapted from [73], with permission from © 2005 American Chemical Society.

Recently, flexible polymers like compressible elastomers with different densities (graded materials) have been described by a worm-like chain model [74]. A self-consistent field theory (SCFT) was developed using thermodynamic calculations of the system based on Helmholtz free energy. There were two types of length scales relevant for a flexible polymer chain: polymer length L and persistence Lp. The ratio of polymer length/persistence (L/Lp) ratio was found to be L/Lp << 1 for rod-like elastomer structures, while L/Lp >> 1 was proposed for coil-like elastomer structures. This combination of thermodynamic and computational modeling paves the way for a new foam material.

Foam processing computational models are useful to predict and estimate the properties of polymer foams. These models are often related to a finite element (FE) method. These empirical models employ time- or temperature-dependent density related to the nucleation and growth of bubbles in the polymer foam. The continuum-level model applies a description of homogeneous nucleation through the density model but does not include the gas model; this type of model has been developed to explore optimum properties of liquid phase/gas bubbles during the self-expansion process [75,76].

Foam rheological property measurements are complicated to carry out, since the foam microstructure usually changes. Thus, the viscosity is separated into two parts dependent on (1) continuous-phase polymer properties and (2) gas bubble volume fraction: these two phases are quite different [77]. The component mass fractions and densities can be utilized to determine the gas volume fraction using the density model of polymer foam. Moreover, the gas volume fraction components of foam heat capacity and thermal conductivity, which can be utilized for the energy equation. The foam heat capacity is calculated by the mixture theory for polymer/gas system [78]. The effect of liquid vaporization can be defined from the density evolution and the mass fraction of polymer/gas system [76,79].

There has been increasing recent interest in applying the concept of graded cellular materials to polymers in order to improve their mechanical properties. Such cellular polymers exhibit a gradient in their properties, for example, cell size/cell density, cell distribution, mechanical properties, etc. [54,62]. Therefore, it is worthy to explore the processing and mechanisms of graded cellular polymers, which can be used to control desirable properties and behaviors. For example, researchers investigated the behaviors of voronoi-type density gradient foams using the finite element (FE) method [80]. The results obtained show that the energy absorption is linked to the profiles of graded cell distribution. The FE simulation can be also utilized to study the effect of temperature gradient on the properties of graded foam [81]. Moreover, the density-graded models can be investigated for the deformation pattern and energy absorption capacity of the resulting materials produced using a temperature gradient. The latest advances of energy absorption (or impact resistance) for functionally graded foams relate to density and temperature gradients during foam processing. For example, an increasing temperature gradient leads to the reducing of energy absorption capacity in functionally graded foam [80,81].

4. Recent Processes to Produce Functionally Graded Foams

New processes have been developed to produce polymer foams with gradients in cell size (gradient density) and properties (functionally graded) [82–85]. A gradient structure imparts the foam with an asymmetric structure of cell size (Figure 4), with cell size at one end being smaller (and denser). This structure provides superior properties, such as mechanical properties, and proposes new functionalities for various applications, including in sound absorption and protective equipment (helmets).

An example process of foaming with cell size gradient was presented in [86], where polymethyl methacrylate (PMMA) absorbed CO_2 at 28 MPa and 50 °C for 1 h in order to form the PMMA foam. Figure 4 shows that the cell size at the near surface is smaller than the cell size faraway from surface. The effect of high CO_2 concentration in the surface increases cell density and reduces cell size.

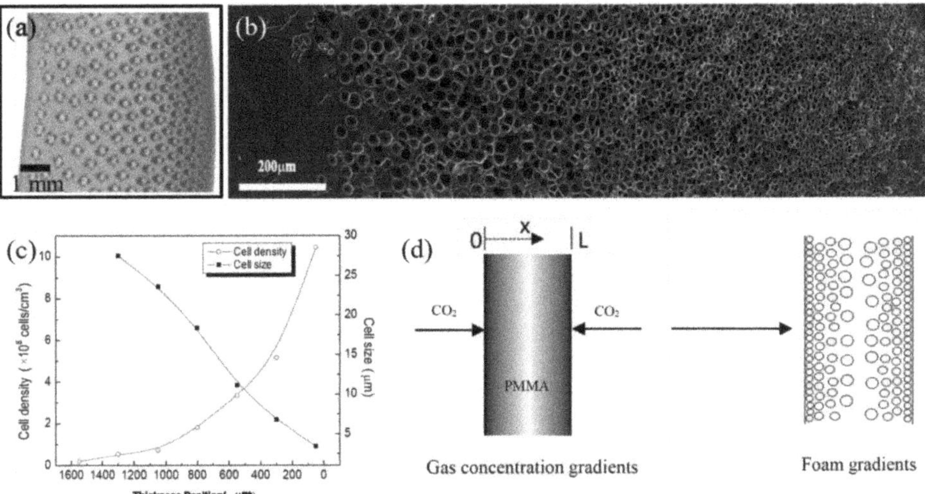

Figure 4. The functionally graded structure of bamboo (**a**) is comparable to the graded microporous foams produced by Yuan et al. [86] (**b**), in which cell size and density are correlated to the location (**b**,**c**), much like in bamboo (**a**). (**d**) The foam gradients are a result of gas concentration gradients during processing. Images (**b**–**d**) by Yuan et al. [86].

Yet another process that produces gradient density polymer foams is when aluminum oxide (AAO) film is used as a surface material for foam preparation [82]. Silane fluoride is used as an agent to change the surface of the substrate. Polystyrene, polymethyl methacrylate, and polyacrylate may be used as the polymer and CO_2 as the blowing agent. For the method of foam preparation (Figure 5), AAO film is modified by fluorinated silane using an impregnation method. The polymer plate is placed on the AAO film and compressed to form a composite structure. The foaming process uses supercritical CO_2 as a blowing agent at 13.8 MPa and 100 °C for 12 h, permitting CO_2 diffusion and reaching an equilibrium state. Figure 6 presents the morphology of polystyrene (PS) foam on anodized aluminum oxide (AAO) film from an SEM image. This method is also successful in producing foam with cell size gradients. The results (Figure 6d) show that near AAO film surface, the cell size is small, while cell density is high. Further away from AAO surface, the cell size is bigger, while cell density is decreased [82].

Gradient density foams with low-density polyethylene (LDPE) have also been reported in literature. Azodicarbonamide is used as a chemical blowing agent, and silicone rubber sheet (SRS) is used as a contact material during the foaming of the polymer. ZnO is used to reduce high-temperature decomposition of azodicarbonamide, and stearic acid is used as an extrusion processing aid. All materials were mixed using a twin-screw extruder, and then LDPE foams were injected inside a mold at temperatures during 200 and 240 °C [26]. The construction of outer solid skin is formed by gas diffusion from the matrix to the surface of mold. The results show that the internal pressure of foam increases when the amount of azodicarbonamide is increased; this consequently affects final density by inducing significant gas absorption in silicone at the surface. X–ray visualization revealed that the properties of foams produced with or without SRS were the same results. However, foams produced with SRS showed the construction of a dense skin of SRS between 1800 and 2990 s, whereas the conventional foam appears to remain free of a solid skin. Increasing foaming temperature and maximum internal pressure induce an increase in the solid skin thickness.

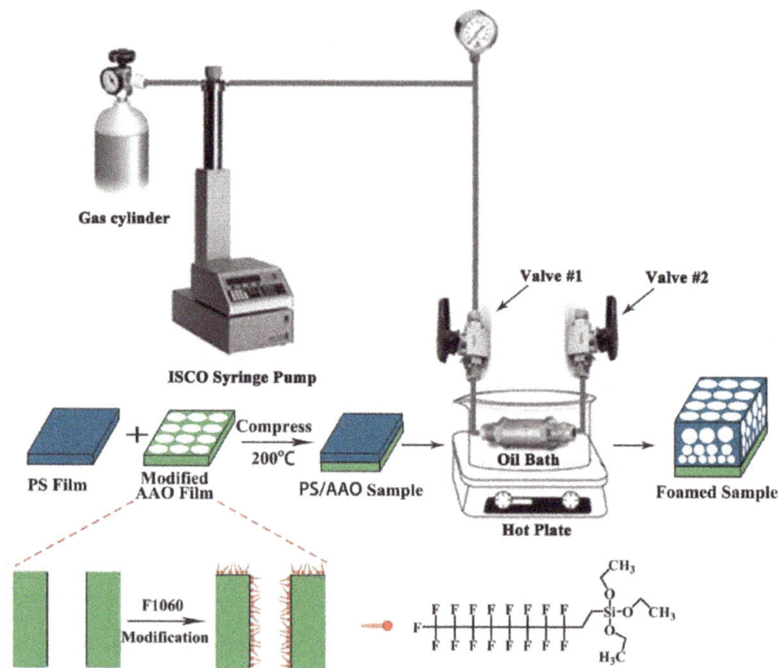

Figure 5. Using a conventional batch foaming method for the processing of graded polymer foams [82]. Adapted from [82], with permission from © 2016 Elsevier.

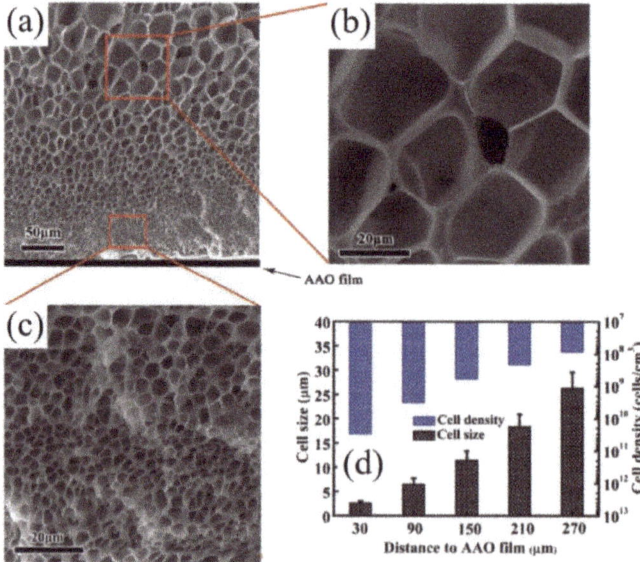

Figure 6. (**a–c**) SEM images of graded polystyrene (PS) foam structure; (**d**) cell size and density at different distance from the surface [82]. Adapted from [82], with permission from © 2016 Elsevier.

LDPE foams produced without SRS presented pores reaching right to the surface of the foam. By contrast, a solid skin at the foam surface with significant thickness is observed when SRS is applied into the mold at the foaming preparation. Based on this new process, two types of foam structure are found: solid skin and porous core. The tensile stress–strain behavior of the skin, core, and structural foam are distinct. From a mechanical properties point of view, the core has the lowest modulus and strength, and the solid skin has the highest modulus and strength, whereas values of the structural foam remain intermediate to the skin and core. The presented process allows the design of gradient density foam structure and related properties [26].

Another interesting example of graded foam is a monodisperse polystyrene (PS) foam [87]: this type of foam can program the pore size and density using different pressures during the foaming process. Concerning this method, one can represent a cell size gradient in the PS foam using the gas pressure variation, in particular, the big cell size of PS foam can be obtained by increasing the gas pressure in the preparation method.

So far we have reviewed functionally graded polymer foams based on PMMA, LDPE, and PS. A range of other thermoplastic polymers have been utilized to produce foams with a controlled structure, including polypropylene (PP) and polycaprolactone (PCL). In a study by Yang et al., hollow molecular-sieve (MS) particles were used as a nucleating agent in supercritical carbon dioxide ($scCO_2$) for polypropylene (PP) foams. In this study, the PP pellets and MS particles were mixed using a twin-screw extruder before they were pressed into sheets with a thickness of 1 mm. They were then foamed inside an autoclave using CO_2 as a blowing agent under pressure of 20 MPa and temperature of 154 °C for 2 h [88]. Yang et al. [88] found that addition of MS particles substantially decreased the distribution in cell sizes of the PP foam, with the foam cell density increasing by an order of magnitude, and doubled the tensile strength. In another study, Llewelyn et al. used a hybrid foaming method by utilizing a physical blowing agent (super critical nitrogen) and a chemical blowing agent to produce polypropylene (PP) foam, with and without talc filler, by low-pressure foam-injection molding (FIM). Through a hybrid foaming method with low pressure (FIM), foams with high cell density and superior homogeneous cell structure were produced [89]. To produce inhomogeneous foams, using chemical blowing was most effective, as a larger skin wall thickness was obtained. Using a thermally induced phase-separation method, Onder et al. produced polycaprolactone (PCL) foams [90]. The PCL solutions were prepared in the tetrahydrofuran/methanol (THF/MeOH) solvent system by slowly heating to 55 °C in a water bath. The homogeneous polymer solutions were quenched at low temperature for 12 h. The PCL foams were warmed up to room temperature and thereafter dried by vacuum drying. PCL foams with larger pores were obtained at lower PCL concentration, lower THF content, and a higher quench temperature [90].

Apart from foams based on thermoplastic polymers, thermoset polymers and elastomers have also been successfully used to produce foams with controlled structure (and therefore properties). Song et al. [89] mixed a biobased epoxy resin (Greenpoxy 56) and an amine-based hardener using a hand-held mixer for 20 min in order to produce an air-in-resin liquid foam. The biobased polymer foams were formed in self-standing tubes. The porosity of biobased polymer foams was increased from 71% to 85% by heating the air-in-resin liquid foam during the curing step [91]. The compressive modulus and compressive strength of the polymer foams were significantly reduced as a result of an increase in porosity. Vahidifar et al. mixed natural rubber (NR) compounds at room temperature using a two-roll mill to produce an elastomeric foam. The one-step foaming process for the natural rubber/carbon black (NR/CB) foam production was performed by compression molding using an electrically heated press at temperature of 160 °C and pressure of 50 kPa for 30 min. Cell density of NR/CB foam was increased around 14 times by increasing the CB content at the same foam density. The morphology of NR/CB foams was divided into three layers: outer (no cells), middle (indeterminate cells), and inner (circular cells) [92].

5. Conclusions and Future Research Outlook

Functionally graded polymer foams are an emerging innovation in polymer foam technology. Their combination of light weight with efficient use of material and enhanced functional properties help the design of multifunctional products. For example, the graded structure may enable efficient performance in low-energy impact as well as high-energy impact applications. Such functionally graded polymer foams may find diverse applications, such as in tissue bioengineering, protective gear (helmets), engineering structures (construction core materials and automotive car bumpers), and filtration and insulation.

A variety of approaches have been employed by scientists to produce functionally graded polymer foams, synthesized in Figure 7. A number of researchers have explored molding processes, such as (reaction) injection or compression molding, and have employed the use of specific contact surface materials (such as aluminum oxide film or silicone rubber sheet) or ingredients (such as supercritical CO_2 as a blowing agent). These have enabled the production of foams with gradients in porosity, cell size, and properties, sometimes with a skin-core structure. Additive processes, such as 3D printing and, more commonly, layer-by-layer lamination techniques (e.g., with thermal bonding), have also been developed to produce functionally graded polymer foams. Each layer may have a distinct pore and cell size, and the layers can range from nano- to micro- to macroscale. However, in such additive manufactured foams, the presence of interlayer interfaces, which are regions of stress transfer and stress concentration, increases issues with delamination and crack propagation. Templating routes have also been examined using solid, liquid, and emulsion foam templates. Solvent-based approaches (including particulate leaching and freeze drying) have also been explored. Mixed success has been achieved using these for functionally graded polymer foam manufacture. While some approaches may enable achieving versatile structures, producing nanoporous to macroporous structures, the processes may hinder the formation of continuously graded structures and pore interconnectivity (open cell and closed cell structures). A number of other processes have been also developed in literature, including those based on microfluidics and the use of ultrasound to produce heterogeneous polymer foams.

The overarching and biggest challenge is control on microstructure. This includes control on porosity and cell size, the gradient (qualitative such as continuously graded or discretely graded structure or skin-core structures, but also more tailored quantitative), and pore interconnectivity (open-cell or closed-cell or mixed and the extent to which they are tailorable). There is still some way to go in understanding these processing-structure-property relations, let alone controlling them, and developing viable generic processes for tailorable functionally graded polymeric foams.

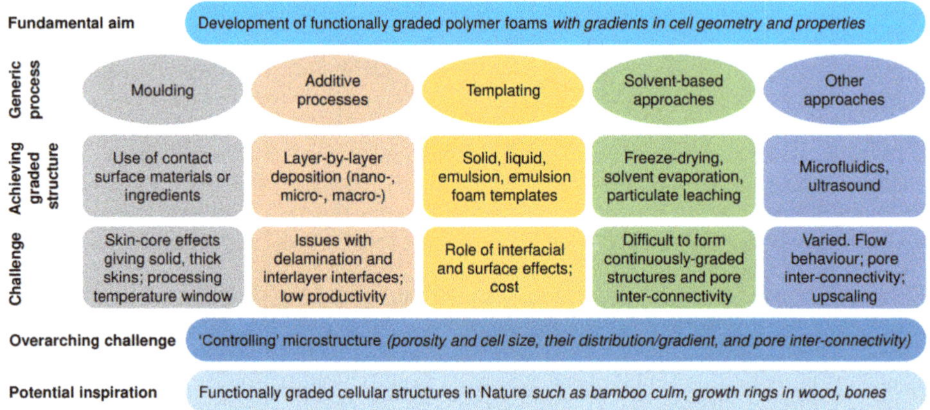

Figure 7. Synthesis of various approaches to process functionally graded polymer foams, and their associated advantages and challenges.

Indeed, better understanding of processing-structure-property relations will require multidisciplinary approaches, including empirical process science and polymer foam technology, but also the physical thermodynamics and chemical kinetics behind the formation of functionally graded polymer foam microstructures, as well as process simulation and computer modeling. The latter may offer more time-saving, cost-effective methods for obtaining insights and optimized routes to fabrication and rapid prototyping for bespoke products (particularly in bioengineering, such as scaffold design and 3D printing for individual patients).

Yet another discipline that may offer insights is the life/biological sciences. An architectural engineering marvel is the Eiffel Tower, which was designed by Gustave Eiffel on the principles of material minimization, through inspirations from the femur bone: a natural, functionally graded cellular structure. Biomimetics and bioinspiration are important sources of design concepts. The intelligent design of functionally graded polymer foams following designs in nature can unveil important insights on structure–property relations. For example, the selection of the polymers—based on their chain length and molecular weight, functional groups and species, and so on)—and how the polymer chains interact with each other will have a notable impact on the morphologies and properties of the foams produced. For instance, in wood and plant stems, other natural functionally graded foam structures, the complex self-assembly and interaction of various polymer species (cellulose, hemicellulose, lignin, pectin, and so on) lead to a beautiful hierarchical cellular structure, making it a model functionally graded material [93–100]. Indeed, the minute changes in the polymer species, and their interactions and formations, lead to a wide variety of wood species in nature with a range of functional properties, including densities, strengths, and hardness [93,101–108].

Moreover, bioinspiration can help produce multifunctional products for specific applications, such as scaffolds for interfacial tissue engineering (e.g., cartilage-bone scaffolds with an order of magnitude difference in modulus between the cartilage and the bone), and protective foam shell of helmets inspired from the functionally graded structures of sheep horn and horn core trabecular bone of bighorn sheep rams. Such studies may also later fuel the design of hierarchical (multiscale porosity distribution) functionally graded polymer foams and composite, fiber-reinforced or polymer-blended, functionally graded polymer foams, taking bamboo as an inspiration, for instance. Furthermore, rather than being based on petrochemical-derivative polymers, exploration of bio-derivative polymers for the fabrication of functionally graded polymer foams may enable improvement in biocompatibility (for bioengineering) and biodegradability (for better end-of-life options) [109–115].

There have been important technological, process-related advancements in functionally graded foams over the past couple of decades, and further understanding and control of the process is key to their inevitable utilization in functional products.

Author Contributions: S.S. wrote the manuscript; D.U.S. wrote and reviewed the manuscript; W.S. provided the original idea to this work, wrote, and reviewed the manuscript. All authors have read and agreed to the published version of the manuscript.

Funding: This research received no external funding.

Acknowledgments: This work was supported by Specialized Center of Rubber and Polymer Materials in Agriculture and Industry (RPM), Department of Materials Science, Faculty of Science, Kasetsart University.

References

1. *CES EduPack 2019*; Ansys (Granta Design Limited): Cambridge, UK, 2020.
2. Fan, D.; Li, M.; Qiu, J.; Xing, H.; Jiang, Z.; Tang, T. Novel method for preparing auxetic foam from closed–cell polymer foam based on the steam penetration and condensation process. *ACS Appl. Mater. Interfaces* **2018**, *10*, 22669–22677. [CrossRef]
3. De Luca Bossa, F.; Verdolotti, L.; Russo, V.; Campaner, P.; Minigher, A.; Lama, G.C.; Boggioni, L.; Tesser, R.; Lavorgna, M. Upgrading sustainable polyurethane foam based on greener polyols: Succinic–based polyol and mannich–based polyol. *Materials* **2020**, *13*, 3170. [CrossRef]
4. Chen, Y.; Luo, Y.; Guo, X.; Chen, L.; Jia, D. The synergistic effect of ionic liquid–modified expandable graphite and intumescent flame–retardant on flame–retardant rigid polyurethane foams. *Materials* **2020**, *13*, 3095. [CrossRef]

5. Suresh, K.I. Rigid polyurethane foams from cardanol: Synthesis, structural characterization, and evaluation of polyol and foam properties. *ACS Sustain. Chem. Eng.* **2013**, *1*, 232–242. [CrossRef]

6. Abbes, B.; Lacoste, C.; Bliard, C.; Maalouf, C.; Simescu-Lazar, F.; Bogard, F.; Polidori, G. Novel extruded starch–beet pulp composites for packaging foams. *Materials* **2020**, *13*, 1571. [CrossRef]

7. Dugad, R.; Radhakrishna, G.; Gandhi, A. Recent advancements in manufacturing technologies of microcellular polymers: A review. *J. Polym. Res.* **2020**, *27*, 182. [CrossRef]

8. Zhang, N.; Cao, H. Enhancement of the antibacterial activity of natural rubber latex foam by blending It with chitin. *Materials* **2020**, *13*, 1039. [CrossRef]

9. Walter, M.; Friess, F.; Krus, M.; Zolanvari, S.M.H.; Grun, G.; Krober, H.; Pretsch, T. Shape memory polymer foam with programmable apertures. *Polymers* **2020**, *12*, 1914. [CrossRef] [PubMed]

10. Jin, F.L.; Zhao, M.; Park, M.; Park, S.J. Recent trends of foaming in polymer processing: A review. *Polymers* **2019**, *11*, 953. [CrossRef] [PubMed]

11. Nalawade, S.P.; Picchioni, F.; Janssen, L.P.B.M. Supercritical carbon dioxide as a green solvent for processing polymer melts: Processing aspects and applications. *Prog. Polym. Sci.* **2006**, *31*, 19–43. [CrossRef]

12. Caballe-Serrano, J.; Zhang, S.; Sculean, A.; Staehli, A.; Bosshardt, D.D. Tissue integration and degradation of a porous collagen–based scaffold used for soft tissue augmentation. *Materials* **2020**, *13*, 2420. [CrossRef] [PubMed]

13. Donnaloja, F.; Jacchetti, E.; Soncini, M.; Raimondi, M.T. Natural and synthetic polymers for bone scaffolds optimization. *Polymers* **2020**, *12*, 905. [CrossRef] [PubMed]

14. Trade Map. *List of Exporters for the Selected Product: Product: 4008 Plates, Sheets, Strip, Rods and Profile Shapes, of Vulcanised Rubber (Excluding Hard Rubber)*; International Trade Centre: Geneva, Switzerland, 2019.

15. Mohebbi, A.; Mighri, F.; Ajji, A.; Rodrigue, D. Current issues and challenges in polypropylene foaming: A review. *Cell. Polym.* **2015**, *34*, 299–338. [CrossRef]

16. Liao, X.; Nawaby, A.V. The sorption behaviors in PLLA–CO_2 system and its effect on foam morphology. *J. Polym. Res.* **2012**, *19*, 9827. [CrossRef]

17. Galakhova, A.; Santiago-Calvo, M.; Tirado-Mediavilla, J.; Villafane, F.; Rodriguez-Perez, M.A.; Riess, G. Identification and quantification of cell gas evolution in rigid polyurethane foams by novel GCMS methodology. *Polymers* **2019**, *11*, 1192. [CrossRef]

18. Zhao, S.; Pan, C.; Xin, Z.; Li, Y.; Qin, W.; Zhou, S. 13X zeolite as difunctional nucleating agent regulating the crystal form and improving the Foamability of blocked copolymerized polypropylene in supercritical CO_2 foaming process. *J. Polym. Res.* **2019**, *26*, 58. [CrossRef]

19. Abbasi, H.; Antunes, M.; Velasco, J.I. Polyetherimide foams filled with low content of graphene nanoplatelets prepared by scCO$_2$ dissolution. *Polymers* **2019**, *11*, 328. [CrossRef]

20. Wang, L.; Jiang, J.; Jiang, P.; Yu, J. Synthesis, characteristic of a novel flame retardant containing phosphorus, silicon and its application in ethylene vinyl–acetate copolymer (EVM) rubber. *J. Polym. Res.* **2010**, *17*, 891–902. [CrossRef]

21. Suksup, R.; Sun, Y.; Sukatta, U.; Smitthipong, W. Foam rubber from centrifuged and creamed latex. *J. Polym. Eng.* **2019**, *39*, 336–342. [CrossRef]

22. Phomrak, S.; Nimpaiboon, A.; Newby, B.Z.; Phisalaphong, M. Natural rubber latex foam reinforced with micro and nanofibrillated cellulose via Dunlop method. *Polymers* **2020**, *12*, 1959. [CrossRef]

23. Lang, X.H.; Wang, D.; Prakashan, K.; Zhang, X.; Zhang, Z.X. Microcellular chlorinated polyethylene (CM) rubber foam by using N_2 as blowing agent. *J. Polym. Res.* **2017**, *24*, 175. [CrossRef]

24. Karim, A.F.A.; Ismail, H.; Ari, Z.M. Properties and characterization of Kenaf–Filled natural rubber latex foam. *Bioresources* **2016**, *11*, 1080–1091.

25. Rathnayake, W.G.I.U.; Ismail, H.; Baharin, A.; Bandara, C.D.; Rajapakse, S. Enhancement of the antibacterial activity of natural rubber latex foam by the incorporation of zinc oxide nanoparticles. *J. Appl. Polym. Sci.* **2013**, *131*, 131. [CrossRef]

26. Pinto, J.; Escudero, J.; Solórzano, E.; Rodriguez-Perez, M.A. A novel route to produce structural polymer foams with a controlled solid skin–porous core structure based on gas diffusion mechanisms. *J. Sandw. Struct. Mater.* **2020**, *22*, 822–832. [CrossRef]

27. Kumar, V.; Suh, N.P. A process for making microcellular thermoplastic parts. *Polym. Eng. Sci.* **1990**, *30*, 1323–1329. [CrossRef]

28. Goel, S.K.; Beckman, E.J. Generation of microcellular polymeric foams using supercritical carbon dioxide. I: Effect of pressure and temperature on nucleation. *Polym. Eng. Sci.* **1994**, *34*, 1137–1147. [CrossRef]

29. Ratcha, A.; Samart, C.; Yoosuk, B.; Sawada, H.; Reubroycharoen, P.; Kongparakul, S. Polyisoprene modified poly(alkyl acrylate) foam as oil sorbent material. *J. Appl. Polym. Sci.* **2015**, *132*. [CrossRef]

30. Ratcha, A.; Yoosuk, B.; Kongparakul, S. Grafted methyl methacrylate and butyl methacrylate onto natural rubber foam for oil sorbent. *Adv. Mater. Res.* **2014**, *844*, 385–390. [CrossRef]

31. Tsivintzelis, I.; Sanxaridou, G.; Pavlidou, E.; Panayiotou, C. Foaming of polymers with supercritical fluids: A thermodynamic investigation. *J. Supercrit. Fluids* **2016**, *110*, 240–250. [CrossRef]

32. Reglero Ruiz, J.A.; Vincent, M.; Agassant, J.-F.; Sadik, T.; Pillon, C.; Carrot, C. Polymer foaming with chemical blowing agents: Experiment and modeling. *Polym. Eng. Sci.* **2015**, *55*, 2018–2029. [CrossRef]

33. Thompson, R.B.; Park, C.B.; Chen, P. Reduction of polymer surface tension by crystallized polymer nanoparticles. *J. Chem. Phys.* **2010**, *133*, 144913. [CrossRef] [PubMed]

34. Huang, D.; Zhang, M.; Guo, C.; Shi, L.; Lin, P. Experimental investigations on the effects of bottom ventilation on the fire behavior of natural rubber latex foam. *Appl. Therm. Eng.* **2018**, *133*, 201–210. [CrossRef]

35. Klempner, D.; Frisch, K.C. *Handbook of Polymeric Foams and Foam Technology*; Hanser Munich etc.: Birmingham, UK, 1991; p. 404.

36. Kudori, S.N.I.; Ismail, H.; Shuib, R.K. Kenaf core and bast loading vs. properties of natural rubber latex foam (NRLF). *BioResources* **2019**, *14*, 1765–1780.

37. Surya, I.; Kudori, S.N.I.; Ismail, H. Effect of partial replacement of kenaf by empty fruit bunch (EFB) on the properties of natural rubber latex foam (NRLF). *BioResources* **2019**, *14*, 9375–9391.

38. Panploo, K.; Chalermsinsuwan, B.; Poompradub, S. Natural rubber latex foam with particulate fillers for carbon dioxide adsorption and regeneration. *RSC Adv.* **2019**, *9*, 28916–28923. [CrossRef]

39. Rathnayake, I.U.; Ismail, H.; De Silva, C.R.; Darsanasiri, N.D.; Bose, I. Antibacterial effect of Ag–doped TiO_2 nanoparticles incorporated natural rubber latex foam under visible light conditions. *Iran. Polym. J.* **2015**, *24*, 1057–1068. [CrossRef]

40. Rathnayake, W.G.I.U.; Ismail, H.; Baharin, A.; Darsanasiri, A.G.N.D.; Rajapakse, S. Synthesis and characterization of nano silver based natural rubber latex foam for imparting antibacterial and anti–fungal properties. *Polym. Test.* **2012**, *31*, 586–592. [CrossRef]

41. Stalder, A.F.; Melchior, T.; Müller, M.; Sage, D.; Blu, T.; Unser, M. Low–bond axisymmetric drop shape analysis for surface tension and contact angle measurements of sessile drops. *Colloids Surf. A* **2010**, *364*, 72–81. [CrossRef]

42. Wulf, M.; Michel, S.; Grungke, K.; Del Rio, O.I.; Kwok, D.Y.; Neumann, A.W. Simultaneous determination of surface tension and density of polymer melts using axisymmetric drop shape analysis. *J. Colloid Interface Sci.* **1999**, *210*, 172–181. [CrossRef]

43. Ramasamy, S.; Ismail, H.; Munusamy, Y. Tensile and morphological properties of rice husk powder filled natural rubber latex foam. *Polym. Technol. Eng.* **2012**, *51*, 1524–1529. [CrossRef]

44. Oliveira-Salmazo, L.; Lopez-Gil, A.; Silva-Bellucci, F.; Job, A.E.; Rodriguez-Perez, M. A Natural rubber foams with anisotropic cellular structures: Mechanical properties and modeling. *Ind. Ind. Crop. Prod.* **2016**, *80*, 26–35. [CrossRef]

45. Sandhu, I.; Kala, M.; Thangadurai, M.; Singh, M.; Alegaonkar, P.; Saroha, D.R. Experimental study of blast wave mitigation in open cell foams. *Mater. Today Proc.* **2018**, *5*, 28170–28179. [CrossRef]

46. Tomasko, D.L.; Burley, A.; Feng, L.; Yeh, S.-K.; Miyazono, K.; Nirmal-Kumar, S.; Kusaka, I.; Koelling, K. Development of CO_2 for polymer foam applications. *J. Supercrit. Fluids* **2009**, *47*, 493–499. [CrossRef]

47. Frerich, S.C. Biopolymer foaming with supercritical CO_2—Thermodynamics, foaming behaviour and mechanical characteristics. *J. Supercrit. Fluids* **2015**, *96*, 349–358. [CrossRef]

48. Tsioptsias, C.; Panayiotou, C. Foaming of chitin hydrogels processed by supercritical carbon dioxide. *J. Supercrit. Fluids* **2008**, *47*, 302–308. [CrossRef]

49. Tsivintzelis, I.; Panayiotou, C. Designing Issues in Polymer Foaming with Supercritical Fluids. *Macromol. Symp.* **2013**, *331–332*, 109–114. [CrossRef]

50. Ma, Z.; Zhang, G.; Yang, Q.; Shi, X.; Shi, A. Fabrication of microcellular polycarbonate foams with unimodal or bimodal cell–size distributions using supercritical carbon dioxide as a blowing agent. *J. Cell. Plast.* **2014**, *50*, 55–79. [CrossRef]

51. Yeh, S.-K.; Liu, W.-H.; Huang, Y.-M. Carbon dioxide–blown expanded polyamide bead foams with bimodal cell structure. *Ind. Eng. Chem. Res.* **2019**, *58*, 2958–2969. [CrossRef]
52. Trofa, M.; Di Maio, E.; Maffettone, P.L. Multi–graded foams upon time–dependent exposition to blowing agent. *Chem. Eng. J.* **2019**, *362*, 812–817. [CrossRef]
53. Sumey, J.L.; Sarver, J.A.; Kiran, E. Foaming of polystyrene and poly(methyl methacrylate) multilayered thin films with supercritical carbon dioxide. *J. Supercrit. Fluids* **2019**, *145*, 243–252. [CrossRef]
54. Cusson, E.; Akbarzadeh, A.H.; Therriault, D.; Rodrigue, D. Density graded polyethylene foams: Effect of processing conditions on mechanical properties. *Cell. Polym.* **2019**, *38*, 3–14. [CrossRef]
55. Bates, S.R.G.; Farrow, I.R.; Trask, R.S. Compressive behaviour of 3D printed thermoplastic polyurethane honeycombs with graded densities. *Mater. Des.* **2019**, *162*, 130–142. [CrossRef]
56. Esmailzadeh, M.; Manesh, H.D.; Zebarjad, S.M. Fabrication and characterization of functional graded polyurethane foam (FGPUF). *Polym. Adv. Technol.* **2018**, *29*, 182–189. [CrossRef]
57. Jahwari, F.A.l.; Huang, Y.; Naguib, H.E.; Lo, J. Relation of impact strength to the microstructure of functionally graded porous structures of acrylonitrile butadiene styrene (ABS) foamed by thermally activated microspheres. *Polymer* **2016**, *98*, 270–281. [CrossRef]
58. Heim, H.-P.; Tromm, M. Injection molded components with functionally graded foam structures–Procedure and essential results. *J. Cell. Plast.* **2016**, *52*, 299–319. [CrossRef]
59. Ghaffari, S.; Naguib, H.E.; Park, C.B.; Atalla, N. Design and development of novel bio–based functionally graded foams for enhanced acoustic capabilities. *J. Mater. Sci.* **2015**, *50*.
60. Zhou, C.; Wang, P.; Li, W. Fabrication of functionally graded porous polymer via supercritical CO_2 foaming. *Compos. B Eng.* **2011**, *42*, 318–325. [CrossRef]
61. Yao, J.; Rodrigue, D. Density graded polyethylene foams produced by compression moulding using a chemical blowing agent. *Cell. Polym.* **2012**, *31*, 189–206. [CrossRef]
62. Stubenrauch, C.; Menner, A.; Bismarck, A.; Drenckhan, W. Emulsion and foam templating—Promising routes to tailor–made porous polymers. *Angew. Chem. Int. Ed.* **2018**, *57*, 10024–10032. [CrossRef]
63. Andrieux, S.; Quell, A.; Stubenrauch, C.; Drenckhan, W. Liquid foam templating—A route to tailor–made polymer foams. *Adv. Colloid Interface* **2018**, *256*, 276–290. [CrossRef]
64. Lee, J.J.; Cho, M.Y.; Kim, B.H.; Lee, S. Development of eco–friendly polymer foam using overcoat technology of deodorant. *Materials* **2018**, *11*, 1898. [CrossRef] [PubMed]
65. Obradovic, J.; Voutilainen, M.; Virtanen, P.; Lassila, L.; Fardim, P. Cellulose fibre–reinforced biofoam for structural applications. *Materials* **2017**, *10*, 619. [CrossRef] [PubMed]
66. Forest, C.; Chaumont, P.; Cassagnau, P.; Swoboda, B.; Sonntag, P. Polymer nano–foams for insulating applications prepared from CO_2 foaming. *Prog. Polym. Sci.* **2015**, *41*, 122–145. [CrossRef]
67. Chollakup, R.; Smitthipong, W.; Chworos, A. Specific interaction of DNA–functionalized polymer colloid. *Polym. Chem.* **2010**, *1*, 658–662. [CrossRef]
68. Chollakup, R.; Smitthipong, W.; Chworos, A. DNA–functionalized polystyrene particles and their controlled self–assembly. *RSC Adv.* **2014**, *4*, 30648–30653. [CrossRef]
69. Strachota, B.; Morand, A.; Dybal, J.; Matejka, L. Control of gelation and properties of reversible Diels–Alder networks: Design of a self–healing network. *Polymers* **2019**, *11*, 930. [CrossRef]
70. Stephanou, P.S.; Tsimouri, I.C.; Mavrantzas, V.G. Simple, Accurate and user–friendly differential constitutive model for the rheology of entangled polymer melts and solutions from nonequilibrium thermodynamics. *Materials* **2020**, *13*, 2867. [CrossRef]
71. Sauceau, M.; Fages, J.; Common, A.; Nikitine, C.; Rodier, E. New challenges in polymer foaming: A review of extrusion processes assisted by supercritical carbon dioxide. *Prog. Polym. Sci.* **2011**, *36*, 749–766. [CrossRef]
72. Pakornpadungsit, P.; Smitthipong, W.; Chworos, A. Self–assembly nucleic acid–based biopolymers: Learn from the nature. *J. Polym. Res.* **2018**, *25*, 45. [CrossRef]
73. Yokoyama, H.; Sugiyama, K. Nanocellular structures in block copolymers with CO_2–philic blocks using CO_2 as a blowing agent: Crossover from micro- to nanocellular structures with depressurization temperature. *Macromolecules* **2005**, *38*, 10516–10522. [CrossRef]
74. Jiang, Y.; Greco, C.; Daoulas, K.h.; Chen, J.Z.Y. Thermodynamics of a compressible Maier–Saupe model based on the self–consistent field theory of wormlike polymer. *Polymers* **2017**, *9*, 48. [CrossRef] [PubMed]

75. Mondy, L.; Rao, R.; Grillet, A.; Adolf, D.; Brotherton, C.; Russick, E.; Cote, R.; Castañeda, J.; Thompson, K.; Bourdon, C.; et al. *Experiments for Foam Model Development and Validation*; Sandia National Laboratories: Albuquerque, NM, USA, 2008.

76. Seo, D.; Youn, J.R.; Tucker III, C.L. Numerical simulation of mold filling in foam reaction injection molding. *Int. J. Numer. Methods Fluids* **2003**, *42*, 1105–1134. [CrossRef]

77. Prud'homme, R.; Khan, S.A. *Foams: Theory, Measurements, and Applications*; Marcel Dekker, Inc.: New York, NY, USA, 1996.

78. Gibson, L.J.; Ashby, M.F. *Cellular solids: Structure and Properties*, 2nd ed.; Cambridge University Press: Cambridge, UK, 1990.

79. May, C. *Epoxy Resins: Chemistry and Technology*, 2nd ed.; CRC Press: Boca Raton, FL, USA, 1987.

80. Lin, H.; Lv, L.; Zhang, J.; Wang, Z. Energy–absorbing performance of graded Voronoi foams. *J. Cell. Plast.* **2019**, *55*, 589–613. [CrossRef]

81. Liu, H.; Ding, S.; Ng, B.F. Impact response and energy absorption of functionally graded foam under temperature gradient environment. *Compos. B Eng.* **2019**, *172*, 516–532. [CrossRef]

82. Yu, J.; Song, L.; Chen, F.; Fan, P.; Sun, L.; Zhong, M.; Yang, J. Preparation of polymer foams with a gradient of cell size: Further exploring the nucleation effect of porous inorganic materials in polymer foaming. *Mater. Today Commun.* **2016**, *9*, 1–6. [CrossRef]

83. Timothy, J.J.; Meschke, G. A cascade continuum micromechanics model for the effective elastic properties of porous materials. *Int. J. Solids Struct.* **2016**, *83*, 1–12. [CrossRef]

84. Bashir, A.S.; Munusamy, Y.; Chew, T.L.; Ismail, H.; Ramasamy, S. Mechanical, thermal, and morphological properties of (eggshell powder)–filled natural rubber latex foam. *J. Vinyl Addit. Technol.* **2015**, *23*, 3–12. [CrossRef]

85. Sadik, T.; Pillon, C.; Carrot, C.; Ruiz, J.A.R.; Vincent, M.; Billon, M. Propylene structural foam: Measurements of the core, skin, and overall mechanical properties with evaluation of predictive models. *J. Cell. Plast.* **2017**, *53*, 25–44. [CrossRef]

86. Yuan, H.; Li, J.; Xiong, Y.; Luo, G.; Shen, Q.; Zhang, L. Preparation and characterization of PMMA graded microporous foams via one–step supercritical carbon dioxide foaming. *J. Phys. Conf. Ser.* **2013**, *419*. [CrossRef]

87. Elsing, J.; Quell, A.; Stubenrauch, C. Toward functionally graded polymer foams using microfluidics. *Adv. Eng. Mater.* **2017**, *19*. [CrossRef]

88. Yang, C.; Wang, M.; Zhao, Z.; Wang, M.; Wu, G. A new promising nucleating agent for polymer foaming: Effects of hollow molecular–sieve particles on polypropylene supercritical CO_2 microcellular foaming. *RSC Adv.* **2018**, *8*, 20061–20067. [CrossRef]

89. Llewelyn, G.; Rees, A.; Griffiths, C.A.; Jacobi, M. A novel hybrid foaming method for low–pressure microcellular foam production of unfilled and talc–filled copolymer polypropylenes. *Polymers* **2019**, *11*, 1896. [CrossRef] [PubMed]

90. Onder, O.C.; Yilgor, E.; Yilgor, I. Preparation of monolithic polycaprolactone foams with controlled morphology. *Polymer* **2018**, *136*, 166–178. [CrossRef]

91. Song, W.; Barber, K.; Lee, K.Y. Heat–induced bubble expansion as a route to increase the porosity of foam–templated bio–based macroporous polymers. *Polymer* **2017**, *118*, 97–106. [CrossRef]

92. Vahidifar, A.; Khorasani, S.N.; Park, C.B.; Naguib, H.E.; Khonakdar, H.A. Fabrication and characterization of closed–cell rubber foams based on natural rubber/carbon black by one–step foam processing. *Ind. Eng. Chem. Res.* **2016**, *55*, 2407–2416. [CrossRef]

93. Ramage, M.H.; Burridge, H.; Busse-Wicher, M.; Fereday, G.; Reynolds, T.; Shah, D.U.; Wu, G.; Yu, L.; Fleming, P.; Densley-Tingley, D.; et al. The wood from the trees: The use of timber in construction. *Renew. Sustain. Energy Rev.* **2017**, *68*, 333–359. [CrossRef]

94. Shah, D.U.; Reynolds, T.P.S.; Ramage, M.H. The strength of plants: Theory and experimental methods to measure the mechanical properties of stems. *J. Exp. Bot.* **2017**, *68*, 4497–4516. [CrossRef]

95. Ciechanska, D. Multifunctional bacterial cellulose/chitosan composite materials for medical applications. *Fibres Text. East. Eur.* **2004**, *12*, 69–72.

96. Dominic CD, M.; Joseph, R.; Begum, P.; Joseph, M.; Padmanabhan, D.; Morris, L.A.; Kumar, A.S.; Formela, K. Cellulose nanofibers isolated from the cuscuta reflexa plant as a green reinforcement of natural rubber. *Polymers* **2020**, *12*, 814. [CrossRef]

97. Abdul Azam, F.A.; Rajendran Royan, N.R.; Yuhana, N.Y.; Mohd Radzuan, N.A.; Ahmad, S.; Sulong, A.B. Fabrication of porous recycled HDPE biocomposites foam: Effect of rice husk filler contents and surface treatments on the mechanical properties. *Polymers* **2020**, *12*, 475. [CrossRef]

98. Phomrak, S.; Phisalaphong, M. Lactic acid modified natural rubber–bacterial cellulose composites. *Appl. Sci.* **2020**, *10*, 3583. [CrossRef]

99. Ismail, H.; Edyham, M.; Wirjosentono, B. Bamboo fibre filled natural rubber composites: The effects of filler loading and bonding agent. *Polym. Test.* **2002**, *21*, 139–144. [CrossRef]

100. Karmarkar, A.; Chauhan, S.; Modak, J.M.; Chanda, M. Mechanical properties of wood–fiber reinforced polypropylene composites: Effect of a novel compatibilizer with isocyanate functional group. *Compos. Part A Appl. Sci. Manuf.* **2007**, *38*, 227–233. [CrossRef]

101. Gao, C.; Xiao, W.; Ji, G.; Zhang, Y.; Cao, Y.; Han, L. Regularity and mechanism of wheat straw properties change in ball milling process at cellular scale. *Bioresour. Technol.* **2017**, *241*, 214–219. [CrossRef]

102. Zhou, L.; He, H.; Li, M.-C.; Song, K.; Cheng, H.; Wu, Q. Morphological influence of cellulose nanoparticles (CNs) from cottonseed hulls on rheological properties of polyvinyl alcohol/CN suspensions. *Carbohydr. Polym.* **2016**, *153*, 445–454. [CrossRef]

103. Ling, Z.; Edwards, J.V.; Guo, Z.; Prevost, N.T.; Nam, S.; Wu, Q.; French, A.D.; Xu, F. Structural variations of cotton cellulose nanocrystals from deep eutectic solvent treatment: Micro and nano scale. *Cellulose* **2019**, *26*, 861–876. [CrossRef]

104. Phisalaphong, M.; Suwanmajo, T.; Sangtherapitikul, P. Novel nanoporous membranes from regenerated bacterial cellulose. *J. Appl. Polym. Sci.* **2008**, *107*, 292–299. [CrossRef]

105. Liu, M.; Wang, H.; Han, J.; Niu, Y. Enhanced hydrogenolysis conversion of cellulose to C2–C3 polyols via alkaline pretreatment. *Carbohydr. Polym.* **2012**, *89*, 607–612. [CrossRef]

106. Nomura, S.; Kugo, Y.; Erata, T. 13C NMR and XRD studies on the enhancement of cellulose II crystallinity with low concentration NaOH post–treatments. *Cellulose* **2020**, *27*, 3553–3563. [CrossRef]

107. Williams, T.; Hosur, M.; Theodore, M.; Netravali, A.; Rangari, V.; Jeelani, S. Time effects on morphology and bonding ability in mercerized natural fibers for composite reinforcement. *Int. J. Polym. Sci.* **2011**, *2011*, 1–9. [CrossRef]

108. Roy, K.; Debnath, S.C.; Tzounis, L.; Pongwisuthiruchte, A.; Potiyaraj, P. Effect of various surface treatments on the performance of jute fibers filled natural rubber (NR) composites. *Polymers* **2020**, *12*, 369. [CrossRef] [PubMed]

109. Pakornpadungsit, P.; Prasopdee, T.; Swainson, N.M.; Chworos, A.; Smitthipong, W. DNA:chitosan complex, known as a drug delivery system, can create a porous scaffold. *Polym. Test.* **2020**, *83*, 106333. [CrossRef]

110. Chollakup, R.; Uttayarat, P.; Chworos, A.; Smitthipong, W. Noncovalent sericin–chitosan scaffold: Physical properties and low cytotoxicity effect. *Int. J. Mol. Sci.* **2020**, *21*, 775. [CrossRef] [PubMed]

111. Deng, C.-M.; He, L.-Z.; Zhao, M.; Yang, D.; Liu, Y. Biological properties of the chitosan–gelatin sponge wound dressing. *Carbohydr. Polym.* **2007**, *69*, 583–589. [CrossRef]

112. Czaja, W.; Krystynowicz, A.; Bielecki, S.; Brown, R.M., Jr. Microbial cellulose—The natural power to heal wounds. *Biomaterials* **2006**, *27*, 145–151. [CrossRef] [PubMed]

113. Bodhibukkana, C.; Srichana, T.; Kaewnopparat, S.; Tangthong, N.; Bouking, P.; Martin, G.P.; Suedee, R. Composite membrane of bacterially–derived cellulose and molecularly imprinted polymer for use as a transdermal enantioselective controlled–release system of racemic propranolol. *J. Control. Release* **2006**, *113*, 43–56. [CrossRef]

114. Klemm, D.; Schumann, D.; Udhardt, U.; Marsch, S. Bacterial synthesized cellulose–Artificial blood vessels for microsurgery. *Prog. Polym. Sci.* **2001**, *26*, 1561–1603. [CrossRef]

115. Ari, Z.; Zakaria, Z.; Tay, L.; Lee, S. Effect of foaming temperature and rubber grades on properties of natural rubber foams. *J. Appl. Polym. Sci.* **2008**, *107*, 2531–2538.

Article

Large Deformation Finite Element Analyses for 3D X-ray CT Scanned Microscopic Structures of Polyurethane Foams

Makoto Iizuka [1,*], **Ryohei Goto** [2], **Petros Siegkas** [1], **Benjamin Simpson** [1] and **Neil Mansfield** [1]

1 Department of Engineering, School of Science and Technology, Nottingham Trent University, Clifton Lane, Nottingham NG11 8NS, UK; petros.siegkas@ntu.ac.uk (P.S.); ben.simpson@ntu.ac.uk (B.S.); neil.mansfield@ntu.ac.uk (N.M.)

2 Bridgestone Corporation, 1, Kashio-Cho, Totsuka-Ku, Yokohama, Kanagawa 244-8510, Japan; ryohei.goto@bridgestone.com

* Correspondence: makoto.iizuka@ntu.ac.uk

Abstract: Polyurethane foams have unique properties that make them suitable for a wide range of applications, including cushioning and seat pads. The foam mechanical properties largely depend on both the parent material and foam cell microstructure. Uniaxial loading experiments, X-ray tomography and finite element analysis can be used to investigate the relationship between the macroscopic mechanical properties and microscopic foam structure. Polyurethane foam specimens were scanned using X-ray computed tomography. The scanned geometries were converted to three-dimensional (3D) CAD models using open source, and commercially available CAD software tools. The models were meshed and used to simulate the compression tests using the implicit finite element method. The calculated uniaxial compression tests were in good agreement with experimental results for strains up to 30%. The presented method would be effective in investigating the effect of polymer foam geometrical features in macroscopic mechanical properties, and guide manufacturing methods for specific applications.

Keywords: polyurethane foam; structure-property relationships; finite element analysis; microscale analysis; X-ray computed tomography

Citation: Iizuka, M.; Goto, R.; Siegkas, P.; Simpson, B.; Mansfield, N. Large Deformation Finite Element Analyses for 3D X-ray CT Scanned Microscopic Structures of Polyurethane Foams. *Materials* **2021**, *14*, 949. https://doi.org/10.3390/ma14040949

Academic Editor: Aleksander Hejna

Received: 31 December 2020
Accepted: 11 February 2021
Published: 17 February 2021

Publisher's Note: MDPI stays neutral with regard to jurisdictional claims in published maps and institutional affiliations.

1. Introduction

Polyurethane foams have many unique properties, such as elasticity, softness, and ease of forming. These properties make polyurethane foams attractive to automotive seat designers since they can effectively support the human body and distribute the body pressure. The improvement of the mechanical properties of the foams is an important challenge. Controlling the mechanical properties of foams would be useful in designing seats that are more comfortable and potentially at lower cost. The mechanical properties of polyurethane foams depend largely on their microstructures (Figure 1). The foam structure consists of a cluster of bubbles and struts at the edges of the cells. Figure 1 shows an example of an open-cell foam in which the bubbles are linked together. The macroscopic stress–strain relationship depends on the mechanical properties of the parent material, of which the struts are made, and the geometrical structure of cells and struts [1]. Understanding the relationships between the microscopic geometrical structures and macroscopic mechanical properties is essential in developing foam products with superior mechanical properties.

Three main regions can be identified in the stress–strain curve for the compressive deformation of elastomeric foams [1]. Figure 2 shows the typical stress–strain curve under the uniaxial compression of foams. Linear elasticity is shown in the small strain region, followed by a collapse plateau, and then densification appears accompanied by a rapid increase in the stress. Firstly, the struts bend and the macroscopically linear elastic behaviour is shown. Next, some of the struts start buckling and the slope of the curve decreases due to the increase of the macroscopic stress. Finally, the slope of the curve increases again up to the same value as the matrix material, because of the contact between struts. The contribution of microstructures to macroscopic properties depends on these deformation mechanisms.

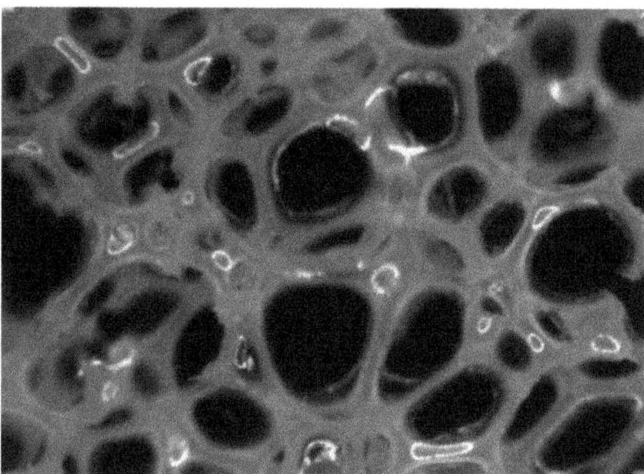

Figure 1. An example of optical microscope images of polyurethane foams.

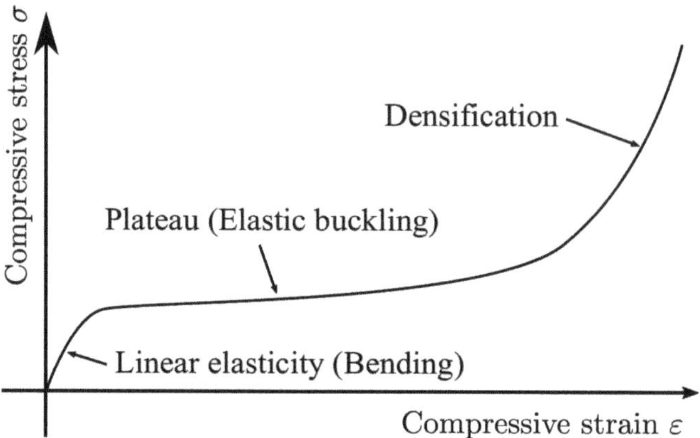

Figure 2. The typical stress–strain relationship of elastomeric foams under the uniaxial compressive stress.

Cell structure geometries are virtually generated and their deformations are analysed to investigate the effect of microstructures on macroscopic properties [2]. The cells were postulated to have same size and the shape of the Kelvin tetrakaidecahedron. The edges of the polyhedron were assumed to be struts that are represented by Euler–Bernoulli beams and the macroscopic elastic properties were analytically calculated. This approach was also expanded to the large compressive strain range up to 70% [3,4] and creep deformations [5]. Other researchers repeated the calculations of Zhu et al. [2], employing a finite element approach, while still making use of Kelvin's cell shape and Euler–Bernoulli beams [6–10]. Okumura et al. [11] and Takahashi et al. [12] analysed the mechanical responses in the [001], [011], and [111] directions, as the Kelvin's cell has anisotropic mechanical properties. Furthermore, closed cell foams have been analysed with shell elements [13]. Modelling the microscopic structures of polyurethane foam materials using the Kelvin's cell is thought to be a simple and effective way of investigating the deformation behaviour.

The Kelvin cell approach assumes that the microstructure is homogeneous; however, in contrast, cell structures are generally heterogeneous. This is a significant disadvantage of the repeated unit cell modelling approach [14]. To model the inhomogeneous structures of foams, the 2D and 3D Voronoi tessellations were employed and the Voronoi edges were regarded as struts [14–17]. Moreover, faces in Voronoi polyhedrons were assumed as cell membranes in closed cell foams [18,19]. The elastic properties in the small strain region and the compressive stress–strain curves on the plateau region were calculated by the finite element method while using beam elements. Furthermore, although the cross-sectional area of a strut is often assumed to be constant, the central parts of struts are thinner than other parts. The effect of this necking can be taken into account using solid elements [11,12,20–27] or beam elements with variable cross-sectional properties [23,28–33]. In addition, the curvature of struts were modelled [34]. Models that consider the heterogeneity of foams are thought to show better results than Kelvin cell models with straight struts. Dynamic crushing behaviour [35,36] and multiaxial crushing [37] were also analysed.

The use of X-ray computed tomography (CT) is one effective method for obtaining a more adequate model that represents actual foam microstructures. The X-ray CT has been performed to observe the microstructures of various kinds of porous materials, for example, biomaterial scaffolds [38,39], soil materials [40], and polyurethane foams [41]. Therefore, the X-ray CT has also been used to generate the geometries for finite element analyses. For example, finite element models for the microstructure of a trabecular bone was generated based on micro-CT [42]. For artificial foam materials, Jeon et al. [43] analysed closed-cell aluminium foams with finite element models meshed with solid tetrahedron elements. The compressive stress–strain curves of the foam were calculated and compared to the experimental results and the 20.86% volume error was shown up to 5.31% strain. Similarly, linear elastic properties under the small strain regions were obtained from X-ray CT scanned finite element models for ceramic foams [44] and a rigid organic foam [45]. Models that were obtained from the X-ray CT have been effectively used to investigate the mechanical properties of foams under small deformations.

For cushioning products, such as automotive seat pads or bed mattresses, the mechanical properties in the plateau regions are more important than the linear elastic regions. As the slope of the stress–strain curve decreases in the plateau region, elastic foams soften and help to distribute body pressure. Most of the studies employ tetrahedron meshing due to the complexity of the geometry; however, this makes analysing large deformations difficult. To analyse the deformation within the plateau region, hexahedron meshing is required, as it is more suitable for large deformation problems.

This study aims to use X-ray CT scans of foam specimens in order to construct validated finite element (FE) models that can be used to study and manipulate the foam microstructure for achieving desirable stress–strain behaviour in the plateau region. The microstructures of elastic polyurethane foams for automotive seat pads are scanned using X-ray computed tomography and converted to STL files. The STL files are smoothed and converted to solid CAD files with commercial CAD software, so that they can be meshed with a hexahedron dominant solid mesh. The uniaxial compressive deformation of the models are analysed with a finite element method and then compared with the experimental results.

2. Materials and Methods

The methodology for analysing the deformation of X-ray CT scanned foam materials and the materials supplied to validate its accuracy are explained here. The specimens were scanned using X-ray CT, converted to CAD models, and then analysed with the implicit finite element method. The tools used for this study are either commercially available CAD or open-source software. Moulded elastic polyurethane foams were investigated using the presented method and physically tested to compare with the result of the analyses.

2.1. Materials

The tested materials were supplied by Bridgestone Corporation in Tokyo, Japan. Polyols, isocyanates, water, and low amounts of other materials were mixed and poured into a $400 \times 400 \times 100$ (mm^3) sized mould and then expanded and polymerized. After demoulding, the foams were crushed between rollers, so that cell membranes were broken and resulted in open-cell foams. The foams were left at least 24 h before proceeding to any other process of the investigation to let the chemical reactions be completed. The foam materials that were investigated in this study are mainly used for automotive seat pads by moulding in product shaped moulds.

2.2. Scanning by the X-ray Computed Tomography

Specimens from the centre of larger samples were cut into $5 \times 5 \times 5$ (mm^3) sized cubes. The X-ray tomography equipment that was employed for this study was the ScanXmate RA150S145/2Be, a product of Comscantecno Co., Ltd. in Kanagawa, Japan. Figure 3, shows an example X-ray CT scan image of the foam. The white parts indicate the foam struts and the black parts are the pores. The size of the pixel was 7.5 (μm). The cross section images were taken by rotating the specimens every 0.18°, so that the cell structures could be observed in three dimensions.

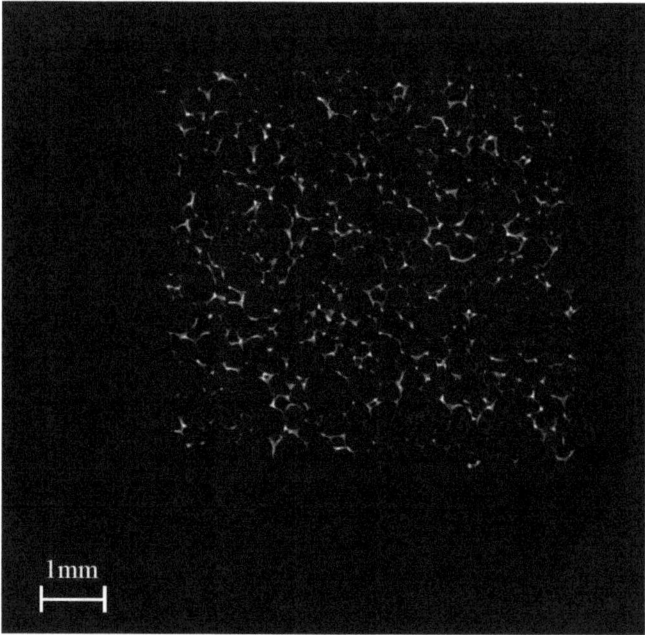

Figure 3. An example of the X-ray computed tomography (CT) scanned images for the polyurethane foams.

2.3. Converting the Scanned Images to 3D STL Files

The cross sectional two-dimensional (2D) images were converted to three-dimensional (3D) STL files by Fiji [46], a distribution of Image J2 [47]. Firstly, the scanned images were binarized to black and white images using a threshold of the brightness. The threshold was determined using Otsu's method [48] and then verified by comparing the relative densities measured with the actual specimen and calculated from the computational models. The borders between the black and white pixels were regarded as the surfaces of the struts. Triangles were then applied to the strut surfaces and the resulting surfaces exported as STL files. Figure 4a shows an example STL file.

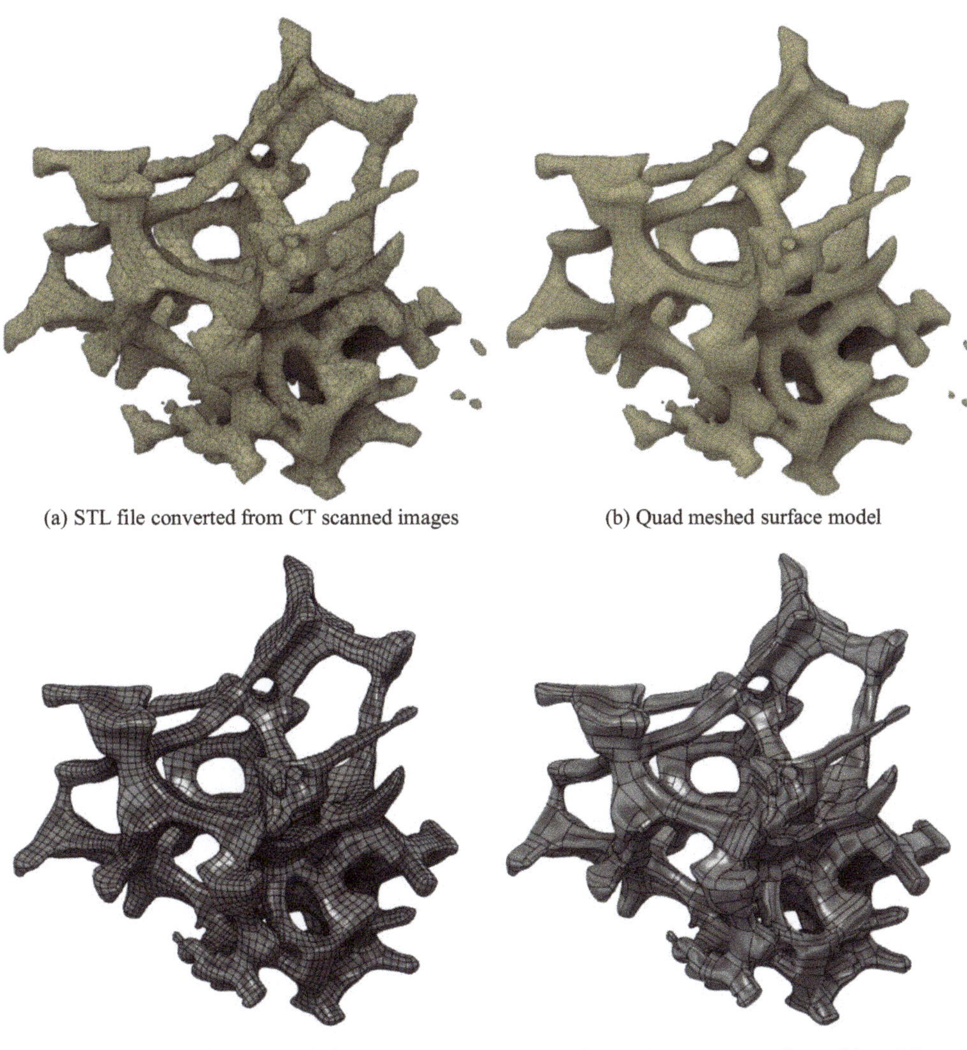

(a) STL file converted from CT scanned images

(b) Quad meshed surface model

(c) T-spline interpolation

(d) Boundary representation solid model

Figure 4. Conversion from the STL files to the boundary representation solid models.

2.4. Converting to Smoothed Solid CAD Models

The STL formatted files consist of only triangle surfaces and the triangle edges are sharp. When dividing STL files to finite elements directly, the triangle surfaces are divided into further small elements that result in a considerable number of nodes and elements. Therefore, the vertices of the triangle surfaces should be interpolated by mathematically smooth surfaces. This smoothing can be performed using Recap® and Fusion 360® software, both products of Autodesk, Inc. in California, CA, USA. Firstly, the STL files with the triangle meshing were converted to surface models with quad meshing (Figure 4b). The quad meshed surfaces were then interpolated and smoothed by T-spline surfaces (Figure 4c). Finally, boundary representation solid models were generated based on the T-spline surface models (Figure 4d). The resulting solid models were then capable of being analysed in commercial FEA software.

2.5. Hexahedron Dominant Meshing

Although the geometries were smoothed by the T-spline interpolation, they were still too complex for hexahedron meshing to be applied. Therefore, mixed hexahedron and tetrahedron meshing was employed. These two kinds of elements were joined by the pyramid mesh elements. The mesh divisions were performed using Ansys® Academic Research Meshing, Release 19.2 [49]. In this study, three representative geometric models were analysed. Figure 5 shows the mesh divisions of these models and Table 1 summarises the numbers of the nodes and the elements in each. Where possible, the models were meshed with hexahedron or pyramid elements.

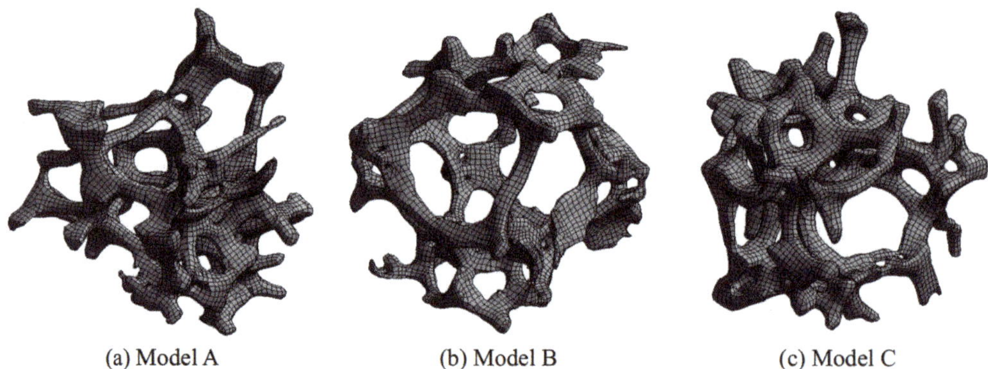

(a) Model A   (b) Model B   (c) Model C

Figure 5. Mesh divisions for the models.

Table 1. Numbers of nodes and elements for the models.

	Model A	Model B	Model C
Nodes	27,730	24,937	30,081
Tetrahedron elements	10,873	11,586	12,041
Pyramid elements	16,202	16,795	17,083
Hexahedron elements	13,231	13,746	15,089

2.6. Finite Element Analyses

The deformation behaviour of three different specimen models was calculated with the commercial FEA software Ansys® Academic Research Mechanical, Release 19.2 [50]. The large deflection was taken into account to analyse the deformations up to the plateau region. As this study focused on the static mechanical properties of polyurethane foams, the static implicit method was employed and the damping or the dynamic characteristics were neglected.

2.7. Strut Material Model

A specimen without pores is needed in order to measure the stress–strain relationship of the matrix material. The diameters of the struts are less than 0.1 mm and form a complex microstructure. Foam was compressed between plates that were heated to 150 °C in order to obtain a parent material specimen without pores. The original thickness of the foam was 50 mm and the compressed specimen had a thickness of 0.7 mm. The measured density of the specimen was 1200 kg/m^3.

Tensile testing was performed to obtain the tensile stress–strain relationship. The test equipment was a universal testing machine AGS-X 10 kN with a 500 N load cell, products of SHIMADZU CORPORATION. The specimen was cut into 50 × 5 mm^2 rectangular shape specimens and then a tensile test was performed under the strain rate 0.01 s^{-1}. The difference between the grippers was regarded as the elongation of the specimen.

Figure 6 shows the measured nominal stress–strain curve. The experimental result is approximated by the neo-Hookean (Equation (1)) and Mooney–Rivlin (Equation (2)) hyper elastic models, respectively.

$$W = C_{10}(\bar{I}_1 - 3) + \frac{1}{d}(J - 1)^2 \tag{1}$$

$$W = C_{10}(\bar{I}_1 - 3) + C_{01}(\bar{I}_2 - 3) + \frac{1}{d}(J - 1)^2 \tag{2}$$

W is the strain energy density, $\bar{I}_1$ and $\bar{I}_2$ are the first and second deviatoric strain invariants, and J is the determinant of the deformation gradient. Table 2 shows the material constants C_{10}, C_{01}, and d. Because the matrix material is thought to be incompressive, d was calculated to let the initial Poisson's ratio ν equal to 0.48. The Mooney–Rivlin model was employed in this study, as it shows better agreement with the experimental result than the neo-Hookean model.

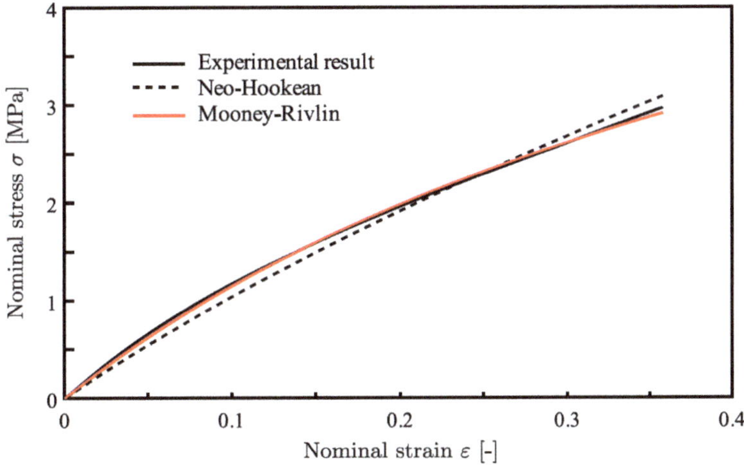

Figure 6. The result of the tensile test for the matrix material and its approximations by hyperelastic models.

Table 2. Material constants for the matrix material.

Hyperelasticity Models	C_{10} [MPa]	C_{01} [MPa]	d [MPa^{-1}]
Neo-Hookean	1.89	-	0.0661
Mooney-Rivlin	0.476	1.78	0.0554

2.8. Boundary Conditions

The foam model specimen was uniaxially compressed between two rigid shell plates, as in Figure 7. The lower plate was fixed preventing any translational or rotational displacements. Translational displacement was applied to the upper plate, whilst all other degrees of freedom were constraint. Frictionless contacts between the foam model and the rigid walls were defined while using the penalty method with a stiffness factor of 0.01. Self-contacts between the struts were not considered, as this study focuses on the buckling behaviour in the transitions to the plateau regions. Finally, remote displacements were used to constraint the specimen lateral boundaries from rigid translational and rotational movement. The average values of the displacements of the nodes on the boundaries that correspond to these directions were fixed. This would allow for deformation, but not rigid body movement.

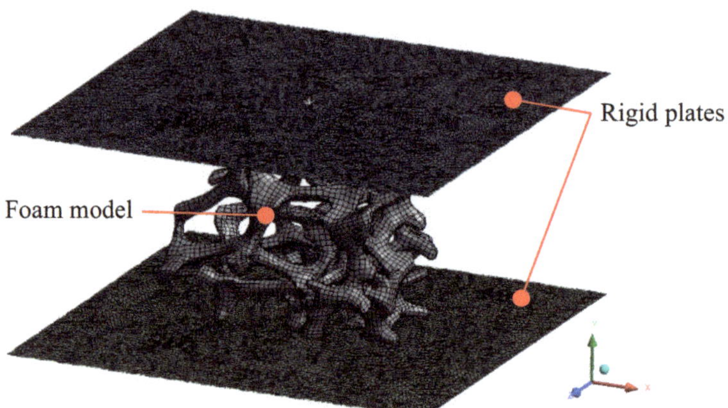

Figure 7. Boundary condition for the uniaxial compression analyses of the foam models.

2.9. Experimental Measurement for the Macroscopic Stress Strain Relationships

The uniaxial compression tests for the actual foam specimens were performed to compare with the FEA results. The testing method was similar to ISO3386-1 [51]. The $25 \times 25 \times 10$ (mm^3) sized specimens were cut from the centre parts of the moulded foams. The specimens were set into the same equipment as seen in Section 2.7 with the compression plates. The lower plate was perforated by 6 mm holes arranged in a latticed pattern with 20 mm distances, so that the air in the foam could be ventilated. Firstly, the foams were compressed to achieve 75% nominal strain with the speed 50 mm/s as the pre-compression. Afterwards, the load was taken off with the same speed and the foams were left for 60 s. After that, the foams were compressed again with the same speed and compressive strain to measure the load and the displacement.

3. Results

3.1. The Deformed Shapes of the Models

Figure 8 shows the deformed shapes of the different specimen models at the macroscopic nominal compressive strains $\varepsilon^c = 0.05$, 0.25 and 0.50 respectively. The coloured contour represents the Von–Mises equivalent strains ε^{eq}. The struts bend in the linear elastic region ($\varepsilon^c = 0.05$), as mentioned in Section 1. After that, some struts start to buckle, which indicates a transition to the plateau region ($\varepsilon^c = 0.25$). Finally, the models gradually become denser and transfer into the densification region ($\varepsilon^c = 0.50$). Because self-contacts were not applied in the foam models, the struts did not touch, but instead overlapped. The results of the analyses enable the microscopic behaviour of the struts to be carefully observed.

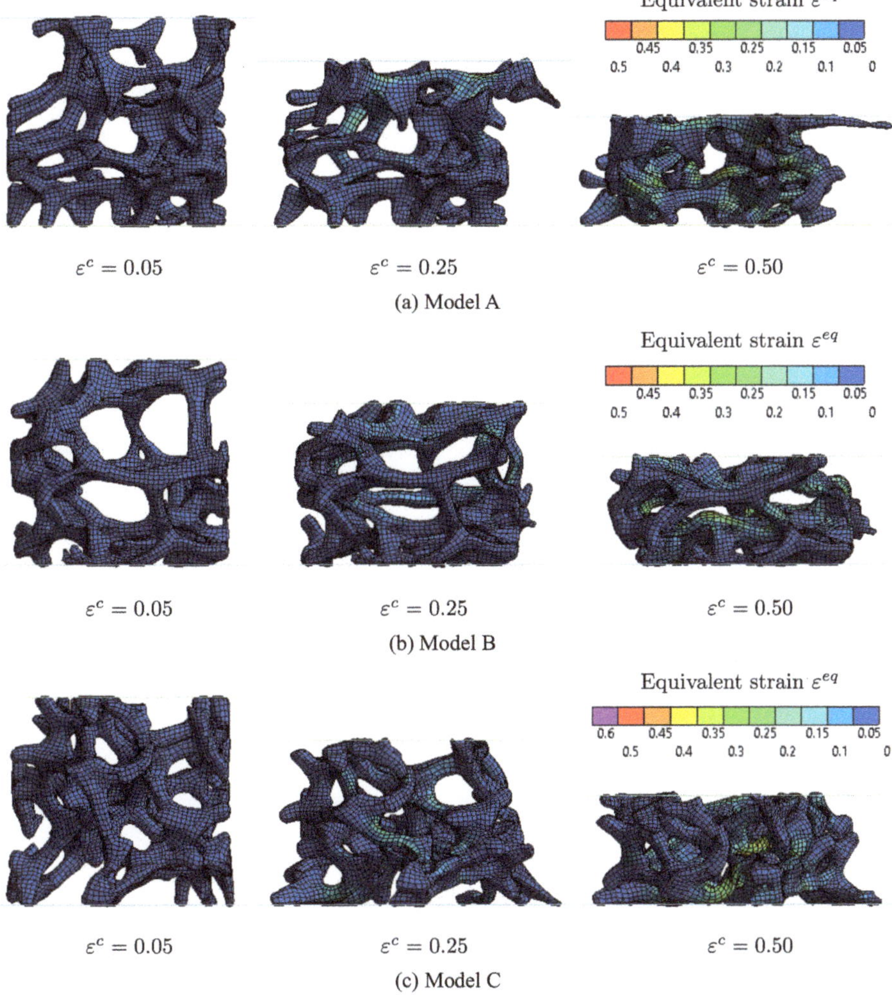

Equivalent strain ε^{eq}

$\varepsilon^c = 0.05$ $\varepsilon^c = 0.25$ $\varepsilon^c = 0.50$

(a) Model A

Equivalent strain ε^{eq}

$\varepsilon^c = 0.05$ $\varepsilon^c = 0.25$ $\varepsilon^c = 0.50$

(b) Model B

Equivalent strain ε^{eq}

$\varepsilon^c = 0.05$ $\varepsilon^c = 0.25$ $\varepsilon^c = 0.50$

(c) Model C

Figure 8. The deformed shapes of the models with the distributions of the equivalent strain ε^{eq}.

3.2. Macroscopic Stress-Strain Relationships

The FEA results were compared with the experiment results to validate the accuracy of the presented analysis method. Figure 9 shows both the experimental and FEA results of the relations between the nominal compressive stress and strain. The slopes of the stress-strain curves for the FEA results start decreasing in the strain region around 0.05 as compared to the smaller strain region. It is thought to mean the transition from the linear elastic regions to the plateau regions.

The models appear to be in good agreement with experiments in the linear elastic and the plateau regions, and up to the strain of 0.30. Differences of the stresses between the experimental and FEA results at the nominal compressive strain of 0.25 were 0.1%, 16.5%, and 6.6% for the models A, B, and C, respectively. In contrast, the FEA results are stiffer than the experimental results in larger strain regions than 0.30. After reaching the strain of 0.30, the slopes of the stress strain curves start increasing again. This behaviour looks similar to the transition to the densification regions; however, self-contacts were not enabled within the model and the stress increase occurs far too early in the strain regions. The presented method should be modified when applied for the densification region.

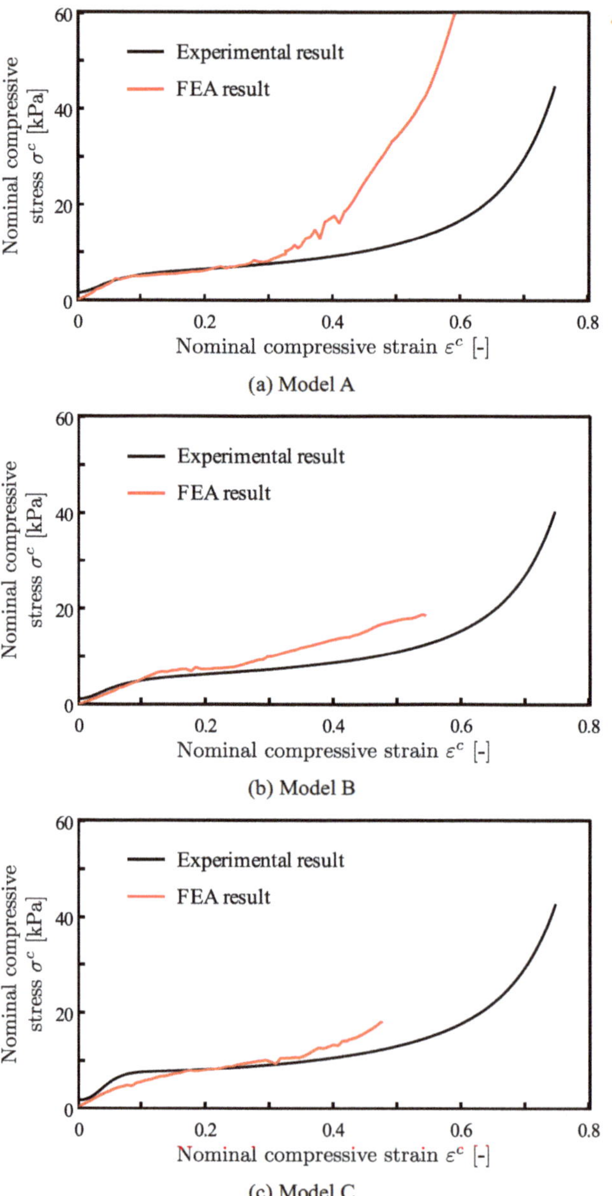

(a) Model A

(b) Model B

(c) Model C

Figure 9. The experimental and FEA results in the relations between the macroscopic compressive stress and strain.

4. Discussions

The compressive response of Polyurethane foam geometries was simulated while using FE methods and then compared with experiments. Foam specimens were scanned using X-ray CT and analysed to obtain geometries for FE simulations. The simulation results were in good agreement with experiments up to 0.3 strain. The finite element model over-predicted stresses, beyond that strain. Three different specimens were scanned and modelled to ensure the repeatability of results.

Elastic buckling appears to be one of the dominant deformation mechanisms. The finite element simulation results seem to have captured the strut deformation behaviour in agreement to relevant literature [1]. Similar deformation mechanisms have been captured with virtually generated cell structures, such as Kelvin's cells [2,4,8,9,11,12,23,29] or Voronoi polyhedrons [24,27,30,31]. Previous models based on X-ray CT scanned foam structures were mostly limited to small strains (up to 5.31%) [43–45].

Hexahedron dominant meshing was used for the large deformation analyses. Struts in foam materials can be long and narrow. Euler–Bernoulli beams have been widely employed for analytical calculations [2,4] and numerical simulations [8,9,24,27,29–31]. However whilst beam models might be beneficial in reducing complexity and calculation time, they might also add stiffness to the structure and result in higher stress predictions in comparison to the experimental values. Hexahedron meshes in large deformation problems have been used for simplified geometries [11,12]. The presented smoothing method and hexahedron dominant meshing are recommended for the complex X-ray scanned geometries.

The foam struts at the lateral specimen boundary were unconstraint. Similarly to other studies [31–33,36,43], compressive loads were applied in the model by using rigid plates. Contact was defined between the foam specimen and rigid plates. The modelled specimens were smaller than those that were used for experiments. However the boundary conditions seem to have been sufficient in capturing the strut behaviour for strains up to 0.3. The effect of surrounding material at the boundary might have been effectively negligible for up to the strain of interest due to the high porosity of the foam. However, more sophisticated boundary conditions might be required for achieving better accuracy beyond 0.3 strain, or for lower porosity foams. The surrounding cell structures could affect the computed region with bending moments, forces, or contacts between the struts, particularly as the foam densifies. These effects could potentially be taken into account by considering periodic boundary conditions [11,12]. However, this type of boundary condition requires the geometry in the model to be periodic and, therefore, might be more difficult to apply in models of stochastic foam geometries.

The finite element model over-predicted stresses, beyond 0.3 strain. Figure 10 shows an example of the deformed modelled specimen at 0.3 strain. As the specimen is compressed, struts that were initially away from the boundary, might then deform and come into contact with the loading plates at the boundary of the specimen. This could cause an increase in the stress response. This could arguably also occur during experiments; however, the model size is considerably smaller than the specimen size in experiments; therefore, the effect of these interactions would be more pronounced in the finite element model simulation. A mitigating approach could be to selectively enable contacts between the loading platens and parts of the foam, i.e., only applying contacts to the nodes on the boundary of the foam rather than the whole specimen. Increasing the model domain size could also improve results. However, a larger model would also increase the computational cost. A damage model was not included in this study. The inclusion of a damage model could potentially improve the accuracy at higher stains.

Analysing the models up to the densification region using implicit FE methods, with the periodic boundary conditions or with larger domains, remains a challenge. Additionally, investigating the effect of strut length and cross-section on the buckling behaviour of struts and the effect of the cell size variation on the linearity of the stress-strain response could inform manufacturing processes for future products. These would be the topics of future work.

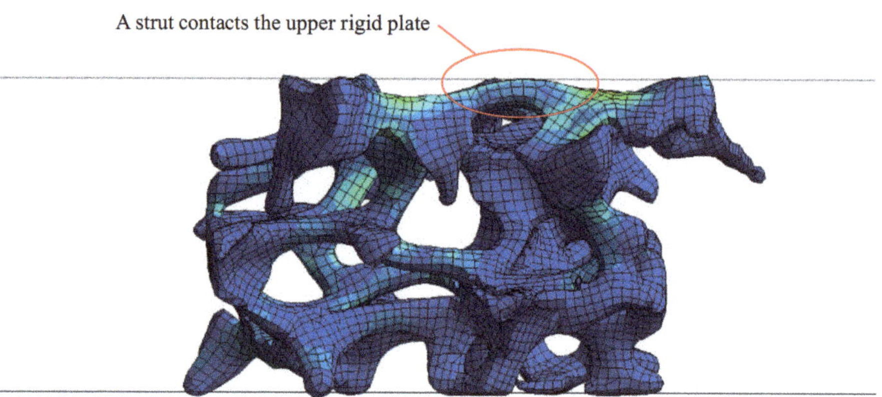

Figure 10. Deformed shape of the model A at the strain of 0.30, when a strut contacts the upper rigid wall.

5. Conclusions

Polyurethane foam specimens, which were intended for automotive seat pads, were scanned using X-ray computed tomography. The scans were converted to 3D CAD models and used to simulate uniaxial compression test using the finite element method. The methodology for the scanning and analyses was described, and the analysis results were compared with the experiments. All three numerical models sufficiently captured the material behaviour in the linear elastic and plateau region of the stress-strain curve. The conclusions for this study are summarised below:

- The investigated foams were scanned by X-ray computed tomography and their structures were captured in 2D cross-section images.
- The observed cross-section images were converted to 3D CAD models using Image J and Autodesk, Inc software products. The smoothed CAD models were analysed with commercial FEA software (Ansys).
- Foam specimens were experimentally tested under uniaxial compression.
- Specimen deformations were analysed by the implicit finite element method with the hexahedron and tetrahedron mixed meshing.
- The mechanical behaviour of foam specimens under compressive loading was sufficiently captured at 0.25 nominal strain and within a reasonable error margin.

The presented method was successfully used to analyse foam structures and it provided a tool in understanding the mechanism of compressive deformations in polyurethane foams. Commercial CAD products and open source software were used for creating a solid mesh for FE analysis from X-ray scans. The chosen approach was perhaps more efficient by comparison to alternative specialised software at a higher cost or in-house development of custom tools. The dependence of the foam macroscopic mechanical behaviour on the microstructural features can now be further investigated to inform manufacturing processes for future polyurethane foam products.

Author Contributions: Conceptualization, M.I.; methodology, M.I. and R.G.; validation, M.I., P.S., B.S. and N.M.; formal analysis, M.I.; investigation, M.I.; writing—original draft preparation, M.I.; writing—review and editing, M.I., R.G., P.S., B.S. and N.M.; supervision, P.S., B.S. and N.M.; project administration, N.M. All authors have read and agreed to the published version of the manuscript.

Funding: This research received no external funding.

Institutional Review Board Statement: Not applicable.

Informed Consent Statement: Not applicable.

Data Availability Statement: Data sharing is not applicable to this article.

Acknowledgments: The authors kindly acknowledge Bridgestone Corporation, which supplied the specimens for the research.

Conflicts of Interest: The authors declare no conflict of interest.

Abbreviations

The following abbreviations are used in this manuscript:

CT	Computed tomography
FEA	Finite element analysis
PUF	Polyurethane foam

References

1. Gibson, L.J.; Ashby, M.F. The mechanics of foams: Basic results. In *Cellular Solids: Structure and Properties*, 2nd ed.; Cambridge Solid State Science Series; Cambridge University Press: Cambridge, UK, 1997; pp. 175–234. [CrossRef]
2. Zhu, H.X.; Knott, J.F.; Mills, N.J. Analysis of the elastic properties of open-cell foams with tetrakaidecahedral cells. *J. Mech. Phys. Solids* **1997**, *45*, 319–343. [CrossRef]
3. Zhu, H.X.; Mills, N.J.; Knott, J.F. Analysis of the high strain compression of open-cell foams. *J. Mech. Phys. Solids* **1997**, *45*, 1875–1899. [CrossRef]
4. Mills, N.J.; Zhu, H.X. The high strain compression of closed-cell polymer foams. *J. Mech. Phys. Solids* **1999**, *47*, 669–695. [CrossRef]
5. Zhu, H.X.; Mills, N.J. Modelling the creep of open-cell polymer foams. *J. Mech. Phys. Solids* **1999**, *47*, 1437–1457. [CrossRef]
6. Laroussi, M.; Sab, K.; Alaoui, A. Foam mechanics: Nonlinear response of an elastic 3D-periodic microstructure. *Int. J. Solids Struct.* **2002**, *39*, 3599–3623. [CrossRef]
7. Gong, L.; Kyriakides, S.; Jang, W.Y. Compressive response of open-cell foams. Part I: Morphology and elastic properties. *Int. J. Solids Struct.* **2005**, *42*, 1355–1379. [CrossRef]
8. Gong, L.; Kyriakides, S. Compressive response of open cell foams part II: Initiation and evolution of crushing. *Int. J. Solids Struct.* **2005**, *42*, 1381–1399. [CrossRef]
9. Gong, L.; Kyriakides, S.; Triantafyllidis, N. On the stability of Kelvin cell foams under compressive loads. *J. Mech. Phys. Solids* **2005**, *53*, 771–794. [CrossRef]
10. Demiray, S.; Becker, W.; Hohe, J. Numerical determination of initial and subsequent yield surfaces of open-celled model foams. *Int. J. Solids Struct.* **2007**, *44*, 2093–2108. [CrossRef]
11. Okumura, D.; Okada, A.; Ohno, N. Buckling behavior of Kelvin open-cell foams under [0 0 1], [0 1 1] and [1 1 1] compressive loads. *Int. J. Solids Struct.* **2008**, *45*, 3807–3820. [CrossRef]
12. Takahashi, Y.; Okumura, D.; Ohno, N. Yield and buckling behavior of Kelvin open-cell foams subjected to uniaxial compression. *Int. J. Mech. Sci.* **2010**, *52*, 377–385. [CrossRef]
13. Ye, W.; Barbier, C.; Zhu, W.; Combescure, A.; Baillis, D. Macroscopic multiaxial yield and failure surfaces for light closed-cell foams. *Int. J. Solids Struct.* **2015**, 60–70. [CrossRef]
14. Zhu, H.X.; Hobdell, J.R.; Windle, A.H. Effects of cell irregularity on the elastic properties of open-cell foams. *Acta Mater.* **2000**, *48*, 4893–4900. [CrossRef]
15. Zhu, H.X.; Hobdell, J.R.; Windle, A.H. EEects of cell irregularity on the elastic properties of 2D Voronoi honeycombs. *J. Mech. Phys. Solids* **2001**, *49*, 857–870. [CrossRef]
16. Zhu, H.X.; Windle, A.H. Effects of cell irregularity on the high strain compression of open-cell foams. *Acta Mater.* **2002**, *50*, 1041–1052. [CrossRef]
17. Zhu, W.; Blal, N.; Cunsolo, S.; Baillis, D. Micromechanical modeling of effective elastic properties of open-cell foam. *Int. J. Solids Struct.* **2017**, *115–116*, 61–72. [CrossRef]
18. Marvi-Mashhadi, M.; Lopes, C.S.; LLorca, J. Effect of anisotropy on the mechanical properties of polyurethane foams: An experimental and numerical study. *Mech. Mater.* **2018**, *124*, 143–154. [CrossRef]
19. Marvi-Mashhadi, M.; Lopes, C.S.; LLorca, J. Surrogate models of the influence of the microstructure on the mechanical properties of closed- and open-cell foams. *J. Mater. Sci.* **2018**, *53*, 12937–12948. [CrossRef]
20. Jang, W.Y.; Kraynik, A.M.; Kyriakides, S. On the microstructure of open-cell foams and its effect on elastic properties. *Int. J. Solids Struct.* **2008**, *45*, 1845–1875. [CrossRef]
21. Storm, J.; Abendroth, M.; Emmel, M.; Liedke, T.; Ballaschk, U.; Voigt, C.; Sieber, T.; Kuna, M. Geometrical modelling of foam structures using implicit functions. *Int. J. Solids Struct.* **2013**, *50*, 548–555. [CrossRef]
22. Storm, J.; Abendroth, M.; Zhang, D.; Kuna, M. Geometry dependent effective elastic properties of open-cell foams based on kelvin cell models. *Adv. Eng. Mater.* **2013**, *15*, 1292–1298. [CrossRef]
23. Zhang, D.; Abendroth, M.; Kuna, M.; Storm, J. Multi-axial brittle failure criterion using Weibull stress for open Kelvin cell foams. *Int. J. Solids Struct.* **2015**, *75–76*, 1–11. [CrossRef]
24. Storm, J.; Abendroth, M.; Kuna, M. Numerical and analytical solutions for anisotropic yield surfaces of the open-cell Kelvin foam. *Int. J. Mech. Sci.* **2016**, *105*, 70–82. [CrossRef]

25. Zhu, W.; Blal, N.; Cunsolo, S.; Baillis, D. Effective elastic properties of periodic irregular open-cell foams. *Int. J. Solids Struct.* **2018**, *143*, 155–166. [CrossRef]

26. Zhu, W.; Blal, N.; Cunsolo, S.; Baillis, D.; Michaud, P.M. Effective elastic behavior of irregular closed-cell foams. *Materials* **2018**, *11*, 2100. [CrossRef]

27. Storm, J.; Abendroth, M.; Kuna, M. Effect of morphology, topology and anisoptropy of open cell foams on their yield surface. *Mech. Mater.* **2019**, *137*, 103145. [CrossRef]

28. Jang, W.Y.; Kyriakides, S. On the crushing of aluminum open-cell foams: Part I. Experiments. *Int. J. Solids Struct.* **2009**, *46*, 617–634. [CrossRef]

29. Jang, W.Y.; Kyriakides, S. On the crushing of aluminum open-cell foams: Part II analysis. *Int. J. Solids Struct.* **2009**, *46*, 635–650. [CrossRef]

30. Jang, W.Y.; Kyriakides, S.; Kraynik, A.M. On the compressive strength of open-cell metal foams with Kelvin and random cell structures. *Int. J. Solids Struct.* **2010**, *47*, 2872–2883. [CrossRef]

31. Gaitanaros, S.; Kyriakides, S.; Kraynik, A.M. On the crushing response of random open-cell foams. *Int. J. Solids Struct.* **2012**, *49*, 2733–2743. [CrossRef]

32. Gaitanaros, S.; Kyriakides, S. On the effect of relative density on the crushing and energy absorption of open-cell foams under impact. *Int. J. Impact Eng.* **2015**, *82*, 3–13. [CrossRef]

33. Gaitanaros, S.; Kyriakides, S.; Kraynik, A.M. On the crushing of polydisperse foams. *Eur. J. Mech. A/Solids* **2018**, *67*, 243–253. [CrossRef]

34. Storm, J.; Abendroth, M.; Kuna, M. Influence of curved struts, anisotropic pores and strut cavities on the effective elastic properties of open-cell foams. *Mech. Mater.* **2015**, *86*, 1–10. [CrossRef]

35. Barnes, A.T.; Ravi-Chandar, K.; Kyriakides, S.; Gaitanaros, S. Dynamic crushing of aluminum foams: Part I—Experiments. *Int. J. Solids Struct.* **2014**, *51*, 1631–1645. [CrossRef]

36. Gaitanaros, S.; Kyriakides, S. Dynamic crushing of aluminum foams: Part II—Analysis. *Int. J. Solids Struct.* **2014**, *51*, 1646–1661. [CrossRef]

37. Yang, C.; Kyriakides, S. Multiaxial crushing of open-cell foams. *Int. J. Solids Struct.* **2019**, *159*, 239–256. [CrossRef]

38. Jones, A.C.; Arns, C.H.; Sheppard, A.P.; Hutmacher, D.W.; Milthorpe, B.K.; Knackstedt, M.A. Assessment of bone ingrowth into porous biomaterials using MICRO-CT. *Biomaterials* **2007**, *28*, 2491–2504. [CrossRef] [PubMed]

39. John, Ł.; Janeta, M.; Rajczakowska, M.; Ejfler, J.; Łydżba, D.; Szafert, S. Synthesis and microstructural properties of the scaffold based on a 3-(trimethoxysilyl)propyl methacrylate–POSS hybrid towards potential tissue engineering applications. *RSC Adv.* **2016**, *6*, 66037–66047. [CrossRef]

40. Munkholm, L.J.; Heck, R.J.; Deen, B. Soil pore characteristics assessed from X-ray micro-CT derived images and correlations to soil friability. *Geoderma* **2012**, *181–182*, 22–29. [CrossRef]

41. Patterson, B.M.; Henderson, K.; Gilbertson, R.D.; Tornga, S.; Cordes, N.L.; Chavez, M.E.; Smith, Z. Morphological and Performance Measures of Polyurethane Foams Using X-ray CT and Mechanical Testing. *Microsc. Microanal.* **2014**, *20*, 1284–1293. [CrossRef]

42. Ulrich, D.; van Rietbergen, B.; Weinans, H.; Rüegsegger, P. Finite element analysis of trabecular bone structure: A comparison of image-based meshing techniques. *J. Biomech.* **1998**, *31*, 1187–1192. [CrossRef]

43. Jeon, I.; Asahina, T.; Kang, K.J.; Im, S.; Lu, T.J. Finite element simulation of the plastic collapse of closed-cell aluminum foams with X-ray computed tomography. *Mech. Mater.* **2010**, *42*, 227–236. [CrossRef]

44. Zhang, L.; Ferreira, J.M.F.; Olhero, S.; Courtois, L.; Zhang, T.; Maire, E.; Rauhe, J.C. Modeling the mechanical properties of optimally processed cordierite–mullite–alumina ceramic foams by X-ray computed tomography and finite element analysis. *Acta Mater.* **2012**, *60*, 4235–4246. [CrossRef]

45. Natesaiyer, K.; Chan, C.; Sinha-Ray, S.; Song, D.; Lin, C.L.; Miller, J.D.; Garboczi, E.J.; Forster, A.M. X-ray CT imaging and finite element computations of the elastic properties of a rigid organic foam compared to experimental measurements: Insights into foam variability. *J. Mater. Sci.* **2015**, *50*, 4012–4024. [CrossRef]

46. Schindelin, J.; Arganda-Carreras, I.; Frise, E.; Kaynig, V.; Longair, M.; Pietzsch, T.; Preibisch, S.; Rueden, C.; Saalfeld, S.; Schmid, B.; et al. Fiji: An open-source platform for biological-image analysis. *Nat. Methods* **2012**, *9*, 676–682. [CrossRef]

47. Rueden, C.T.; Schindelin, J.; Hiner, M.C.; DeZonia, B.E.; Walter, A.E.; Arena, E.T.; Eliceiri, K.W. ImageJ2: ImageJ for the next generation of scientific image data. *BMC Bioinform.* **2017**, *18*, 529. [CrossRef] [PubMed]

48. Otsu, N. A Threshold Selection Method from Gray-Level Histograms. *IEEE Trans. Syst. Man Cybern.* **1979**, *9*, 62–66. [CrossRef]

49. *Ansys® Academic Research Meshing, Release 19.2, Help System, Meshing User's Guide*; ANSYS, Inc.: Canonsburg, PA, USA, 2010.

50. *Ansys® Academic Research Mechanical, Release 19.2, Help System, Mechanical User's Guide*; ANSYS, Inc: Canonsburg, PA, USA, 2010.

51. *Polymeric Materials, Cellular Flexible—Determination of Stress-Strain Characteristics in Compression—Part 1: Low-Density Materials (ISO Standard No. 3386-1:1986)*; International Organization for Standardization: Geneva, Switzerland, 1986.

materials

MDPI

Article

Impact of Different Epoxidation Approaches of Tall Oil Fatty Acids on Rigid Polyurethane Foam Thermal Insulation

Arnis Abolins [1], Ralfs Pomilovskis [1,2], Edgars Vanags [1], Inese Mierina [2], Slawomir Michalowski [3], Anda Fridrihsone [1] and Mikelis Kirpluks [1,*]

[1] Polymer Laboratory, Latvian State Institute of Wood Chemistry, Dzerbenes St. 27, LV-1006 Riga, Latvia; arnisaabolins@gmail.com (A.A.); ralfs.pomilovskis@gmail.com (R.P.); edgars.vanags6@gmail.com (E.V.); anda.fridrihsone@edi.lv (A.F.)

[2] Institute of Technology of Organic Chemistry, Faculty of Materials Science and Applied Chemistry, Riga Technical University, P. Valdena 3/7 St., LV-1048 Riga, Latvia; inese.mierina@rtu.lv

[3] Department of Chemistry and Technology of Polymers, Cracow University of Technology, Warszawska 24, 31-155 Cracow, Poland; slawomir.michalowski@pk.edu.pl

* Correspondence: mkirpluks@gmail.com

Citation: Abolins, A.; Pomilovskis, R.; Vanags, E.; Mierina, I.; Michalowski, S.; Fridrihsone, A.; Kirpluks, M. Impact of Different Epoxidation Approaches of Tall Oil Fatty Acids on Rigid Polyurethane Foam Thermal Insulation. *Materials* **2021**, *14*, 894. https://doi.org/10.3390/ma14040894

Academic Editor: Andrea Spagnoli

Received: 30 December 2020
Accepted: 10 February 2021
Published: 13 February 2021

Publisher's Note: MDPI stays neutral with regard to jurisdictional claims in published maps and institutional affiliations.

Abstract: A second-generation bio-based feedstock—tall oil fatty acids—was epoxidised via two pathways. Oxirane rings were introduced into the fatty acid carbon backbone using a heterogeneous epoxidation catalyst-ion exchange resin Amberlite IR-120 H or enzyme catalyst *Candida antarctica* lipase B under the trade name Novozym® 435. High functionality bio-polyols were synthesised from the obtained epoxidated tall oil fatty acids by oxirane ring-opening and subsequent esterification reactions with different polyfunctional alcohols: trimethylolpropane and triethanolamine. The synthesised epoxidised tall oil fatty acids (ETOFA) were studied by proton nuclear magnetic resonance. The chemical structure of obtained polyols was studied by Fourier-transform infrared spectroscopy and size exclusion chromatography. Average molecular weight and polydispersity of polyols were determined from size exclusion chromatography data. The obtained polyols were used to develop rigid polyurethane (PU) foam thermal insulation material with an approximate density of 40 kg/m^3. Thermal conductivity, apparent density and compression strength of the rigid PU foams were determined. The rigid PU foams obtained from polyols synthesised using Novozym® 435 catalyst had superior properties in comparison to rigid PU foams obtained from polyols synthesised using Amberlite IR-120 H. The developed rigid PU foams had an excellent thermal conductivity of 21.2–25.9 mW/(m·K).

Keywords: tall oil fatty acids; ion-exchange resin; lipase enzyme catalyst; high functionality polyols; rigid polyurethane foam

1. Introduction

Due to environmental concerns, new regulatory policies and a shift in consumer requirements, renewable resources for polymer materials have been widely studied in the last decade [1]. In the past, as well as nowadays, fatty acids present in plants in the form of triglycerides are one of the most appropriate raw products for the manufacturing of bio-based materials [2]. Various chemical modification methods, such as epoxidation and ring-opening, hydroformylation, transesterification, ozonolysis, amidation and thiol-ene coupling [3], have been used to obtain various monomers and polymers with low toxicity, convenient availability and relatively low price [4–7]. Epoxides are one of the most versatile intermediates to be further used for the synthesis of different compounds and has a wide commercial use because of its high reactivity [8].

Numerous epoxidation methods of different kind of vegetable oils, such as canola-rapeseed [9,10], palm [11], cottonseed [12], soybean [13], castor [14], linseed [15], mahua [16] and grape seed [17] among others, have been reported. In addition, non-edible plant oils, such as tall oil [18–20] and jatropha [21,22], have been successfully epoxidised previously. However, the use of vegetable oil for industrial uses is in direct competition with food and feed production. As mentioned, there are sources of plant-derived fatty acids that do not compete, such as tall oil [23,24].

Tall oil fatty acids (TOFA) are important renewable feedstock, which is obtained as a side stream from the softwood Kraft pulping process. After fractional distillation of crude tall oil, a more pure form of TOFA is obtained containing at least 97% free fatty acids (mainly a mixture of 48–52% of oleic and 43–48% of linoleic acid) and less than 3% other components, such as rosin acids and unsaponifiables [25–28]. The relatively high level of unsaturation makes TOFA a suitable raw material for the introduction of reactive functional group using the unsaturated hydrocarbon C=C double bonds, making them suitable for further processing into polymers [29].

The most widely used epoxidation method is the well-known Prilezhaev reaction where peracid is reacted with olefins [20,30–32]. Peracids are conventionally formed in situ from hydrogen peroxide and short-chain carboxylic acid in the presence of highly acidic catalysts [32,33]. However, during epoxidation, acidic catalysts, such as sulphuric acid and acidic ion exchange resins, and carboxyl acids as oxygen carriers, such as formic or acetic acid, exacerbate undesirable side product formation through oxirane rings [18,20,34]. Moreover, the higher the temperature of the epoxidation reaction, the greater the frequency of side reactions [20]. Studies previously carried out by our group showed that if free fatty acids containing unprotected carboxyl groups, as they are in TOFA, are epoxidised, the use of acidic components has even more significance for side reactions occurrence [18,20]. To avoid the formation of side products, other epoxidation routes have to be explored.

Some studies indicate that chemo-enzymatic epoxidation, where acidic catalysts are replaced by lipase, overall is a milder route to free fatty acid conversion into epoxides [19] than the well-known Prilezhaev rection. The advantages of chemo-enzymatic epoxidation are lower reaction temperature [35], the absence of acidic catalysts [36,37] and even lipase reusability [36], which if all taken in to account can result in substantially higher oxirane ring introduction into the substrate. In addition, lipases are highly selective to limit the frequency of side reactions [38]. Moreover, it is possible to prevent the use of additional oxygen carriers if free fatty acids are chemo-enzymatically epoxidised. Lipases can turn free fatty acids into highly reactive peroxy fatty acids, which subsequently epoxidise unsaturated bonds [39], thus improving the feasibility of reaction as there are no acidic catalysts and additional oxygen carriers needed. However, different factors, such as solvent, temperature, pH and presence of activators or deactivators etc., can influence the activity of lipases [40].

After epoxidation, one of the most easily obtainable functional groups are hydroxyl groups, which are essential for polyurethane (PU) production. PUs are a class of polymers that are commonly used in a wide variety of applications to produce high-performance materials. The primary uses for PUs are flexible and rigid foams, sealants, elastomers, adhesives, and coatings [30,41–47]. Usually, PUs are obtained by polycondensation reaction between isocyanates and hydroxyl group containing compounds [48–50]. Hydroxyl group compounds can be polyols with low, medium or high functionality with low, medium or high hydroxyl values (OH value), respectively. Polyols with high average hydroxyl group functionality are needed for the production of rigid PU foams to ensure high dimensional, mechanical and thermal stability of the material [51]. A combination of epoxy ring-opening and transesterification or transamidation of fatty acids with polyfunctional alcohols can lead to such polyols, which would contain primary OH groups to ensure high cross-link density of obtained PU polymer matrix [52–54].

The goal of this study was to compare two different TOFA epoxidation catalysts—ion exchange resin Amberlite IR-120 H and enzymatic catalyst *Candida antarctica* lipase B with a trade name Novozym® 435—and their influence on the properties of resulting polyols and rigid PU foams. In this study, a second-generation bio-based feedstock—TOFA—was epoxidised via two pathways resulting in two different epoxidised tall oil fatty acids (ETOFA). Afterwards, two different polyols were developed using the two different ETOFA and employing oxirane ring-opening and subsequent esterification reactions with two different polyfunctional alcohols (trimethylolpropane (TMP) and triethanolamine (TEOA)). The four developed polyols were used to obtain rigid PU foam thermal insulation material. Its common characteristics, such as thermal conductivity, apparent density and compression strength, were analysed and compared.

2. Materials and Methods

2.1. Materials

TOFA (trade name "FOR2") with a high content of fatty acids (>96%), low content of rosin acids (1.9%) and unsaponifiables (1.8%) was ordered from Forchem Oyj (Rauma, Finland). Glacial acetic acid (AcOH), puriss, $\geq$99.8%; hydrogen peroxide (H_2O_2), purum p.a., $\geq$35%; acetanhydride, puriss, $\geq$99%; 4-(dimethylamino)pyridine (DMAP), reagent plus, $\geq$99%; N,N-dimethylformamide (DMF), ACS reagent, $\geq$99.8%, water content $\leq$150 ppm; potassium hydroxide, puriss, $\geq$85%; potassium iodide, ACS reagent, $\geq$99%; tetraethylammonium bromide, reagent grade, 98%; perchloric acid, ACS reagent, 70%; dichloromethane, puriss p.a., ACS reagent; anhydrous sodium sulphate, puriss; TMP, reagent grade, 97%, were ordered from Sigma-Aldrich (Schnelldorf, Germany). Amberlite IR-120 H, strongly acidic, hydrogen form and sodium thiosulphate fixanals 0.1 M and Hanus solution, volumetric 0.1 M IBr were ordered from Fluka (Steinheim, Germany). Lipase Novozym® 435 (immobilised on acrylic resin) was kindly supplied by Novozymes A/S (Bagsvaerd, Denmark). Tetrafluoroboric acid solution, 48 wt.% in H_2O (HBF_4), was ordered from Alfa Aesar (Kandel, Germany). TEOA, 99.2%, was ordered from Huntsman (Rotterdam, The Netherlands), and was used as purchased.

For the development of rigid PU foams, the following materials were used as purchased: two tertiary amine-based catalysts Polycat® 5, Polycat® NP10 as well as 30 wt.% of potassium acetate in diethylene glycol (PC CAT TKA 30) (Air Products and Chemicals Inc., Halfweg, The Netherlands); Niax Silicone L-6915 as a surfactant (Momentive Performance Materials Inc., Rotterdam, Germany); tris (1-chloro-2-propyl phosphate 99% (TCPP) as a flame retardant (Albermarle, Louvain-la-Neuve, Belgium)) and cyclopentane as a physical blowing agent (Sigma-Aldrich, Schnelldorf, Germany). Desmodur 44V20 L was purchased from (Covestro, Krefeld, Germany), and was used as the isocyanate component for all PU materials. It is a solvent-free product based on 4,4′-diphenylmethane diisocyanate (pMDI) and contains oligomers of high functionality. The average functionality is 2.8–2.9 and the isocyanate group (–NCO) content of 30.5–32.5 wt.%.

2.2. Epoxidation of TOFA with Ion Exchange Resin Amberlite IR-120 H

The epoxidation of TOFA was carried out in a four-necked round bottom flask. A thermocouple, mechanical stirrer, dropping funnel and a reflux condenser were attached to the necks of the flask. The epoxidation of TOFA was achieved by in-situ generated peroxyacetic acid, which forms from acetic acid and hydrogen peroxide in the presence of an acidic catalyst. During the epoxidation, the molar ratio of TOFA (double bonds-155 g $I_2/100$ g) to H_2O_2 and AcOH was 1.0:1.5:0.5. At first, the calculated amount of TOFA (700.0 g), acetic acid (128.5 g) and ion exchange resin Amberlite IR-120 H (140.0 g, 20 wt.% of TOFA weight), as the catalyst, was added to the flask. The flask was immersed in a thermostatic water bath (preheated to 40 °C). The speed of the mechanical stirrer was set to 600 rpm, and the mixture was started to stir. A hydrogen peroxide/water (35%/65%) solution (638.5 g) was poured into a dropping funnel. When the content of the flask reached 40 °C, hydrogen peroxide solution was added dropwise to the round bottom flask in a time interval of 30 min. Meanwhile, the temperature of the reaction medium was slowly increased to 60 °C. After the complete addition of hydrogen peroxide, the reaction medium was continued to stir for 6 h at 60 °C and 600 rpm [18]. Afterwards, the reaction mixture was poured into a separating funnel and washed four times with warm (T = 60 °C) distilled water. The product was dried using a rotatory vacuum evaporator to remove water residues. As a result, ETOFA were obtained exhibiting the acid value of 144 mg KOH/g, oxirane content of 2.30 mmol/g and iodine value of 27.0 g I_2/100 g and characterised by pomegranate red colour. An acronym ETOFA_IR is used for the ETOFA synthesised via TOFA epoxidation with ion exchange resin Amberlite IR-120 H.

2.3. Epoxidation of TOFA with Novozym® 435

The epoxidation of TOFA was carried out in a four-necked round bottom flask. The flask was immersed in a water bath and equipped with a mechanical stirrer, a reflux condenser, a thermocouple and a dropping funnel. Using the data that was obtained from the previous study [55], the optimal epoxidation parameters were determined and used for TOFA epoxidation.

During the epoxidation, the molar ratio of double bond present in TOFA and H_2O_2 was 1.0:2.0. The required amount of TOFA (700.0 g) was poured in the flask, and necessary amount of *Candida antarctica* lipase B (Novozym® 435) (22.4 g, 3.2 wt.% of TOFA weight) was added. The mixture of TOFA and catalyst was heated to 44 $\pm$ 0.1 °C. Afterwards, the necessary amount of 32% H_2O_2 water solution (1038.2 g) was added to the reactants dropwise through the dropping funnel. The rate of addition speed was adjusted so that the whole peroxide was added within 30 min. After the complete addition of hydrogen peroxide, the reaction medium was continued to stir for 5.5 h at 44 $\pm$ 0.1 °C and 500 rpm. Afterwards, the reaction mixture was poured into a separating funnel and washed four times with warm (T = 60 °C) distilled water. The product was dried using a rotatory vacuum evaporator to remove water residues. As a result, ETOFA were obtained exhibiting the acid value of 146 mg KOH/g, oxirane content of 3.28 mmol/g and iodine value of 17.0 g I_2/100 g and characterised by pomegranate red colour. An acronym ETOFA_E is used for the ETOFA obtained from TOFA epoxidation with lipase catalyst Novozym® 435.

2.4. Synthesis of Polyols Using Two Different ETOFA

High functionality polyols were synthesised by functionalising ETOFA_IR or ETOFA_E, which were obtained by epoxidising TOFA using two epoxidation catalysts either Amberlite IR-120 H or Novozym® 435. Bio-based polyols were synthesised by opening the oxirane ring of ETOFA_IR or ETOFA_E and subsequent esterification with various polyfunctional alcohols, such as TMP and TEOA. For bio-polyols synthesised from intermediate ETOFA_IR, the following acronyms were used ETOFA_TMP_IR and ETOFA_TEOA_IR depending on the used polyfunctional alcohol for epoxide ring-opening. For polyols obtained from TOFA epoxidation with lipase catalyst Novozym® 435, the following acronyms were used ETOFA_TMP_E and ETOFA_TEOA_E. The general scheme of bio-polyol development is depicted in Figure 1.

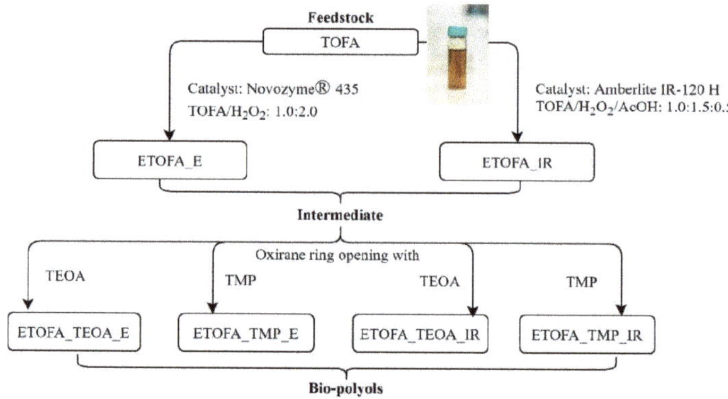

Figure 1. Scheme of developed tall oil fatty acids (TOFA)-based bio-polyols and intermediates.

Due to the two different chemical processes that are carried out simultaneously, namely, oxirane ring-opening reaction and esterification reaction, the molar amount of the polyfunctional alcohol needed for polyol synthesis is calculated using the following equation:

$$n_{MP} = n_{KG} + n_{EG},\qquad(1)$$

where n_{MP} is the molar amount of the polyfunctional alcohol used for oxirane ring-opening and esterification reaction (TMP or TEOA), in mol; n_{KG} is the molar amount of the ETOFA carboxylic groups, in mol and n_{EG} is the molar amount of the ETOFA oxirane groups, in mol.

To obtain TOFA-based bio-polyols, the oxirane ring-opening was first carried out in the four-necked round bottom flask. The multifunctional alcohol/amine and tetrafluoroboric acid solution, 48 wt.% in H_2O and as a catalyst, 0.4 wt.% of ETOFA mass, was added, see Table 1 for corresponding mass for each type of polyols. The flask was immersed into an oil thermobath, and a mechanical stirrer was inserted into the central neck. A purge gas tube, Liebig condenser, and a dropping funnel were attached to the vacant necks. The mixer was set to 400 rpm, and the flow of purge gas (argon) through the flask was provided, while the content of the flask was heated up to 120 °C. When the required temperature of 120 °C for oxirane ring-opening was reached, 200 g of ETOFA were added dropwise to the flask in a time interval of 20 min. After the complete addition of ETOFA, the reaction medium was continued to stir for 30 min at the temperature of 120 °C to open the oxirane rings completely. Afterwards, the synthesis temperature was increased to carry out the esterification reactions. The ETOFA_TMP_IR and ETOFA_TMP_E polyol synthesis were carried out at 200 °C, whereas ETOFA_TEOA_IR and ETOFA_TEOA_E synthesis were carried out at 180 °C (Figure 2). The stirring of the reaction medium and the argon gas flow was retained until the acid value of the product decreased below 5 mg KOH/g [54]. After which, the synthesis was considered to be finished, and the bio-polyol was obtained for further rigid PU foam development.

Table 1. Multifunctional alcohols/amines used in the bio-polyol syntheses.

Multifunctional Alcohol for Polyol Synthesis	Mass (g) Added if Polyol is Synthesised from ETOFA_IR	Mass (g) Added if Polyol is Synthesised from ETOFA_E
TEOA	145.0	175.4
TMP	130.4	157.8

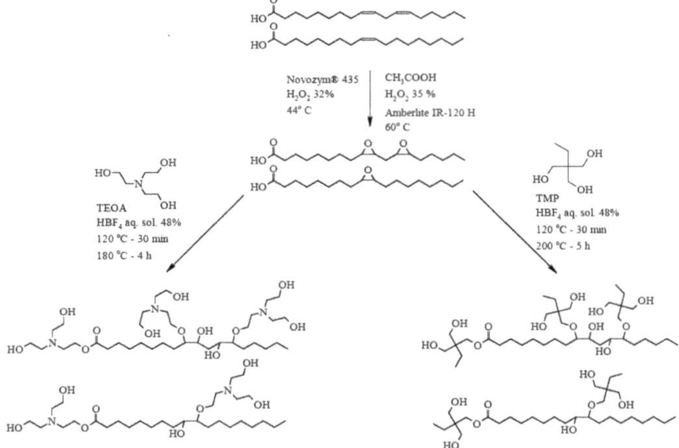

Figure 2. Idealised chemical structure of developed bio-polyols.

2.5. Characterisation of Products and Precursors

The obtained bio-polyols were characterised by hydroxyl and acid values calculated according to ISO 4629-2:2016 and ISO 2114:2000 testing standards using titrimetric methods. The epoxy content was calculated in accordance with ASTM D1652-04:2004. The viscosity of polyols was measured at 25 °C using the Thermo Science HAAKE (Medium-High Range Rotational Viscometer, Thermo Fisher Scientific, Waltham, MA, USA). Polyol density was determined using a series of hydrometers. In a thermostatic bath at 20 °C, a graduated cylinder filled with polyol was immersed for 20 min, and afterwards, the density was measured. The moisture content was measured using the Denver Instrument Model 275KF automatic titrator (Denver Instrument, Bohemia, NY, USA) using Karl Fisher titration.

Polyol structure was analysed using Fourier-transform infrared spectrometry data (FTIR), which were obtained with a Thermo Scientific Nicolet iS50 spectrometer (Thermo Fisher Scientific, Waltham, MA, USA) at a resolution of 4 cm^{-1} (32 scans). The FTIR data were collected using attenuated total reflectance technique with ZnSe and diamond crystals. Moreover, ^{1}H NMR spectra for the samples were recorded on a Bruker spectrometer (Bruker BioSpin AG, Fällanden, Switzerland) at 500 MHz. The chemical shifts (δ) are reported in ppm. The residual chloroform peak was used as an internal reference (δ = 7.26 ppm). Size exclusion chromatography from Knauer equipped with refractive index detector (Detector RI) and polystyrene/divinylbenzene matrix gel column with a measurement range up to 30,000 Da at tetrahydrofuran (THF) eluent flow of 1.0 mL/min was used to analyse the number-average molecular weight (M_n) and number-average functionality (f_n) of the synthesised bio-polyols. The polyols f_n was calculated based on hydroxyl values and M_n as seen from Equation (2) [52].

$$f_n = M_n \times OH_{value} \times 56{,}110^{-1}, \tag{2}$$

where f_n is the number-average functionality, OH groups/mol; M_n is the number-average molecular weight, g/mol; OH_{valuel} is the hydroxyl value of the polyol, mg KOH/g and 56,100 is equivalent weight of KOH, in milligrams.

2.6. Rigid PU foam Preparation and Characterisation

For rigid PU foam development, ETOFA-based bio-polyols were used. For the PU foams obtained from bio-based polyols synthesised from ETOFA using Amberlite IR-120 H and polyfunctional alcohols like TMP and TEOA, the following acronyms PU_TMP_IR and PU_TEOA_IR were used. Similarly, for the PU foams obtained from bio-based polyols synthesised from ETOFA using Novozym® 435 and polyfunctional alcohols like TMP and TEOA, the following acronyms PU_TMP_E and PU_TEOA_E were used. In Table 2, the developed rigid PU foam formulations are depicted. In order to obtain rigid PU foams using newly synthesised bio-polyols a previously synthesised bio-polyol with lower functionality based on tall oil (TO) esterification with TEOA was also used (TO_TEOA with OH value of 334 mg KOH/g, the water content of 0.45 wt.%, the viscosity of 280 mPa·s at 25 °C, f_n = 2.4 and M_n = 391 g/mol). The polyol component was obtained by weighing all the required components presented in Table 2 (polyols, blowing agent, flame retardant, catalysts and surfactant) and stirring them with a mechanical stirrer at 2000 rpm for 1 min. The polyol system was then conditioned in a sealed container at room temperature for at least 2 h to de-gas the mixed air, and afterwards, PU foams were prepared. The isocyanate index was chosen to be 150 for all PU foams. To produce rigid PU foams, isocyanate (pMDI) and a polyol portion were weighed and mixed at 2000 rpm for 15 s with a mechanical stirrer. After that, the reactive mixture was poured into an open-top mould [54].

Table 2. Polyurethane (PU) foam formulations and renewable material content in PU foams.

Components		Reagents, Parts by Weight			
		PU_TEOA_IR	PU_TEOA_E	PU_TMP_IR	PU_TMP_E
Bio-polyols	ETOFA_TEOA_IR	85.0	–	–	–
	ETOFA_TEOA_E	–	85.0	–	–
	ETOFA_TMP_IR	–	–	85.0	–
	ETOFA_TMP_E	–	–	–	85.0
	TO_TEOA	10.0	10.0	10.0	10.0
	Glycerol	5.0	5.0	5.0	5.0
Blowing agents	c-pentane	12.0	12.0	12.0	10.0
	Water	2.0	2.0	2.0	2.0
Catalysts	Polycat$^®$ 5	–	–	0.5	0.5
	Polycat$^®$ NP10	3.0	3.0	3.0	3.0
	PC CAT TKA 30	1.5	1.5	1.5	1.5
Surfactant	L-6915	2.5	2.5	2.5	2.5
Flame retardant	TCPP	31 (8 wt.%)	30 (8 wt.%)	32 (8 wt.%)	27 (8 wt.%)
Isocyanate	pMDI	243.4	218.3	246.2	190.5
Isocyanate index	–	150	150	150	150
Renewable materials in PU foam, %		16.3	18.1	16.2	19.8

The content of renewable materials was determined based on the mass of renewable materials used in the formulation of the PU foams. The rigid PU foam formulations were designed to obtain foams with an apparent density of ~40 kg/m^3. The physical and mechanical properties of the foams were measured in accordance with the following standards: foam density—ISO 845:2009, closed cell content—ISO 4590:2003, compression strength—ISO 844:2009 and thermal conductivity—ISO 8301:1991. The compression strength of the PU foams was tested parallel and perpendicular to foam rise with one offset from ISO 844:2009 standard—sample size; cylinders with a diameter of 20 mm and a height of 22 mm were tested. The mechanical testing of PU foams was done using Zwick/Roell 1000 N testing machines (Zwick Roell Group, Ulm, Germany)

3. Results and Discussion

3.1. Characteristics of Synthesised High Functionality Polyols

The key characteristics of synthesised bio-polyols, such as OH value, acid value, moisture content, viscosity, density, f_n, M_n and polydispersity (p_d) are summarised in Table 3.

Table 3. Common characteristics of developed tall oil fatty acids (TOFA)-based bio-polyols.

Polyol	OH Value, mg KOH/g	Acid Value, mg KOH/g	Moisture, %	Viscosity (20 °C), mPa·s	Density (20 °C), g/cm^3	f_n	M_n, g/mol	p_d
ETOFA_TEOA_IR	510	<5	0.14	7400	1.047	8.1	893	1.78
ETOFA_TEOA_E	519	<5	0.05	10,800	1.048	8.3	899	1.88
ETOFA_TMP_IR	427	<5	0.24	77,000	1.056	9.7	1279	1.58
ETOFA_TMP_E	335	<5	0.06	278,300	1.058	12.6	2112	1.57

The synthesised TOFA bio-polyols have high OH values and low acid value, which is typical and necessary for polyols to be used for rigid PU foam production. Unfortunately, the bio-polyols that were synthesised using TMP reached very high viscosity. It can be explained by the intermolecular hydrogen bonding of the polyol chemical structure as well as the higher molecular weight of the polyols. The f_n of synthesised bio-polyols is high as it ranges from 8.1 to almost 12.6. Polyols with such high f_n are rarely used as the only polyol in rigid PU foam formulation. Therefore, such polyols' main application is their use as cross-linking reagents in rigid PU foam development.

3.2. FTIR Analysis of Synthesised Bio-Polyols from ETOFA

The FTIR spectra of TOFA, ETOFA_IR and ETOFA_E are depicted in Figure 3a.

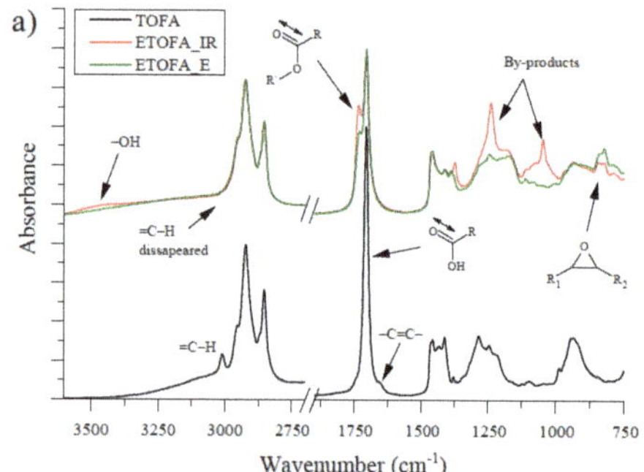
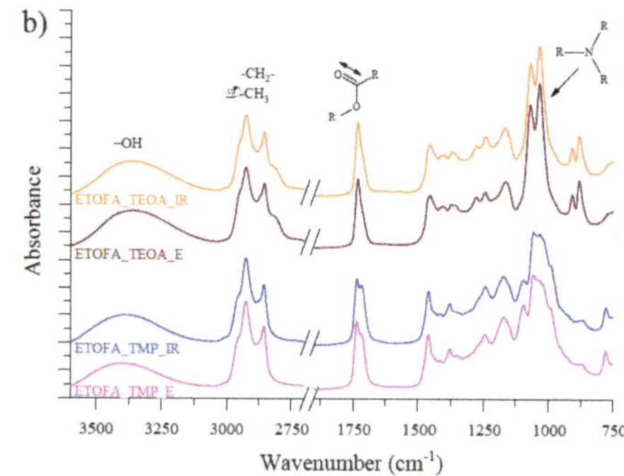

Figure 3. FTIR spectra: (**a**) spectra of TOFA, epoxidised tall oil fatty acids synthesised via TOFA epoxidation with ion exchange resin Amberlite IR-120 H (ETOFA_IR) and ETOFA obtained from TOFA epoxidation with lipase catalyst Novozym® 435 (ETOFA_E) and (**b**) spectrum of the resulting TOFA-based bio-polyols.

The oxirane ring stretching vibration peak appeared at 823 cm^{-1} after epoxidation of TOFA. Furthermore, for ETOFA_E, its intensity is higher when compared with ETOFA_IR, which correlates with the determined oxirane content from titrimetric analysis data of 3.28 and 2.30 mmol/g, respectively. The introduction of oxirane rings coincides with the disappearance of =C–H and –C=C– absorption peaks at 3009 and 1659 cm^{-1}, respectively. The side reactions of oxirane ring cleavage during TOFA epoxidation occurred for both epoxidation methods and are confirmed by the occurrence of a C=O ester stretching shoulder peak for the both ETOFA intermediate products at 1736 cm^{-1}. In case of ETOFA_IR, its intensity is higher as the oxirane ring cleavage side reactions could occur with COOH groups of TOFA as well as acetic acid present in reaction medium among other reactants. The side reaction occurrence was also confirmed by the appearance of the broad –OH stretching vibration peak between 3600 and 3100 cm^{-1}, which for ETOFA_IR was more intensive than ETOFA_E. In the case of ETOFA_IR, the C–O bond absorption peaks at 1240 and 1046 cm^{-1} of various by-products were also more distinguishable. It can be observed that the Novozym® 435 is more selective epoxidation catalyst, and ETOFA_E could be better suited for polyol development.

When TOFA-based bio-polyols were obtained (Figure 3b), the oxirane ring stretching vibration peak at 823 cm^{-1} disappeared. This correlates with a titrimetric analysis of oxirane content in all TOFA-based bio-polyols, which was lower than 0.01 mmol/g. These kinds of changes in FTIR spectra confirm conversion towards the desired bio-polyol products. A very noticeable distinction between bio-polyols synthesised from TMP and TEOA is the C–N stretching peak of a tertiary amine group that appeared at ~1036 cm^{-1} when oxirane ring was opened with TEOA. The tertiary amine group introduction in chemical structure could provide the developed bio-polyol with autocatalytic properties when rigid PU foam is developed. For all four polyols, the carboxylic group C=O stretching vibration at 1707 cm^{-1} of TOFA shifted towards C=O stretching ester vibration at ~1733 cm^{-1}. For all feedstock, TOFA and intermediates, both ETOFAs and four obtained polyols a typical symmetric and asymmetric –CH$_2$– stretching peaks were observed at ~2930 and 2860 cm^{-1}. The broad absorption peak between 3600 and 3100 cm^{-1} of all spectra of synthesised bio-polyols was identified as characteristic stretching vibrations of the O–H group. The intensity of peak between 3600 and 3100 cm^{-1} of synthesised bio-polyols correlated with the increase in OH value obtained from the titrimetric analysis of the polyols presented in Table 3.

3.3. NMR Analysis of Epoxidated TOFA Using Different Catalysts

Tall oil mainly consists of linoleic acid and oleic acid. According to ^{1}H NMR spectra, the ratio of both these fatty acids is about 1:1: characteristic signal for linoleic acid is *bis*-allylic protons at 2.8 ppm (Figure 4). The amount of oleic acid was calculated from the amount of allylic protons at 1.9–2.1 ppm.

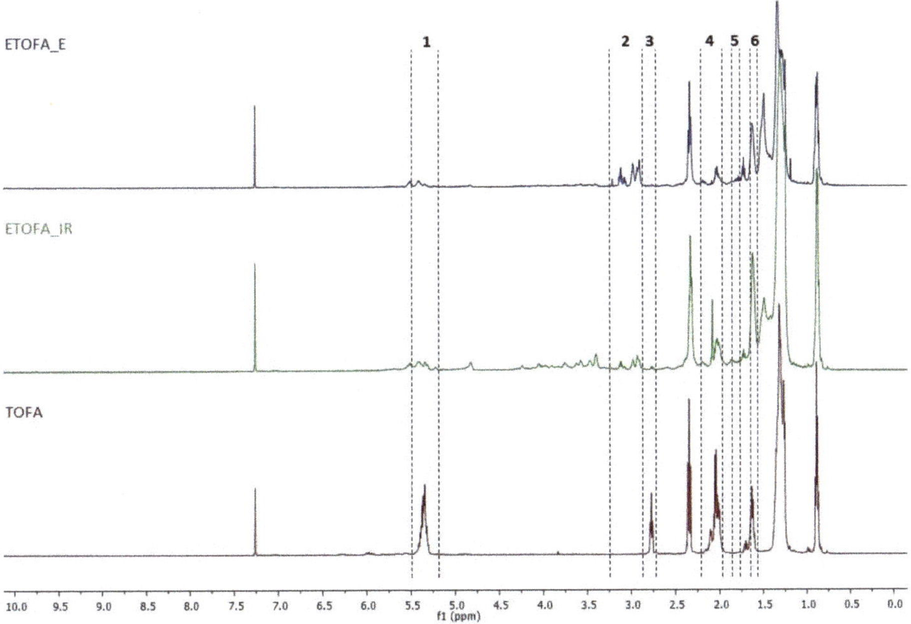

Figure 4. ^{1}H-NMR of TOFA and TOFA-catalysed epoxidation by ion-exchange resin (ETOFA_IR) and TOFA-catalysed epoxidation by lipase catalyst Novozym® 435 (ETOFA_E) (500 MHz, CDCl$_3$). Abbreviations: 1—protons of double bond; 2—protons of oxirane moiety; 3—protons of *bis*-allylic position; 4—protons of allylic position; 5—CH$_2$ protons of oxiran-2-yl moiety; 6—CH$_2$ protons of *bis*-(oxiran-2-yl) moiety.

^{1}H NMR spectra demonstrate that enzyme-catalysed reaction leads to a rather clean formation of epoxides (characteristic signals at 2.9–3.1 ppm), without cleavage of the oxirane ring in ETOFA_E product. The ETOFA_E contained up to 60% of epoxide moieties (calculated per number of double bonds in the starting material) and nearly 20% of untreated double bond moieties (signals at 5.1–5.7 ppm).

The yield of oxiranes was significantly smaller when ion exchange resin was used as the epoxidation catalyst. Although the oxirane moiety was formed, a significant amount of oxirane cleavage products is observed (characteristic signals in the range of 3.3–4.5 ppm). According to the ^{1}H NMR spectra, the product contained less than 20% of epoxide moieties (calculated per number of double bonds). Besides, the conversion of double bond was not full (signals at 5.1–5.7 ppm).

3.4. SEC Analysis of Synthesised High Functionality Bio-Polyols

The SEC chromatograms of different ETOFA and synthesised bio-polyols are depicted in Figure 5. The oxirane cleavage products discussed in previous paragraphs were also visible in SEC chromatogram of ETOFA_IR as a shoulder of epoxidised fatty acid products at the retention time of ~29 min. The oxirane ring-opening dimerisation, trimerisation and oligomerisation products were identified at retention times of ~27, 26 and 25 min, respectively, in ETOFA of both types of epoxidation methods. Although, the higher selectivity of Novozym® 435 catalyst ensured a lower amount of dimerisation products.

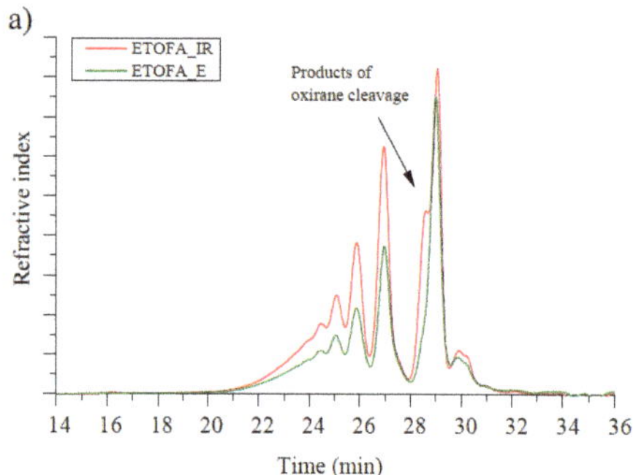
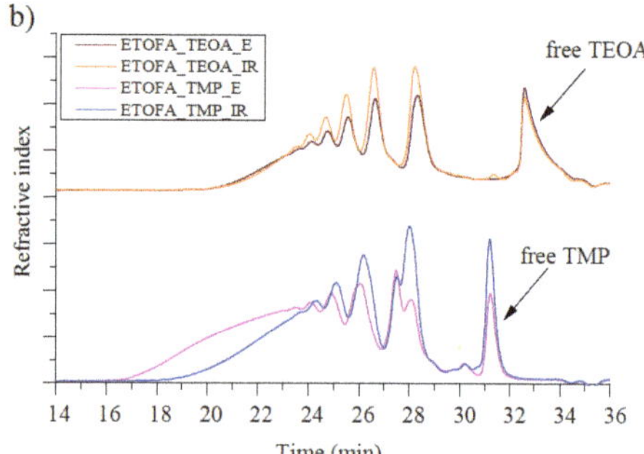

Figure 5. SEC chromatograms of (**a**) different ETOFA and (**b**) developed TOFA bio-polyols.

The obtained polyols had relatively similar characteristics when compared by used oxirane opening polyfunctional amine/alcohol despite ETOFA_E having fewer products of oxirane cleavage side reactions. The SEC chromatograms of ETOFA_TEOA_E and ETOFA_TEOA_IR are almost the same. Unfortunately, in the case of ETOFA_TMP_E polyol, an even larger amount of oligomerisation products has been identified. The oligomerisation of ETOFA_TMP_E polyol led to increased M_n of 2112 g/mol compared to ETOFA_TMP_IR polyol with M_n of 1279 g/mol (See Table 3). The oligomerisation could be explained due to the lower reactivity of TMP towards oxirane ring-opening. The method of oxirane ring-opening has to be further optimised to reduce the amount of oligomerisation products as they increase the viscosity of the bio-polyols, which is already considerable. High viscosity is undesirable, as it complicates the industrial upscale of the technology and production of rigid PU foams. Moreover, peaks of unreacted TEOA and TMP have been identified at the retention time of ~33 and 31 min, respectively. The presence of unreacted polyfunctional amines/alcohols is undesired but not critical, as they contain OH groups and can react with isocyanate in the formation of PU material. Obtained bio-polyols are a mixture of different TOFA derivatives with high OH group functionality and are suitable for rigid PU foam development.

3.5. Characteristics of Rigid PU Foams Developed from Synthesised High Functional Bio-Polyols

The main goal of the study was to obtain good-quality rigid PU foams using two different types of bio-polyols obtained from ETOFA_IR and ETOFA_E and to compare their common physical and mechanical properties. Synthesised TOFA-based bio-polyols were used to prepare rigid PU foams by formulations shown in Table 2. Typical characteristics of obtained rigid PU foams are summarised in Table 4.

Table 4. The common characteristic of developed rigid PU foams based on TOFA-type bio-polyols.

PU Material	PU_TEOA_IR	PU_TEOA_E	PU_TMP_IR	PU_TMP_E
Technological parameters		–		
Foaming start time, s	28	29	32	23
Foaming rise time, s	61	61	78	67
PU foam apparent density, kg/m³	36.4	35.4	35.8	38.5
Closed cell content, %	90	91	91	95
Thermal conductivity, mW/(m·K)	23.0	23.3	21.3	21.2

The foaming start time of rigid PU foam formulations developed from TEOA-type bio-polyols is on par with formulations from TMP-type bio-polyol at ~30 s. The presence of tertiary amine groups in TEOA-type bio-polyols expresses autocatalytic properties. Thus, this allowed to bypass the use of Polycat® 5 catalyst. This could be beneficial because Polycat® 5 is an additive catalyst with low boiling temperature, and it could leak out from the foamed material over time. This is not desirable as amine catalysts tend to have a strong odour which could influence the air quality in a rigid PU foam application environment, such as inside of buildings. The TEOA-type bio-polyol autocatalytic activity also influenced a faster foaming rise time of 61 s when compared to the foaming rise time of ~70 s of TMP-type bio-polyol formulations. The apparent density of developed rigid PU foams was almost the same and varied between 35.4 and 38.5 kg/m^3 range. The slightly higher apparent density of PU_TMP_E foam can be explained by higher viscosity of ETOFA_TMP_E polyol. Nevertheless, the closed cell content of all samples was 90% or above it, which is typical for rigid PU foams intended for thermal insulation application. The thermal conductivity for all developed rigid PU foams was quite low, reaching values between 21.2 and 23.3 mW/(m·K), which is considered appropriate for industrial use. The type of ETOFA used for polyols synthesis did not influence the thermal conductivity of obtained rigid PU foams. The thermal conductivity of rigid PU foams mostly depends on the gas composition inside the material's closed cells. The slight difference in the cell size (see Figure 6) of developed rigid PU foams did not influence the thermal conductivity.

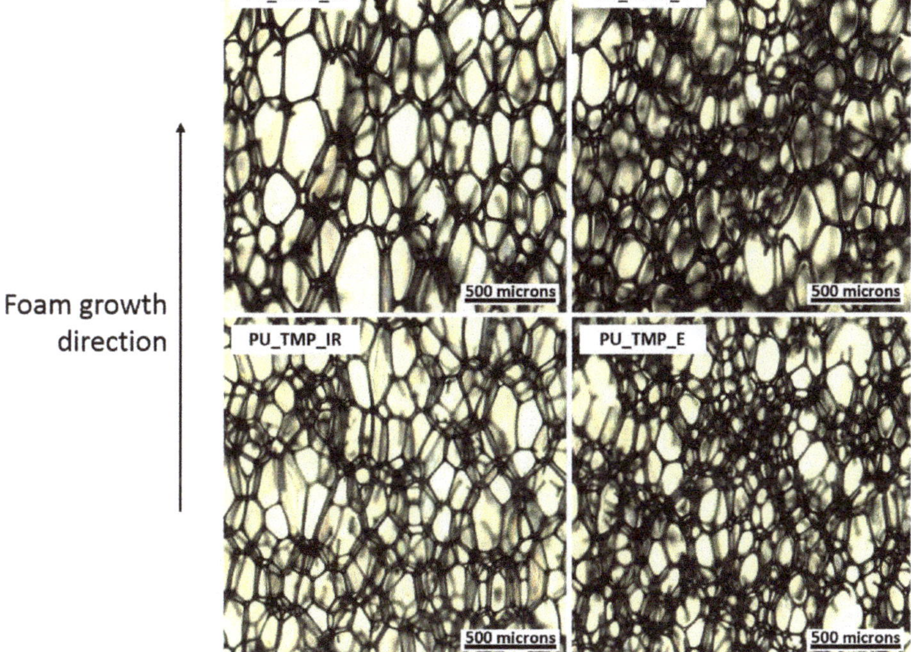

Figure 6. Microscope images of developed rigid polyurethane (PU) foams.

The compression strength and compression modulus of ETOFA_TEOA_IR-, ETOFA_TEOA_E-, ETOFA_TMP_IR- and ETOFA_TMP_E-based rigid PU foams were measured in parallel and perpendicular to foaming direction and are depicted in Figure 7. The data have been normalised to the average apparent density of 40 kg/m^3, according to Hawkins et al. [56], to compare the mechanical properties. Developed rigid PU foams had significant anisotropicity as the mechanical properties varied depending on the testing direction. Anisotropicity is typical for rigid PU foams that are obtained in open-type mould. Rigid PU foam cells tend to elongate with the foaming direction, as seen in Figure 6. Nevertheless, the compression strength was ~0.2 MPa, which is typical for this type of material applied for civil engineering [57].

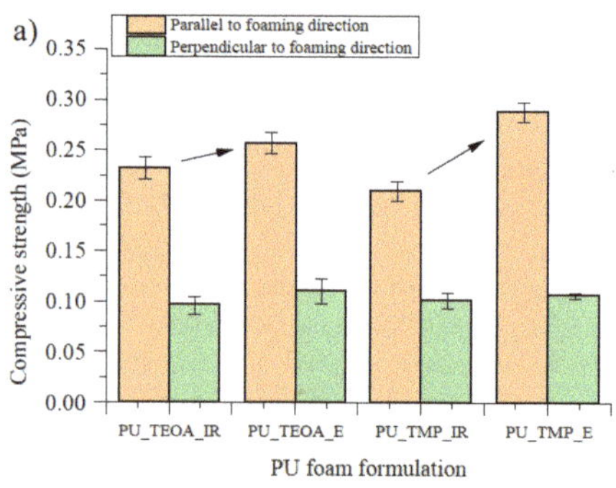

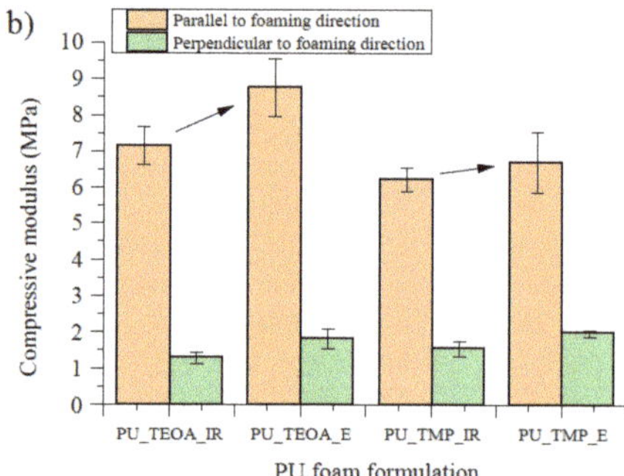

Figure 7. Compression properties (**a**) compressive strength and (**b**) compressive modulus of developed rigid PU foams normalised to the average apparent density of 40 kg/m^3.

A slight difference can be distinguished when the mechanical properties are compared regarding the used epoxidation method (ETOFA_IR and ETOFA_E). Rigid PU foams obtained from ETOFA_E showed a slight increase in compression properties due to the higher functionality of the derived polyols compared to polyols obtained from ETOFA_IR. A higher functionality of bio-polyol will result in the higher cross-link density of the obtained PU polymer matrix, which will increase the mechanical properties of rigid PU foam. In addition, the increase in mechanical properties can also be explained by slightly smaller cell size rigid PU foams obtained from ETOFA_E as depicted in Figure 5.

The common properties of the developed rigid PU foams were compared to other rigid PU foams that were obtained using other bio-based polyols (Table 5). The rigid PU foams used for comparison were produced from different bio-based polyols, such as polyols from waste cooking oil (PU 60 PWCO) [30], pumpkin seed oil (PU-PSO) [47], soybean oil-derived polyol (PU-0) [46] as well as commercially used polyether type polyols with the trade name Rokopol® RF-551 (PU-BMC/0) [45] and Raypol® 4218 [47]. The chosen rigid PU foams from literature data were selected because they had somewhat similar apparent density with closed cell structure. Some differences in-between foams can be distinguished, namely, the isocyanate index, which was 110, whereas the present study describes rigid PU foams with the isocyanate index of 150. Nevertheless, a comparison can be made to assess developed materials suitability as a thermal insulation. The compression properties of TOFA-based rigid PU foams were similar to materials described in the most recent literature; the slightly lower values can be explained by the lower apparent density. The thermal conductivity between the compared rigid PU foams is similar. In the case of TOFA-based foams, it is even slightly better than in most of the compared foams. It is crucial to mention that the TOFA-based rigid PU foams were developed using only bio-based polyols in the formulation. The developed TOFA-based high functionality polyols can be used as a cross-linking reagent and might be used as a replacement for sorbitol-based polyether type polyols.

Table 5. The common characteristic of developed rigid PU foams comparison with other materials.

PU Material	Compressive Strength, MPa	Compressive Modulus, MPa,	PU Foam Apparent Density, kg/m^3	Closed Cell Content, %	Thermal Conductivity, mW/(m·K)
PU_TEOA_IR	0.191	5.6	36.4	90.0	23.0
PU_TEOA_E	0.195	6.7	35.4	91.0	23.3
PU_TMP_IR	0.198	5.9	35.8	91.0	21.3
PU_TMP_E	0.309	7.2	38.5	95.0	21.2
PU_60_PWCO [1]	0.275	n.a	41.3	85.3	26.4
PU-BMG/0 [2]	0.248	n.a	42.1	87.5	22.5
PU-0 [3]	0.250	6.4	37.0	88.0	n.a
PU-PSO [4]	0.238	n.a	40.8	n.a	33.9
Raypol® 4218 [4]	0.303	n.a	44.1	n.a	36.8

[1] Data were taken from M. Kurańska et al. [30]; [2] Data were taken from M. Barczewski et al. [45]; [3] Data were taken from S. Czlonka et al. [46]; [4] Data were taken from P. Ekkaphan et al. [47].

4. Conclusions

Two different epoxidation catalyst have been applied for tall oil fatty acid epoxidation and the obtained epoxidised oil have been used for high functionality bio-polyol synthesis. Application of lipase-based catalyst allowed to obtain epoxidised tall oil fatty acids with a lower amount of side reaction by-products and higher content of oxirane rings when compared to oil epoxidated with ion exchange resin type catalyst (3.28 and 2.30 mmol/g, respectively). The higher oxirane content of epoxidised oil yielded bio-polyols with higher number average functionality. Unfortunately, a significant amount of oligomerisation products formation was observed during bio-polyol synthesis, which increased the viscosity of bio-polyols. To decrease the amount of oligomerisation products, the bio-polyol synthesis process has to be further optimised. The synthesised bio-polyols were used to obtain rigid PU foams. The type of chosen epoxidation route, ion-exchange resin or enzyme catalyst, for tall oil fatty acid epoxidation did not significantly influence the properties of foams. Only mechanical properties had a slight increase for rigid PU foams obtained from feedstock synthesised via chemo-enzymatic epoxidation due to the higher number average functionality of bio-polyols. Overall, developed rigid PU foams had typical properties as for material applied as thermal insulation in civil engineering.

Author Contributions: Conceptualisation, M.K.; formal analysis, M.K. and A.A.; investigation, A.A., E.V., R.P., I.M., A.F., and S.M.; writing—original draft preparation, A.A. and M.K.; writing—review and editing, A.F., I.M., S.M., and M.K.; visualisation, M.K.; supervision, M.K. All authors have read and agreed to the published version of the manuscript.

Funding: This research is supported/funded by the Latvian Council of Science, project Forest industry by-product transformation into valuable bio-polyols using advanced heterogeneous phase biocatalyst and characterisation of the process kinetics (FORinPOL), Grant No. lzp-2018/2-0020.

Institutional Review Board Statement: Not applicable.

Informed Consent Statement: Not applicable.

Data Availability Statement: Data Sharing is not applicable.

Acknowledgments: Special acknowledgement to Novozymes A/S, Denmark, who kindly supplied a sample of Novozym® 435 (on acrylic resin immobilised lipase enzyme catalyst).

Conflicts of Interest: The authors declare no conflict of interest.

Abbreviations

ETOFA	Epoxidised tall oil fatty acids
ETOFA_E	Epoxidised tall oil fatty acids obtained from tall oil fatty acids epoxidation with lipase catalyst Novozym® 435
ETOFA_IR	Epoxidised tall oil fatty acids obtained from tall oil fatty acids epoxidation with ion exchange resin Amberlite IR-120 H
ETOFA_TEOA_E	Polyol that is obtained from epoxidised tall oil, enzyme is used as a catalyst, oxirane ring opened with triethanolamine
ETOFA_TEOA_IR	Polyol that is obtained from epoxidised tall oil, ion exchange resin is used as a catalyst, oxirane ring opened with triethanolamine
ETOFA_TMP_E	Polyol that is obtained from epoxidised tall oil if enzyme is used as a catalyst, oxirane ring opened with trimethylolpropane
ETOFA_TMP_IR	Polyol that is obtained from epoxidised tall oil if ion exchange resin is used as a catalyst, oxirane ring opened with trimethylolpropane
–NCO	Isocyanate group
OH value	Hydroxyl value
pMDI	4,4′-diphenylmethane diisocyanate
PU	Polyurethane
PU_TEOA_E	Polyurethane formulated using polyol that is obtained from epoxidised tall oil, enzyme is used as a catalyst, oxirane ring opened with triethanolamine
PU_TEOA_IR	Polyurethane formulated using polyol that is obtained from epoxidised tall oil, ion exchange resin is used as a catalyst, oxirane ring opened with triethanolamine
PU_TMP_E	Polyurethane formulated using polyol that is obtained from epoxidised tall oil, enzyme is used as a catalyst, oxirane ring opened with trimethylolpropane
PU_TMP_IR	Polyurethane formulated using polyol that is obtained from epoxidised tall oil, ion exchange resin is used as a catalyst, oxirane ring opened with trimethylolpropane
TEOA	Triethanolamine
TMP	Trimethylolpropane
TO_TEOA	Polyol that is obtained by tall oil tall oil esterification with trietanolamine
TOFA	Tall oil fatty acids

References

1. Zhang, C.; Garrison, T.F.; Madbouly, S.A.; Kessler, M.R. Recent advances in vegetable oil-based polymers and their composites. *Prog. Polym. Sci.* **2017**, *71*, 91–143. [CrossRef]
2. Biermann, U.; Bornscheuer, U.; Meier, M.A.R.; Metzger, J.O.; Schäfer, H.J. Oils and Fats as Renewable Raw Materials in Chemistry. *Angew. Chem. Int. Ed.* **2011**, *50*, 3854–3871. [CrossRef]
3. Desroches, M.; Escouvois, M.; Auvergne, R.; Caillol, S.; Boutevin, B. From vegetable oils to polyurethanes: Synthetic routes to polyols and main industrial products. *Polym. Rev.* **2012**, *52*, 38–79. [CrossRef]
4. Li, Y.; Luo, X.; Hu, S. *Bio-Based Polyols and Polyurethanes*, 1st ed.; Springer International Publishing: Berlin/Heidelberg, Germany, 2015; ISBN 9783319215389.
5. Montero De Espinosa, L.; Meier, M.A.R.R. Plant oils: The perfect renewable resource for polymer science?! *Eur. Polym. J.* **2011**, *47*, 837–852. [CrossRef]
6. Lomège, J.; Negrell, C.; Robin, J.J.; Lapinte, V.; Caillol, S. Oleic acid-based poly(alkyl methacrylate) as bio-based viscosity control additive for mineral and vegetable oils. *Polym. Eng. Sci.* **2019**, *59*, E164–E170. [CrossRef]
7. Noreen, A.; Zia, K.M.; Zuber, M.; Tabasum, S.; Zahoor, A.F. Bio-based polyurethane: An efficient and environment friendly coating systems: A review. *Prog. Org. Coat.* **2016**, *91*, 25–32. [CrossRef]
8. Tan, S.G.; Chow, W.S. Biobased epoxidized vegetable oils and its greener epoxy blends: A review. *Polym. Plast. Technol. Eng.* **2010**, *49*, 1581–1590. [CrossRef]
9. Mungroo, R.; Pradhan, N.C.; Goud, V.V.; Dalai, A.K. Epoxidation of canola oil with hydrogen peroxide catalyzed by acidic ion exchange resin. *J. Am. Oil Chem. Soc.* **2008**, *85*, 887–896. [CrossRef]
10. Kirpluks, M.; Kalnbunde, D.; Walterova, Z.; Cabulis, U. Rapeseed Oil as Feedstock for High Functionality Polyol Synthesis. *J. Renew. Mater.* **2017**, *5*, 1–23. [CrossRef]
11. Lee, P.L.; Wan Yunus, W.M.Z.; Yeong, S.K.; Abdullah, D.K.; Lim, W.H. Optimization of the epoxidation of methyl ester of palm fatty acid distillate. *J. Oil Palm Res.* **2009**, *21*, 675–682.
12. Dinda, S.; Patwardhan, A.V.; Goud, V.V.; Pradhan, N.C. Epoxidation of cottonseed oil by aqueous hydrogen peroxide catalysed by liquid inorganic acids. *Bioresour. Technol.* **2008**, *99*, 3737–3744. [CrossRef]
13. Patel, M.M.; Patel, B.P.; Patel, N.K. Utilization of soya-based polyol for High solid PU-coating application. *Int. J. Plast. Technol.* **2012**, *16*, 67–79. [CrossRef]

14. Sinadinović-Fišer, S.; Janković, M.; Borota, O. Epoxidation of castor oil with peracetic acid formed in situ in the presence of an ion exchange resin. *Chem. Eng. Process. Process Intensif.* **2012**, *62*, 106–113. [CrossRef]

15. Abolins, A.; Yakushin, V.; Vilsone, D. Properties of polyurethane coatings based on linseed oil phosphate ester polyol. *J. Renew. Mater.* **2018**, *6*, 737–745. [CrossRef]

16. Goud, V.V.; Patwardhan, A.V.; Pradhan, N.C. Studies on the epoxidation of mahua oil (Madhumica indica) by hydrogen peroxide. *Bioresour. Technol.* **2006**, *97*, 1365–1371. [CrossRef]

17. de Haro, J.C.; Izarra, I.; Rodríguez, J.F.; Pérez, Á.; Carmona, M. Modelling the epoxidation reaction of grape seed oil by peracetic acid. *J. Clean. Prod.* **2016**, *138*, 70–76. [CrossRef]

18. Abolins, A.; Kirpluks, M.; Vanags, E.; Fridrihsone, A.; Cabulis, U. Tall Oil Fatty Acid Epoxidation Using Homogenous and Heterogeneous Phase Catalysts. *J. Polym. Environ.* **2020**, *28*, 1822–1831. [CrossRef]

19. Kirpluks, M.; Vanags, E.; Abolins, A.; Fridrihsone, A.; Cabulis, U. Chemo-enzymatic oxidation of tall oil fatty acids as a precursor for further polyol production. *J. Clean. Prod.* **2019**, *215*, 390–398. [CrossRef]

20. Vanags, E.; Kirpluks, M.; Cabulis, U.; Walterova, Z. Highly functional polyol synthesis from epoxidized tall oil fatty acids. *J. Renew. Mater.* **2018**, *6*, 764–771. [CrossRef]

21. Goud, V.V.; Patwardhan, A.V.; Dinda, S.; Pradhan, N.C. Kinetics of epoxidation of jatropha oil with peroxyacetic and peroxyformic acid catalysed by acidic ion exchange resin. *Chem. Eng. Sci.* **2007**, *62*, 4065–4076. [CrossRef]

22. Hazmi, A.S.A.; Aung, M.M.; Abdullah, L.C.; Salleh, M.Z.; Mahmood, M.H. Producing Jatropha oil-based polyol via epoxidation and ring opening. *Ind. Crops Prod.* **2013**, *50*, 563–567. [CrossRef]

23. Maisonneuve, L.; Chollet, G.; Grau, E.; Cramail, H. Vegetable oils: A source of polyols for polyurethane materials. *OCL* **2016**, *23*, D508. [CrossRef]

24. Hachemi, I.; Kumar, N.; Mäki-Arvela, P.; Roine, J.; Peurla, M.; Hemming, J.; Salonen, J.; Murzin, D.Y. Sulfur-free Ni catalyst for production of green diesel by hydrodeoxygenation. *J. Catal.* **2017**, *347*, 205–221. [CrossRef]

25. Chemistry, C. Preperation of biodiesel and separation of hemicellulose. *Cellul. Chem. Technol.* **2016**, *50*, 247–255.

26. Aro, T.; Fatehi, P. Tall oil production from black liquor: Challenges and opportunities. *Sep. Purif. Technol.* **2017**, *175*, 469–480. [CrossRef]

27. Panda, H. *Handbook on Tall Oil Rosin Production, Processing and Utilization*; Asia Pacific Business Press Inc.: Delhi, India, 2013.

28. Demirbas, A. Methylation of wood fatty and resin acids for production of biodiesel. *Fuel* **2011**, *90*, 2273–2279. [CrossRef]

29. Lubguban, A.A.; Ruda, R.J.G.; Aquiatan, R.H.; Paclijan, S. Soy-Based Polyols and Polyurethanes. *KIMIKA* **2017**, *1*, 1–19. [CrossRef]

30. Kurańska, M.; Leszczyńska, M.; Kubacka, J.; Prociak, A.; Ryszkowska, J. Effects of Modified Used Cooking Oil on Structure and Properties of Closed-Cell Polyurethane foams. *J. Polym. Environ.* **2020**, *28*, 2780–2788. [CrossRef]

31. Omonov, T.S.; Kharraz, E.; Curtis, J.M. The epoxidation of canola oil and its derivatives. *RSC Adv.* **2016**, *6*, 92874–92886. [CrossRef]

32. Ranganathan, S.; Sieber, V. Development of semi-continuous chemo-enzymatic terpene epoxidation: Combination of anthraquinone autooxidation and the lipase-mediated epoxidation process. *React. Chem. Eng.* **2017**, *2*, 885–895. [CrossRef]

33. Re, R.N.; Proessdorf, J.C.; La Clair, J.J.; Subileau, M.; Burkart, M.D. Tailoring chemoenzymatic oxidation: Via in situ peracids. *Org. Biomol. Chem.* **2019**, *17*, 9418–9424. [CrossRef]

34. Cai, X.; Zheng, J.L.; Aguilera, A.F.; Vernières-Hassimi, L.; Tolvanen, P.; Salmi, T.; Leveneur, S. Influence of ring-opening reactions on the kinetics of cottonseed oil epoxidation. *Int. J. Chem. Kinet.* **2018**, *50*, 726–741. [CrossRef]

35. Zhang, X.; Wan, X.; Cao, H.; Dewil, R.; Deng, L.; Wang, F.; Tan, T.; Nie, K. Chemo-enzymatic epoxidation of Sapindus mukurossi fatty acids catalyzed with Candida sp. 99–125 lipase in a solvent-free system. *Ind. Crops Prod.* **2017**, *98*, 10–18. [CrossRef]

36. Mashhadi, F.; Habibi, A.; Varmira, K. Determination of Activation Energy and Ping-Pong Kinetic Model Constants of Enzyme-Catalyzed Self-Epoxidation of Free Fatty Acids using Micro-reactor. *Catal. Lett.* **2018**, *148*, 3236–3247. [CrossRef]

37. Danov, S.M.; Kazantsev, O.A.; Esipovich, A.L.; Belousov, A.S.; Rogozhin, A.E.; Kanakov, E.A. Recent advances in the field of selective epoxidation of vegetable oils and their derivatives: A review and perspective. *Catal. Sci. Technol.* **2017**, *7*, 3659–3675. [CrossRef]

38. Kumar, A.; Dhar, K.; Kanwar, S.S.; Arora, P.K. Lipase catalysis in organic solvents: Advantages and applications. *Biol. Proced. Online* **2016**, *18*, 1–11. [CrossRef]

39. Milchert, E.; Malarczyk-Matusiak, K.; Musik, M. Technological aspects of vegetable oils epoxidation in the presence of ion exchange resins: A review. *Polish J. Chem. Technol.* **2016**, *18*, 128–133. [CrossRef]

40. Mateo, C.; Palomo, J.M.; Fernandez-Lorente, G.; Guisan, J.M.; Fernandez-Lafuente, R. Improvement of enzyme activity, stability and selectivity via immobilization techniques. *Enzyme Microb. Technol.* **2007**, *40*, 1451–1463. [CrossRef]

41. Ivdre, A.; Soto, G.D.; Cabulis, U. Polyols Based on Poly(ethylene terephthalate) and Tall Oil: Perspectives for Synthesis and Production of Rigid Polyurethane Foams. *J. Renew. Mater.* **2016**, *4*, 285–293. [CrossRef]

42. Zhang, C.; Ding, R.; Kessler, M.R. Reduction of epoxidized vegetable oils: A novel method to prepare bio-based polyols for polyurethanes. *Macromol. Rapid Commun.* **2014**, *35*, 1068–1074. [CrossRef]

43. Zeltins, V.; Yakushin, V.; Cabulis, U.; Kirpluks, M. Crude Tall Oil as Raw Material for Rigid Polyurethane Foams with Low Water Absorption. *Solid State Phenom.* **2017**, *267*, 17–22. [CrossRef]

44. Pfister, D.P.; Xia, Y.; Larock, R.C. Recent advances in vegetable oil-based polyurethanes. *ChemSusChem* **2011**, *4*, 703–717. [CrossRef] [PubMed]

45. Barczewski, M.; Kurańska, M.; Sałasińska, K.; Michałowski, S.; Prociak, A.; Uram, K.; Lewandowski, K. Rigid polyurethane foams modified with thermoset polyester-glass fiber composite waste. *Polym. Test.* **2020**, *81*. [CrossRef]

46. Członka, S.; Strakowska, A.; Strzelec, K.; Kairyte, A.; Kremensas, A. Bio-based polyurethane composite foams with improved mechanical, thermal, and antibacterial properties. *Materials* **2020**, *13*, 1108. [CrossRef]

47. Ekkaphan, P.; Sooksai, S.; Chantarasiri, N.; Petsom, A. Bio-Based Polyols from Seed Oils for Water-Blown Rigid Polyurethane Foam Preparation. *Int. J. Polym. Sci.* **2016**, *2016*. [CrossRef]

48. Alagi, P.; Ghorpade, R.; Jang, J.H.; Patil, C.; Jirimali, H.; Gite, V.; Hong, S.C. Functional soybean oil-based polyols as sustainable feedstocks for polyurethane coatings. *Ind. Crops Prod.* **2018**, *113*, 249–258. [CrossRef]

49. Omrani, I.; Farhadian, A.; Babanejad, N.; Shendi, H.K.; Ahmadi, A.; Nabid, M.R. Synthesis of novel high primary hydroxyl functionality polyol from sunflower oil using thiol-yne reaction and their application in polyurethane coating. *Eur. Polym. J.* **2016**, *82*, 220–231. [CrossRef]

50. Fridrihsone-Girone, A.; Stirna, U.; Misane, M.; Lazdiņa, B.; Deme, L. Spray-applied 100% volatile organic compounds free two component polyurethane coatings based on rapeseed oil polyols. *Prog. Org. Coat.* **2016**, *94*, 90–97. [CrossRef]

51. Considine, D.M.; Considine, G.D. *Van Nostrand's Scientific Encyclopedia*; Springer US: Boston, MA, USA, 1995; ISBN 978-1-4757-6920-3.

52. Kurańska, M.; Prociak, A. The influence of rapeseed oil-based polyols on the foaming process of rigid polyurethane foams. *Ind. Crops Prod.* **2016**, *89*, 182–187. [CrossRef]

53. Zieleniewska, M.; Leszczyński, M.K.; Kurańska, M.; Prociak, A.; Szczepkowski, L.; Krzyżowska, M.; Ryszkowska, J. Preparation and characterisation of rigid polyurethane foams using a rapeseed oil-based polyol. *Ind. Crops Prod.* **2015**, *74*, 887–897. [CrossRef]

54. Kirpluks, M.; Vanags, E.; Abolins, A.; Michalowski, S.; Fridrihsone, A.; Cabulis, U. High Functionality Bio-Polyols from Tall Oil and Rigid Polyurethane Foams Formulated Solely Using Bio-Polyols. *Materials* **2020**, *13*, 1985. [CrossRef] [PubMed]

55. Kirpluks, M.; Pomilovskis, R.; Vanags, E.; Abolins, A.; Mierina, I.; Fridrihsone, A. Optimisation of the chemo-enzymatic epoxidation of tall oil fatty acids using response surface methodology. *J. Clean. Prod.* **2020**. Manuscript submitted for publication.

56. Hawkins, M.C.; O'Toole, B.; Jackovich, D. Cell Morphology and Mechanical Properties of Rigid Polyurethane Foam. *J. Cell. Plast.* **2005**, *41*, 267–285. [CrossRef]

57. Szycher, M. *Szycher's Handbook of Polyurethanes*, 1st ed.; CRC Press: Boca Raton, FL, USA, 1999.

 *materials*

Article

Study on the Structure-Property Dependences of Rigid PUR-PIR Foams Obtained from Marine Biomass-Based Biopolyol

Paulina Kosmela [1,*], Aleksander Hejna [1] **, Jan Suchorzewski [2,3], Łukasz Piszczyk [1] and Józef Tadeusz Haponiuk [1]**

[1] Department of Polymer Technology, Chemical Faculty, Gdansk University of Technology, Narutowicza Str. 11/12, 80-233 Gdansk, Poland; aleksander.hejna@pg.edu.pl (A.H.); lukpiszc@pg.gda.pl (Ł.P.); jozhapon@pg.gda.pl (J.T.H.)

[2] RISE Research Institutes of Sweden, Infrastructure and Concrete Technology, Material Design, 501-15 Borås, Sweden; jan.suchorzewski@ri.se

[3] Department of Civil and Material Engineering, Faculty of Civil and Environmental Engineering, Gdansk University of Technology, Narutowicza Str.11/12, 80-233 Gdansk, Poland; jansucho@pg.edu.pl

* Correspondence: paulina.kosmela@pg.gda.pl

Received: 23 January 2020; Accepted: 4 March 2020; Published: 10 March 2020

Abstract: The paper describes the preparation and characterization of rigid polyurethane-polyisocyanurate (PUR-PIR) foams obtained with biopolyol synthesized in the process of liquefaction of biomass from the Baltic Sea. The obtained foams differed in the content of biopolyol in polyol mixture (0–30 wt%) and the isocyanate index (I_{ISO} = 200, 250, and 300). The prepared foams were characterized in terms of processing parameters (processing times, synthesis temperature), physical (sol fraction content, apparent density) and chemical structure (Fourier transform infrared spectroscopy), microstructure (computer microtomography), as well as mechanical (compressive strength, dynamic mechanical analysis), and thermal properties (thermogravimetric analysis, thermal conductivity coefficient). The influence of biopolyol and I_{ISO} content on the above properties was determined. The addition of up to 30 wt% of biopolyol increased the reactivity of the polyol mixture, and the obtained foams showed enhanced mechanical, thermal, and insulating properties compared to foams prepared solely with petrochemical polyol. The addition of up to 30 wt% of biopolyol did not significantly affect the chemical structure and average cell size. With the increase in I_{ISO}, a slight decrease in processing times and mechanical properties was observed. As expected, foams with higher I_{ISO} exhibited a higher relative concentration of polyisocyanurate groups in their chemical structure, which was confirmed using principal component analysis (PCA).

Keywords: bio-based polyol; biomass liquefaction; rigid polyurethane-polyisocyanurate foams; structure-property relationship; isocyanate index

1. Introduction

Algae biomass is commonly used around the world on an industrial scale, mainly for consumer, industrial, pharmaceutical, and cosmetology purposes. Algae have been used in the food industry for many years as a rich source of low-calorie nutrients [1–3]. In the cosmetics industry, mainly dried, freeze-dried algae, and algae extracts are used [4]. Algae biomass is characterized by a high content of various biologically active compounds, among others with antibacterial, antiviral, antifungal, and anti-inflammatory effects [5–8]. Due to its renewable nature, interest in using algae has increased in recent years, including in the energy sector [9] and the plastics industry [10].

Plastics with algae biomass can be obtained in many ways. The main methods of preparation include direct and indirect methods. Direct methods rely on mixing algae biomass with a polymer matrix to obtain biocomposites. Much research has been done on the use of algae biomass as a reinforcement in composites [11–13]. Jang et al. [14] received biocomposites using polypropylene (PP) and pre-processed algae Laminaria japonica (BA) and Enteromorpha (GA). Algae pre-treatment included treatment with ethanol at 78 °C for 3 h, followed by washing with ethanol or acetone. Other samples were treated with 3 wt% of sulfuric acid at 121 °C for 2 h. In the last stage of preparation, BA or GA was washed with distilled water, dried at 100 °C for 24 h, and milled. Biocomposites were obtained by the compression molding method. PP was evenly mixed with BA or GA and dried in a convection oven at 100 °C for 24 h. PP/GA biocomposites showed higher thermal stability and impact resistance than PP/BA ones. It was also found that sulfuric acid-treated GA biocomponents showed the best thermal and mechanical properties.

Indirect methods consist of obtaining plastics from the algae biomass used as an intermediate in different processes. An example of the indirect use of algae is the extraction of alginate, well described in the literature, and its use in the synthesis of polyurethane [15–18]. Another indirect method is the use of algae oil in the synthesis of polyurethane foams [19–21]. Pawar et al. [22] modified oil from chlorella microalgae and used it to obtain rigid polyurethane foams. The modification consisted of oil epoxidation and opening of the oxirane ring using lactic acid and ethylene glycol. Oil epoxidation was carried out in solution using peracetic acid and hydrogen peroxide for 6–8 h with a yield of 90–94%. In the next stage, the epoxidized oil was mixed with lactic acid (1:6 molar ratio) or ethylene glycol (1:10 molar ratio) to obtain biocomponents with active hydroxyl groups. Rigid polyurethane foams were obtained using polyol solely from algae oil and compared with their petroleum-based analogs. Foams obtained with the use of algae oil are characterized by a higher content of closed cells and similar thermal stability compared to petrochemical foam. Although the presented procedure resulted in foams showing satisfactory properties, the synthesis of applied biopolyol consisted of several steps including extraction of oils from microalgae, its epoxidation and reaction with lactic acid, as well as purification steps. Furthermore, it required the use of significant amounts of solvents, which cannot be considered environmentally friendly.

In our previous work [23], we proposed a method of direct liquefaction of marine biomass from the Baltic Sea, which was only dried before processing. Moreover, to enhance the eco-friendly character of the process, we applied waste material (crude glycerol) as one of the solvents in the process. The developed procedure provided a method of utilization of marine biomass from the Baltic Sea, which especially during the summer is a troublesome waste. This paper presents the method of preparation and the characteristics of rigid polyurethane-polyisocyanurate foams obtained using biopolyol resulting from the abovementioned liquefaction process. Rigid polyurethane-polyisocyanurate (PUR-PIR) foams were obtained by a one-step method, by replacing up to 30 wt% petrochemical polyol with biopolyol. Processing, chemical structure, and morphology, as well as mechanical and thermal properties were characterized.

2. Materials and Methods

2.1. Materials

For the preparation of rigid PUR-PIR foams, LB biopolyol obtained via marine biomass (10 wt%) liquefaction with a mixture of crude glycerol and poly(ethylene glycol) (90 wt%) was applied. The applied biomass consisted of Enteromorpha macroalgae and Zostera marina seagrass in the ratio of 1 to 9. The proposed structure of biopolyol is presented in Figure 1, while its synthesis and properties were described in another work [23]. Rokopol®RF551 (polyoxyalkylene polyhydric alcohol), obtained from the PCC Group (, Brzeg Dolny, Poland), was used as the petrochemical polyol. The selected properties of both polyols are shown in Table 1. The isocyanate component was polymeric methylene diphenyl-4,4′-diisocyanate (pMDI) with a free isocyanate groups (NCO) content of 31.5% from BASF

(Ludwigshafen, Germany). The solution of potassium acetate in ethylene glycol (AC) (PC CAT® TKA30 from Performance Chemicals (Buchholz in der Nordheide, Germany)), 75 wt% solution of potassium octoate in diethylene glycol (Dabco K15), 33 wt% solution of triethyl diamine in dipropylene glycol (Dabco33LV from Air Products (Allentown, PA, USA)), and dibutyltin dilaurate (DBTDL) from Sigma Aldrich (Saint Louis, MO, USA) were applied as catalysts. Tegostab B8465 from Evonik Industries AG (Essen, Germany) was applied as a surfactant (SPC), and n-pentane from Lachner was used as a blowing agent. Trichloropropyl phosphate (TCPP) was also added as a flame retardant, which also reduced the viscosity of the polyol mixture (TCPP was characterized with a viscosity of 61–89 mPa·s) from LANXESS Deutschland GmbH (Koln, Germany).

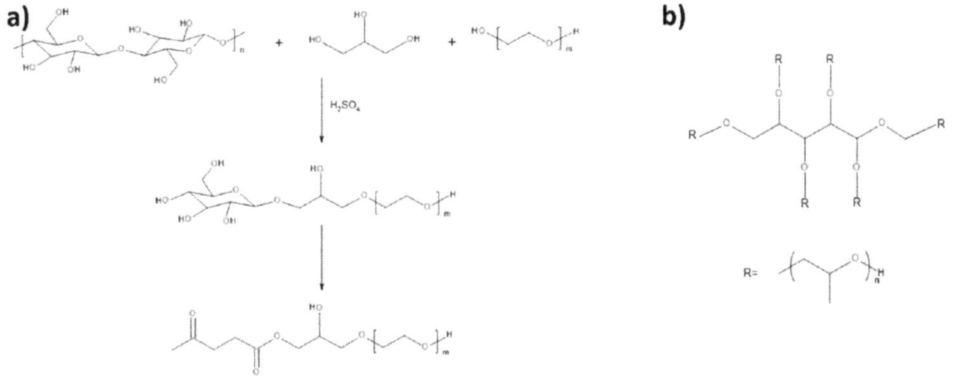

Figure 1. (**a**) proposed course of the main reaction occurring as a result of liquefaction of biomass; (**b**) chemical structure of Rokopol®RF551.

Table 1. Selected properties of polyols applied for the preparation of PUR-PIR foams.

Polyol	L_{OH}, mg KOH/g	η, mPa s	ρ, g/cm^3	H_2O Content, wt%
LB	650 *	2236	1.21	0.2
Rokopol®RF551	440 *	3000–5000	1.06	0.1

*-determined experimentally.

2.2. Preparation of Rigid PUR-PIR Foams

Rigid PUR-PIR foams were obtained on a laboratory scale using the single-step method from the two-component system. The impact of the isocyanate index (I_{ISO}) on the processing of polyurethane systems and the performance of the resulting PUR-PIR foams was investigated. Three values of I_{ISO} were applied, basing on literature reports and our previous works: 200, 250, and 300 [24,25]. Component A consisted of the proper amounts of Rokopol®RF551 and biopolyol LB, catalysts, surfactant, flame retardant, and blowing agent. Component B was isocyanate. Both components were mixed in a polypropylene cup with a mechanical stirrer at 2000 rpm and poured into the open mold. Foams were cured for 24 h at 60 °C and then conditioned at room temperature for 24 h. Table 2 contains the details of foam formulations.

The performance of rigid PUR-PIR foams, as well as all other porous materials, is closely related to their apparent density. Therefore, it is essential to analyze the materials characterized by a similar level of this parameter. In Table 2, there are also presented values of apparent density, which for all samples was in the range of 49.2–53.2 kg/m^3. It was possible by the adjustment of foams' formulations and varying content of potassium catalyst and foaming agent. The rise of the isocyanate index increased the amount of these components required for providing the desired level of apparent density, which was also observed in other works [26].

Table 2. Formulations of prepared polyurethane-polyisocyanurate (PUR-PIR) foams. DBTDL, dibutyltin dilaurate; TCPP, Trichloropropyl phosphate; pMDI, polymeric methylene diphenyl-4,4′-diisocyanate.

Component	Foam Symbol											
	200_LB0	200_LB10	200_LB20	200_LB30	250_LB0	250_LB10	250_LB20	250_LB30	300_LB0	300_LB10	300_LB20	300_LB30
	Content, pbw											
RF551	100	90	80	70	100	90	80	70	100	90	80	70
LB	0	10	20	30	0	10	20	30	0	10	20	30
AC	0.5	0.5	0.5	0.5	0.5	0.5	0.5	0.5	0.5	0.5	0.5	0.5
Dabco 15K	0.5	0.5	0.5	0.5	0.75	0.75	0.75	0.75	1.00	1.00	1.00	1.00
Dabco 33LV	0.5	0.5	0.5	0.5	0.5	0.5	0.5	0.5	0.5	0.5	0.5	0.5
DBTDL	0.5	0.5	0.5	0.5	0.5	0.5	0.5	0.5	0.5	0.5	0.5	0.5
SPC	6	6	6	6	6	6	6	6	6	6	6	6
TCPP	10	10	10	10	10	10	10	10	10	10	10	10
n-pentane	12.5	12.5	12.5	12.5	15.0	15.0	15.0	15.0	20.0	20.0	20.0	20.0
Gram equivalents of OH groups	0.784	0.822	0.859	0.897	0.784	0.822	0.859	0.897	0.784	0.822	0.859	0.897
pMDI	203.6	214.7	225.8	237.0	254.5	268.4	282.3	296.2	305.3	322.0	338.7	355.4
I_{ISO}	200				250				300			
Apparent density, kg/m^3	49.2 ± 1.8	50.1 ± 1.1	50.8 ± 1.9	49.6 ± 1.3	53.2 ± 2.0	50.5 ± 1.8	52.4 ± 1.5	51.7 ± 1.4	49.9 ± 2.2	50.6 ± 1.9	51.9 ± 1.4	52.1 ± 1.2

2.3. Characterization of Polyurethanes

2.3.1. Evaluation of the Foaming Process

For all the samples, the following processing times were determined: start time (from the beginning of mixing to the start of volumetric expansion), rise time (time of volumetric expansion), and the tack-free time (from the end of volumetric expansion to the point when the surface stopped being tacky to the touch). After pouring the reaction mixture into the mold, the temperature inside the foam was measured by a thermocouple.

2.3.2. Fourier-Transform Infrared Spectroscopy

FT-IR spectrophotometric analysis was performed to determine the structure of the bio-based polyol and rigid PU-PIR foams. The analysis was performed at a resolution of 4 cm^{-1} using a Nicolet 8700 apparatus (Thermo Electron Corporation, Waltham, MA, USA) equipped with a snap-Gold State II, which allowed for making measurements in the reflection configuration mode. Principal component analysis (PCA) was applied to analyze the results with MATLAB software (MathWorks, Natic, MA, USA).

2.3.3. Apparent Density

The apparent density of samples was calculated in accordance with EN ISO 845:2000 [27], as a ratio of the sample weight to the sample volume (g/cm^3). The cylindrical samples were measured with a slide caliper with an accuracy of 0.1 mm and weighed using an electronic analytical balance with an accuracy of 0.0001 g.

2.3.4. Sol Fraction Content

To determine the sol fraction content in prepared foams, 0.2 g samples were placed in xylene for 72 h at room temperature. Then, samples were taken out and dried until constant weight. Sol fraction content was calculated according to the following Formula (1):

$$X = \frac{m_1 - m_2}{m_1} \cdot 100\% \tag{1}$$

where: m_1 is the initial mass of the sample and m_2 is the mass after extraction and drying.

2.3.5. Compressive Strength

The compressive strength of rigid foams was estimated in accordance with EN ISO 844:2007 [28]. The cylindrical samples with dimensions of 20 mm × 20 mm (height and diameter) were measured with a slide caliper with an accuracy of 0.1 mm. The compression test was performed on a Zwick/Roell 1000 N testing machine (ZwickRoell GmbH & Co. KG., Ulm, Germany) at a constant speed of 10%/min until reaching 15% deformation.

2.3.6. Thermal Conductivity Coefficient

The thermal conductivity of foams was tested with Lambda 2300 heat flow meter (Holometrix, Bedford, MA, USA) according to the ASTM C518 standard [29]. The average temperature of analysis was 10 °C and the temperatures of the lower and upper plates 0 and 20 °C, respectively. Three samples with a size of 300 mm × 300 mm × 50 mm were tested for each composition.

2.3.7. Thermogravimetric Analysis

To evaluate the thermal stability of materials, thermogravimetric analysis (TGA) was performed on 5 mg samples using a NETZSCH TG 209 F3 apparatus (NETZSCH-Gruppe, Selb, Germany) under a nitrogen atmosphere in the temperature range from 40 to 800 °C and at a heating rate of 20 °C/min.

2.3.8. Dynamic Mechanical Analysis

The dynamic mechanical analysis (DMA) was performed using the DMA Q800 TA Instruments apparatus (TA Instruments Inc., New Castle, DE, USA). Samples were analyzed in compression mode with a frequency of 1 Hz and an amplitude of 20 μm. Measurements were performed for the temperature range from 35 to 270 °C with a heating rate of 4 °C/min. Samples were cylindrically shaped with dimensions of 6 mm × 18 mm.

2.3.9. Micro-Computed Tomography

Foams were analyzed with a 3D Skyscan 1173 X-ray microscope (Bruker, Kontich, Belgium) with a resolution of 9 μm and a rotation of 180°. The energy of X-ray radiation was 35 kV, a current of 175 μA, and the sample exposition time was 500 ms. A rotation step of 0.2° was applied for high precision. Skyscan Nrecon software (6.0 Skyscan, Kontich, Belgium) was used to obtain 3D images, which were then analyzed with CTAn software. The X-ray microscope used in this study represented the new generation of apparatus with high resolution [30,31].

3. Results and Discussion

3.1. Evaluation of the Foaming Process

The values of the start time and tack-free time for all samples equaled 10 and 0 s, respectively. this means that when foams reached their maximum volume, their surface was already not sticky. Therefore, in Table 3 are presented only the values of the rise time and maximum temperatures reached by the foams during the reaction. It can be seen that the rise time was slightly reduced by the increase of biopolyol content in foams' formulations, which could be related to the lower viscosity of biopolyol LB compared to Rokopol®RF551. Furthermore, a significant increase of the maximum temperature reached during foaming was noted, which could be associated with the enhanced reactivity of the systems with a higher content of biopolyol. The increased reactivity of biopolyol was associated with its chemical structure (higher hydroxyl number L_{OH} = 650 mg KOH/g) and higher viscosity polyols (η = 2236 mPa·s) compared to petrochemical polyol. Lower viscosity (greater mobility of chains, functional groups) and the higher concentration of reactive groups caused acceleration of the reaction of allophanate crosslinks, urethane, and urea linkages. A similar effect was noted in other works [32].

Table 3. Rise times and maximum temperatures recorded during the synthesis of PUR-PIR foams.

I_{ISO}	Biopolyol Content, wt%	Rise Time, s	T_{MAX} during Foaming, °C
	0	40.2 ± 0.4	86.8 ± 2.3
200	10	40.3 ± 0.3	87.9 ± 3.5
	20	39.5 ± 0.1	138.2 ± 2.8
	30	38.6 ± 0.4	141.6 ± 3.1
	0	38.3 ± 0.3	82.4 ± 2.0
250	10	37.7 ± 0.2	86.3 ± 2.4
	20	37.5 ± 0.5	133.1 ± 3.4
	30	35.7 ± 0.5	134.6 ± 2.1
	0	37.4 ± 0.3	71.4 ± 2.0
300	10	35.8 ± 0.6	85.2 ± 2.5
	20	34.5 ± 0.4	128.0 ± 3.4
	30	33.3 ± 0.5	130.7 ± 3.2

The values of rise time and T_{MAX} were also affected by the isocyanate index. It could be seen that shortening of the rise time was noted, but the maximum temperature was reduced, when I_{ISO} was increased. Such an effect was probably associated with the increased loading of the blowing agent for higher isocyanate content, which was necessary to provide a similar level of the foams' apparent density. The applied *n*-pentane was the physical blowing agent, so it required heat to evaporate and generate a porous structure of PUR-PIR foam. Moreover, the reduction of T_{MAX} could be related to the lower enthalpy for the isocyanate trimerization reaction ($\Delta H = 80$ kJ/kmol), compared to the generation of urethane bonds ($\Delta H = 105$ kJ/kmol) [33].

3.2. Microstructure of Rigid PUR-PIR Foams

Micro-computed tomography (micro-CT or µCT) is a technique that enables three-dimensional imaging of a sample and the generation of virtual 3D models with the use of X-ray radiation. In Figure 2 are presented two-dimensional cross-sections of PUR-PIR foams prepared solely from petrochemical polyol and with a 30 wt% share of biopolyol. A slight decrease of the average cell size was noted, which could be associated with the presence of solid particles in the applied biopolyol. The applied biopolyol was obtained with 78% biomass conversion, as mentioned in our previous work [23]. Therefore, considering the applied formulations, foams with an isocyanate index of 200, 250, and 300 contained 1.8, 1.5, and 1.4 wt% of solid particles, which might act as nucleating agents, simultaneously reducing the average cell diameter. A similar effect was noted by Zhang et al. [34]. Moreover, the values of porosity are presented, which were calculated as the ratio of the total cell volume and the volume of the measured sample. It could be seen that the porosity was slightly decreasing with the addition of biopolyol, which was probably associated with the lower viscosity of biopolyol LB compared to commercial Rokopol®RF551 (see Table 1). Such an effect was related to the lower resistance of the foaming mixture to the cells' coalescence. The more viscous polyol mixture showed a higher ability to trap gas inside [35]. The effect was very significant in the case of the work presented by Fan et al. [36]; however, they substituted conventional polyol with a viscosity of 9000 cP with a soy-based one, which showed a value of 31,351 cP. In the case of presented work, differences in the polyols' viscosity were not so significant. A similar effect was observed in the work of Ciecierska et al. [37], who analyzed the impact of the viscosity of a polyol mixture on the cellular structure of rigid PUR foams. Their results showed that the increase of the viscosity caused an increase in the porosity, without significant changes in cell diameters. Nevertheless, such changes in morphology caused the deterioration of the mechanical performance.

Table 4. Morphological properties of PUR-PIR foams.

Foam Symbol	Average Cell Diameter, µm	Average Cell Volume, mm³·10⁻³	Porosity, %
200_LB0	117 ± 21	8.9 ± 2.1	85
200_LB30	108 ± 22	6.5 ± 1.8	81
250_LB0	115 ± 23	8.3 ± 1.6	83
250_LB30	95 ± 16	5.4 ± 2.0	80
300_LB0	131 ± 25	11.0 ± 1.2	84
300_LB30	124 ± 17	7.2 ± 1.6	83

In Table 4 are presented the parameters of the foams' cellular structure. It could be seen that the incorporation of biopolyol resulted in the reduction of the average cell diameter and volume. Moreover, the increase of I_{ISO} resulted in the enhancement of the microstructure homogeneity, which was confirmed by the decrease of the standard deviation and the sharper shape of the histograms presented in Figure 3. For all samples containing 30 wt% of biopolyol, a significant increase in the content of cells with a diameter of ~60 µm was observed. Regarding the influence of I_{ISO}, the highest homogeneity of structure and the lowest values of cell size and volume were noted for the ratio of isocyanate to hydroxyl groups equal to 250. Moreover, in Table 5 are presented 3D images showing the distribution of cells with a particular volume inside foams.

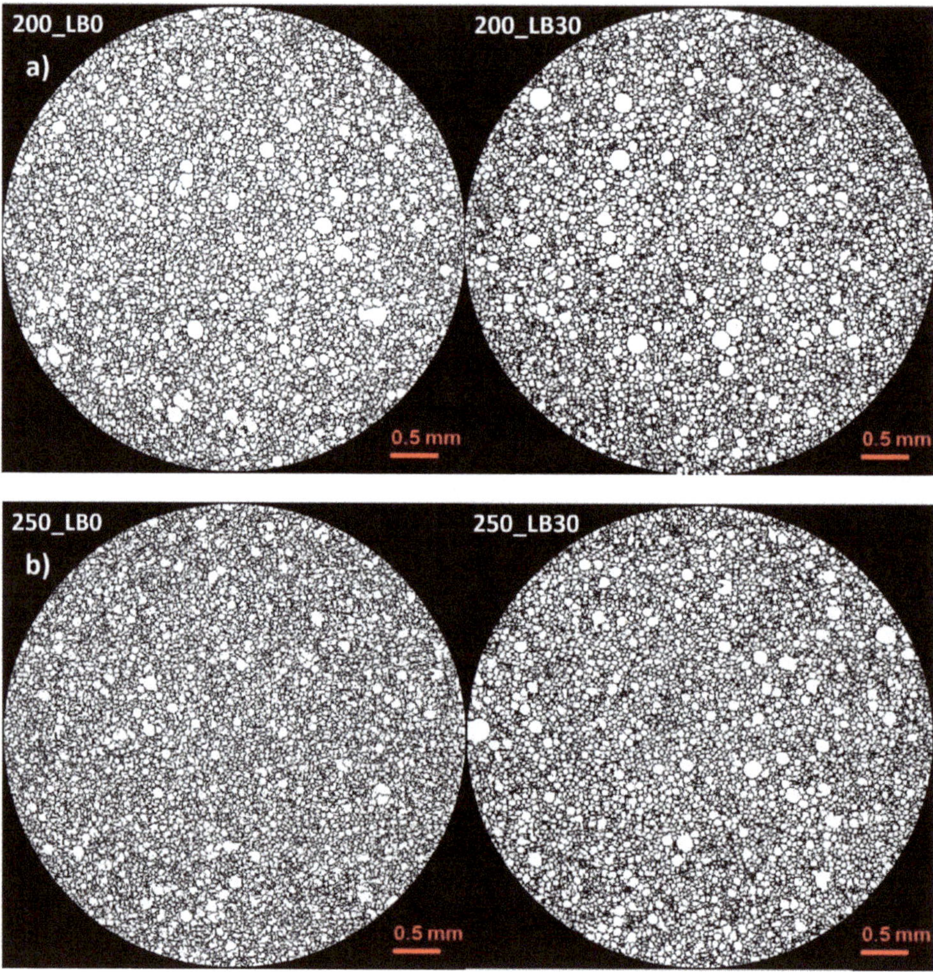

Figure 2. *Cont.*

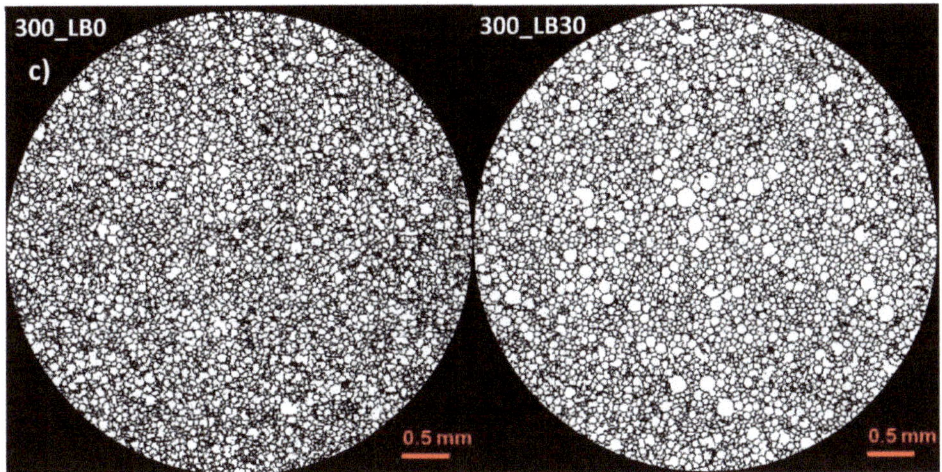

Figure 2. 2D images of PUR-PIR foams: (**a**) I_{ISO} = 200, (**b**) I_{ISO} = 250, (**c**) I_{ISO} = 300.

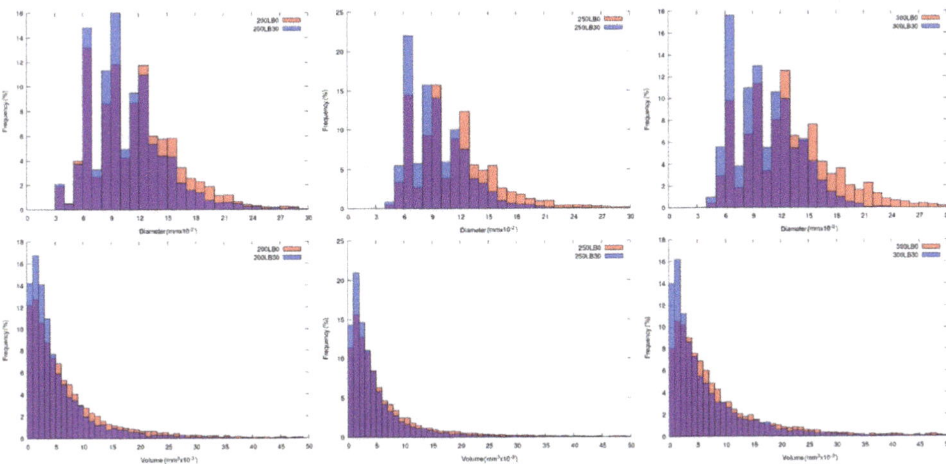

Figure 3. Histograms showing the distribution of the pore size (upper line) and pore volume (lower line) of PUR-PIR foams for different isocyanate indexes.

In Table 6 are presented the thermal insulation properties of prepared foams. The thermal conductivity coefficient (λ) and thermal resistance (R) are crucial for determining the possibility of the application of PUR-PIR foams as insulation materials. For these materials, the λ coefficient was composed of four components λ_{gas}, $\lambda_{PUR-PIR}$, $\lambda_{radiation}$, and $\lambda_{convection}$ [38]. The share of these components was strictly correlated with the cellular structure of the foams. With the increase of apparent density, the thermal conductivity of a solid part of the foam would have a more significant influence on the total λ value. Therefore, a low density was very beneficial for the performance of thermal insulation materials since the λ values for solid PUR-PIR, CO_2, and air were 220.0, 15.3, and 24.9, respectively [25]. Other structural parameters, such as the cell size, also showed a significant impact on the thermal conductivity of foamed materials. According to Glicksman [39], $\lambda_{radiation}$ was proportional to cell diameter and inversely proportional to the density of the foam and solid polymer. Randall and Lee [40] proved that the decrease in average cell diameter from 600 to 250 μm resulted in a drop in the thermal conductivity coefficient by almost 50%. Moreover, to show the lowest possible value of the λ parameter, foams should be characterized by a high content of closed cells, which reduced $\lambda_{convection}$ by trapping gas inside the foam. Volatile hydrocarbons, commonly applied as blowing agents for PUR-PIR foams, showed thermal conductivity coefficient ~1.5 and ~2.5 lower than CO_2 and air.

Table 5. 3D images of PUR-PIR foams.

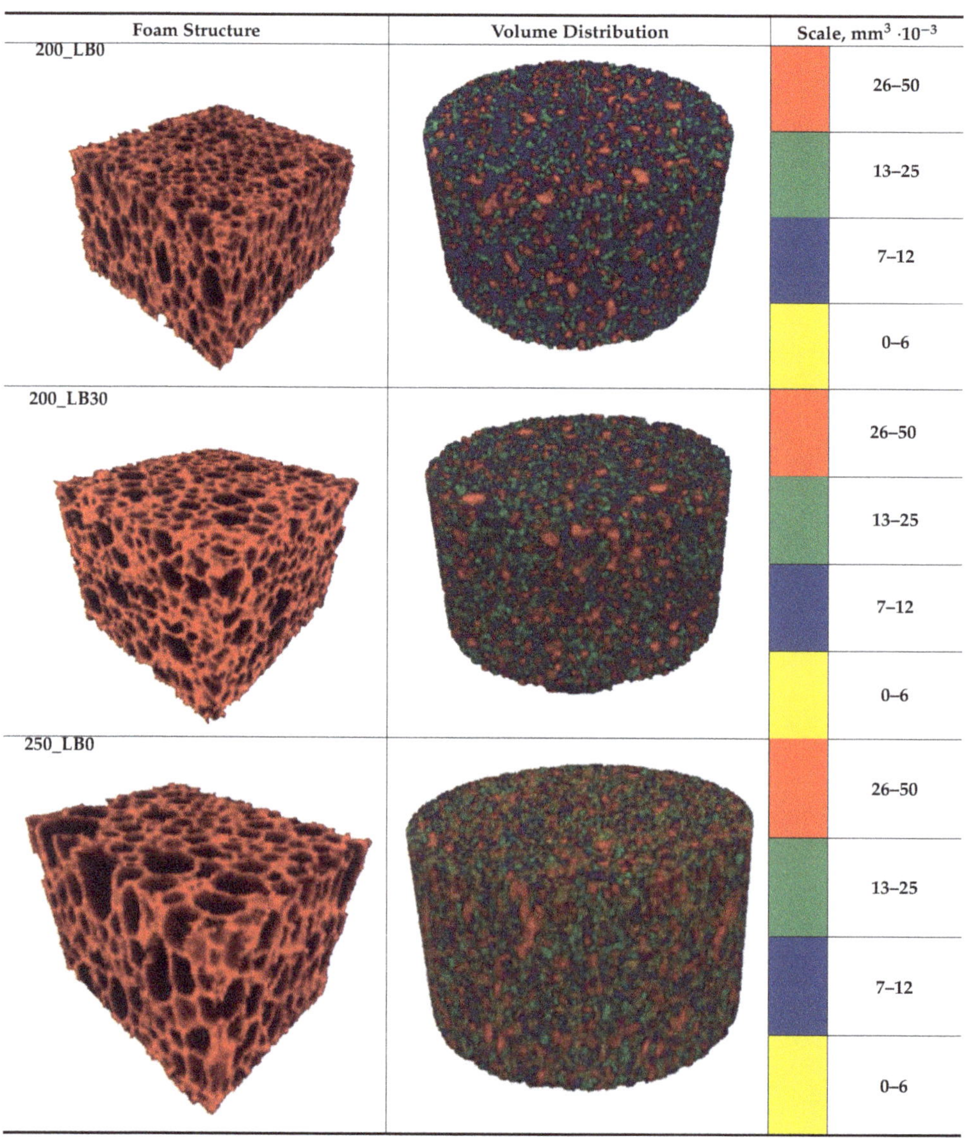

Foam Structure	Volume Distribution	Scale, mm^3 ·10^{-3}
200_LB0		26–50
		13–25
		7–12
		0–6
200_LB30		26–50
		13–25
		7–12
		0–6
250_LB0		26–50
		13–25
		7–12
		0–6

Table 5. *Cont.*

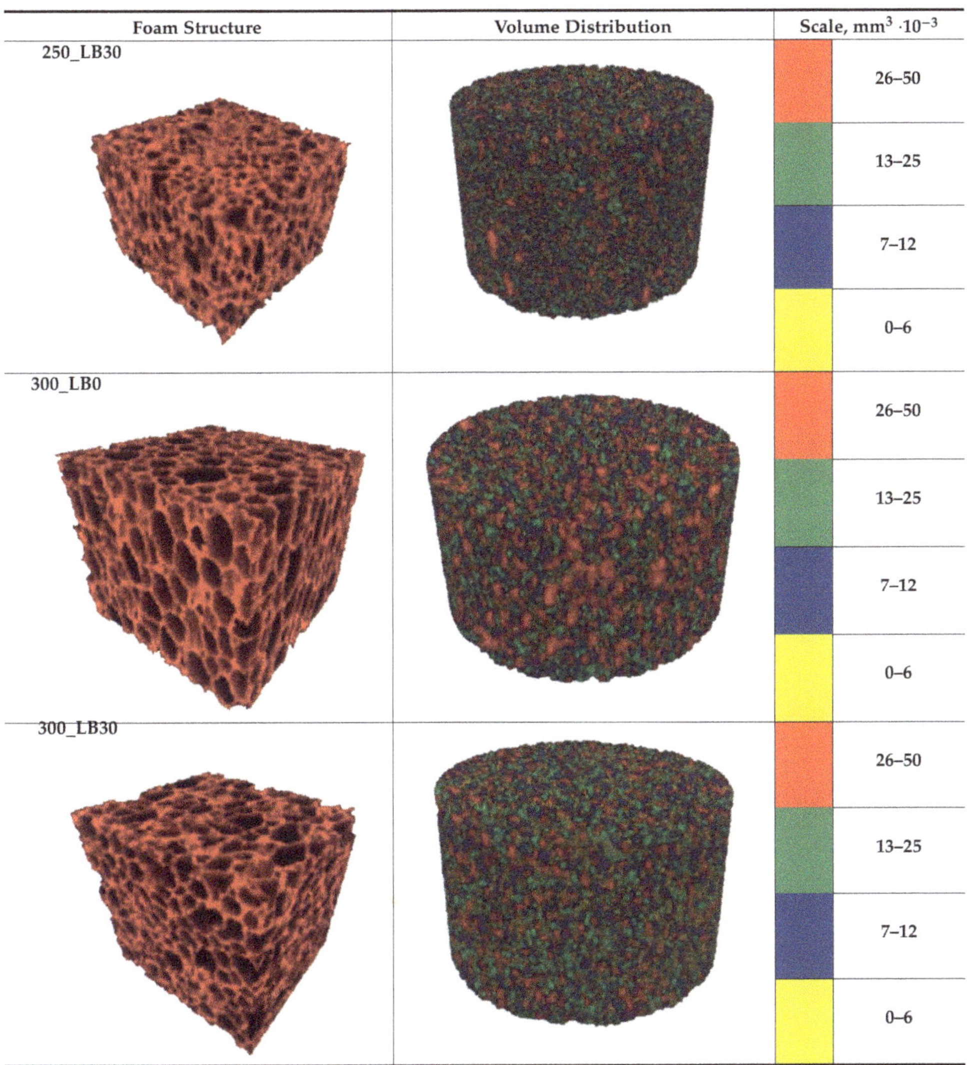

Foam Structure	Volume Distribution	Scale, mm$^3 \cdot 10^{-3}$
250_LB30		26–50 / 13–25 / 7–12 / 0–6
300_LB0		26–50 / 13–25 / 7–12 / 0–6
300_LB30		26–50 / 13–25 / 7–12 / 0–6

Table 6. Thermal insulation properties of the obtained foams.

Foam Symbol	Thermal Conductivity Coefficient, mW/m·K	Thermal Resistance (d = 0.02m), m^2·K/W
200_LB0	26.98 ± 0.14	0.741
200_LB10	25.33 ± 0.15	0.790
200_LB20	24.74 ± 0.21	0.808
200_LB30	24.69 ± 0.12	0.810
250_LB0	26.15 ± 0.20	0.765
250_LB10	25.38 ± 0.18	0.788

Table 6. *Cont.*

Foam Symbol	Thermal Conductivity Coefficient, mW/m·K	Thermal Resistance (d = 0.02m), m^2·K/W
250_LB20	25.25 ± 0.17	0.792
250_LB30	24.91 ± 0.19	0.803
300_LB0	26.09 ± 0.22	0.767
300_LB10	25.44 ± 0.20	0.786
300_LB20	25.11 ± 0.21	0.796
300_LB30	24.83 ± 0.19	0.806

It could be seen that the incorporation of biopolyol into the foams' formulation resulted in the reduction of the average cell size, which was followed by a drop of the thermal conductivity coefficient. A similar effect was observed in our other works [24,25]. Changes of the isocyanate index did not result in significant differences in the thermal conductivity of prepared foams.

3.3. Physico-Mechanical and Thermal Properties Analysis

In Table 7 are presented the values of the sol fraction, compressive strength, and glass transition temperature (T_g) of foams determined from the temperature dependence of the loss tangent. The addition of biopolyol into the formulations of PUR-PIR foams reduced the content of the sol fraction. This was associated with a decrease of the non-crosslinked fraction content, which suggested a higher crosslink density of prepared foams. The decrease of the sol fraction content was also affected by the increase of I_{ISO}, which confirmed other research works [41]. Except for the primary condensation reaction, which led to the generation of urethane groups, at higher values of I_{ISO}, additional reactions may take place. They included the generation of allophanate and biuret groups or trimerization of isocyanates leading to the generation of polyisocyanurates. All of these groups could noticeably affect the level of crosslink density of PUR-PIR foams [42].

Table 7. Physical and mechanical properties of PUR-PIR foams.

Foam Symbol	Sol Fraction Content, wt%	Compressive Strength, kPa			T_g, °C
		Perpendicular to the Rise Direction	Parallel to the Rise Direction	Anisotropy, %	
200_LB0	2.7 ± 0.7	163 ± 9	394 ± 12	2.42	154
200_LB10	2.2 ± 0.2	176 ± 10	410 ± 8	2.33	200
200_LB20	1.5 ± 0.3	229 ± 8	430 ± 11	1.88	204
200_LB30	1.3 ± 0.7	287 ± 11	446 ± 8	1.55	210
250_LB0	1.2 ± 0.2	168 ± 7	368 ± 6	2.19	213
250_LB10	1.4 ± 1.1	215 ± 12	400 ± 7	1.86	220
250_LB20	0.6 ± 0.1	222 ± 8	414 ± 9	1.86	222
250_LB30	1.5 ± 0.3	243 ± 9	440 ± 8	1.81	225
300_LB0	1.2 ± 1.1	151 ± 7	342 ± 9	2.26	221
300_LB10	0.9 ± 0.2	186 ± 8	366 ± 11	1.97	230
300_LB20	0.8 ± 0.2	190 ± 13	382 ± 14	2.01	232
300_LB30	0.7 ± 0.1	200 ± 7	426 ± 10	2.13	233

In Table 7 are also presented the values of the compressive strength of rigid PUR-PIR foams, measured in two directions, parallel and perpendicular to the rise direction. It was noted that the mechanical performance of the foams was significantly affected by their cellular structure, which was also proven in other works related to this type of material [43,44]. For the analyzed materials, the anisotropy of the mechanical properties was associated with the elongation of cells in the direction of the foam rise (see the 3D images in Table 5). More significant differences between compressive strength parallel and perpendicular to the rise direction were noted for lower contents of biopolyol. This effect was not observed for the highest value of I_{ISO}, which could be related to the increased stiffness of the structure. Moreover, the increase of the isocyanate index caused a slight reduction of compressive strength, which was also observed by other researchers [45]. Such an effect was associated with the higher friability of the foams' structure, which was already noted by Modesti and Lorenzetti [42].

Except for the decrease of anisotropy in the mechanical properties, a general enhancement of compressive strength was observed for the increase of biopolyol content. Such an effect was related to the reduced values of the sol fraction, suggesting the enhanced crosslink density of modified foams, as well as changes in the cellular structure (a decrease of the cell size and porosity). Generally, a decrease of the cell size with a similar level of the materials' porosity resulted in the enhancement of the mechanical performance [46].

The glass transition temperature of the foams was significantly affected by the changes in the foams' formulations. The increase of the biopolyol content and isocyanate index led to the rise of T_g, which indicated the stiffening of the material, and together with a drop of the sol fraction content, indicated an increase of crosslink density. Ivani et al. [47] stated that for rigid PUR-PIR foams, an increase of T_g with the addition of biopolyol was associated with the lower flexibility of macromolecular chains compared to the conventionally applied petrochemical polyols.

3.4. Thermogravimetric Analysis

In Figure 4 and Table 8 are presented the results of the thermogravimetric analysis. The thermal decomposition of polyurethanes is a very complicated process, due to its complex structure, especially when also polyisocyanurate groups are present. Generally, it could be seen that the increase of biopolyol content shifted the onset of thermal degradation, measured by the temperature of 2 wt% mass loss, towards higher temperatures. The only exception was observed for I_{ISO} of 200 and the highest content of biopolyol. It could be considered as very beneficial because numerous literature reports pointed to the deterioration of thermal stability caused by the application of biopolyols [48].

Table 8. Results of thermogravimetric analysis of PUR-PIR foams.

Foam Symbol	Mass Loss, wt%			T_{max}, °C
	2	5	10	
	Temperature, °C			
200_LB0	217.0	250.8	289.2	342.1
200_LB10	224.3	259.7	290.7	347.8
200_LB20	235.8	260.7	290.7	348.8
200_LB30	218.0	251.6	284.8	347.5
250_LB0	198.9	255.9	295.4	344.1
250_LB10	220.4	254.6	289.8	345.2
250_LB20	228.6	264.0	298.5	350.1
250_LB30	239.7	268.8	298.6	349.8
300_LB0	214.6	269.0	300.9	344.2
300_LB10	245.0	275.3	303.7	350.6
300_LB20	242.5	270.5	299	348.1
300_LB30	245.6	273.2	301.5	348.9

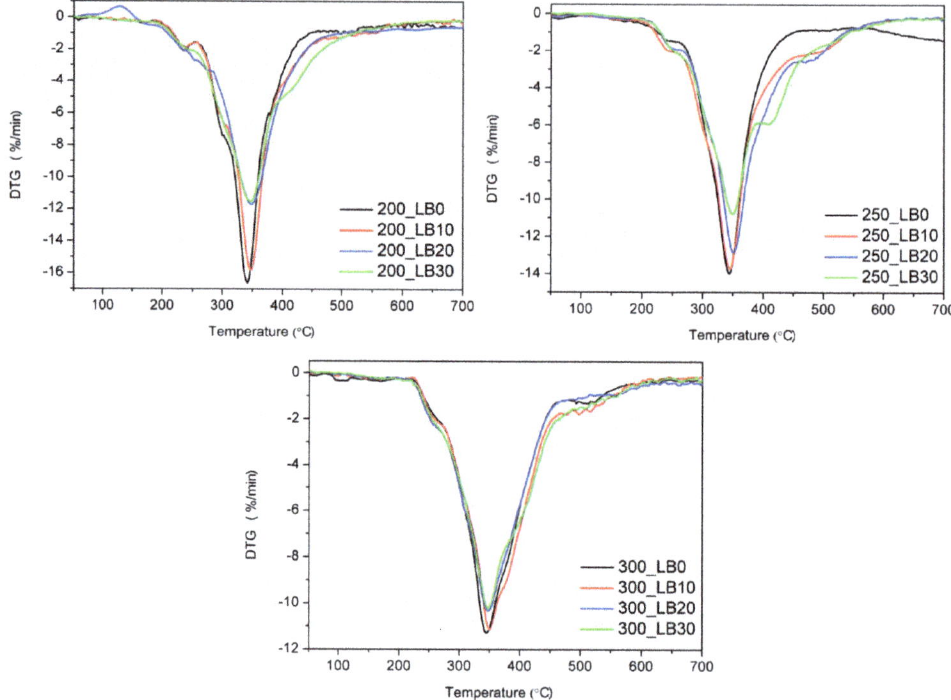

Figure 4. DTG curves of rigid PUR-PIR foams.

Differential thermogravimetric curves indicated the three-step decomposition process for all samples. The first step, observed at 220–280 °C, was associated with the decomposition of TCPP (degradation temperature of 224 °C), as well as the decomposition of soft segments. During the second step, with the temperature of the maximum degradation rate of 350 °C, hard segments of polyurethane were decomposed, and amines and carbon dioxide were formed [49]. For sample 250_LB30, an additional peak around 410 °C was observed, which was associated with biopolyol decomposition [32]. The third step of degradation, in the range of 430–570 °C, was related to the thermolysis of organic residues from previous steps [50]. The incorporation of synthesized biopolyol into foams' formulations caused a shift of peaks on DTG curves towards higher temperatures, which confirmed the enhancement of thermal stability.

3.5. Infrared Spectroscopy and Principal Component Analysis

3.5.1. FTIR Spectra Analysis

The analysis of FTIR spectra presented in Figure 5 did not indicate qualitative changes in the chemical structure of prepared PUR-PIR foams after the addition of biopolyol. Absorption bands observed in the range of 3290–3320 cm^{-1} were attributed to the stretching vibrations of N–H bonds in the urethane groups. Signals for bending vibrations of these bonds were noted at 1510–1520 cm^{-1} [51]. Bands at 1200–1215 and 1705–1715 cm^{-1} related to stretching vibrations of C–N and C=O bonds, respectively, confirmed the presence of urethane groups [52]. Signals present in the range of 1410–1415 cm^{-1} were attributed to the presence of isocyanurate rings generated during the trimerization of isocyanate groups [53]. Around 2260–2280 cm^{-1}, there were noted peaks, which indicated the presence of free isocyanate groups, also associated with relatively high values of applied I$_{ISO}$ [54].

Moreover, around 2860–2870 and 2960–2975 cm^{-1}, there were observed signals related to the symmetric and asymmetric stretching vibrations of C–H bonds in macromolecular chains of applied polyols. Multiplet bands in the range of 1000–1090 cm^{-1} were attributed to the vibrations of C–O bonds in the ester and ether groups present in the structure of polyols [55].

In Figure 6 is presented the impact of I_{ISO} on the FTIR spectra of prepared PUR-PIR foams. The increase of the peaks' magnitude for vibrations of isocyanate groups, carbonyl bonds, and polyisocyanurate rings, marked as I, II, and III, respectively, was noted. As mentioned above, their intensity was noticeably increasing with the rise of I_{ISO} [56].

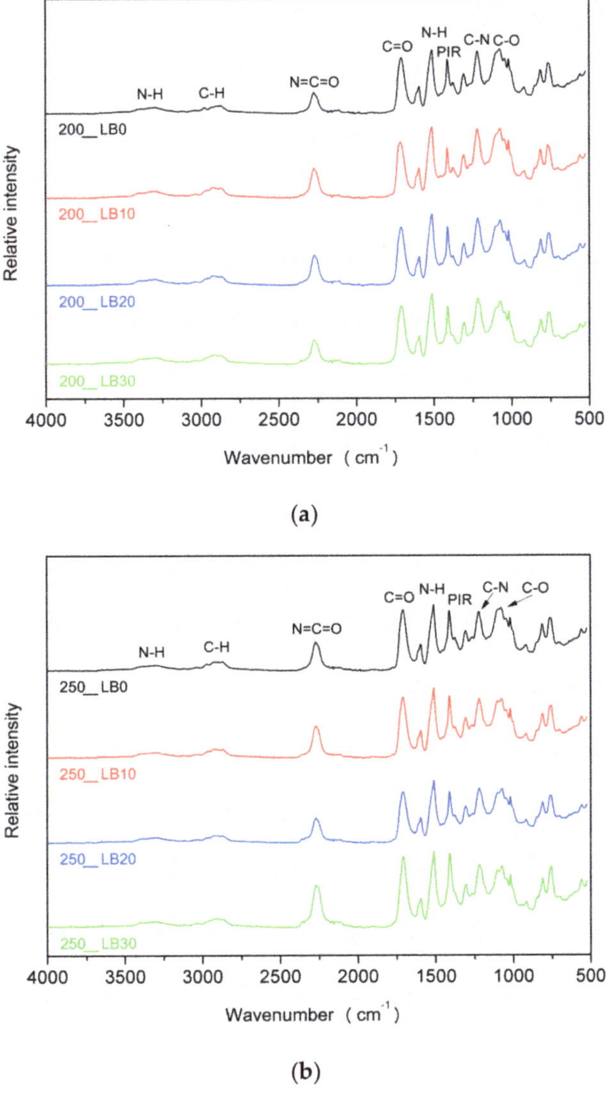

(a)

(b)

Figure 5. *Cont.*

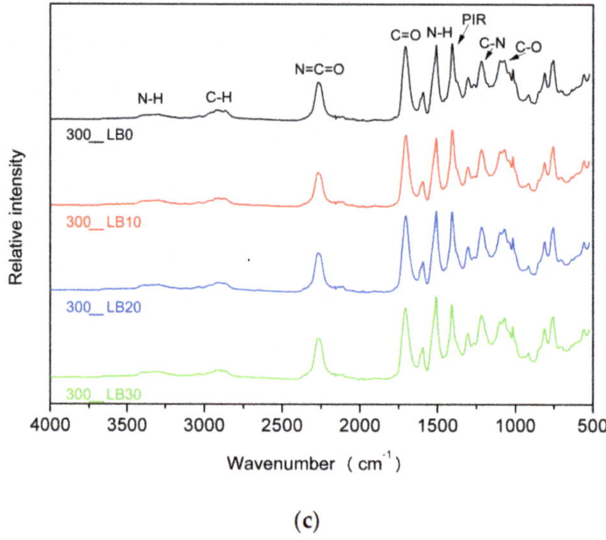

(**c**)

Figure 5. FTIR spectra of PUR-PIR foams: (**a**) I_{ISO} = 200, (**b**) I_{ISO} = 250, (**c**) I_{ISO} = 300.

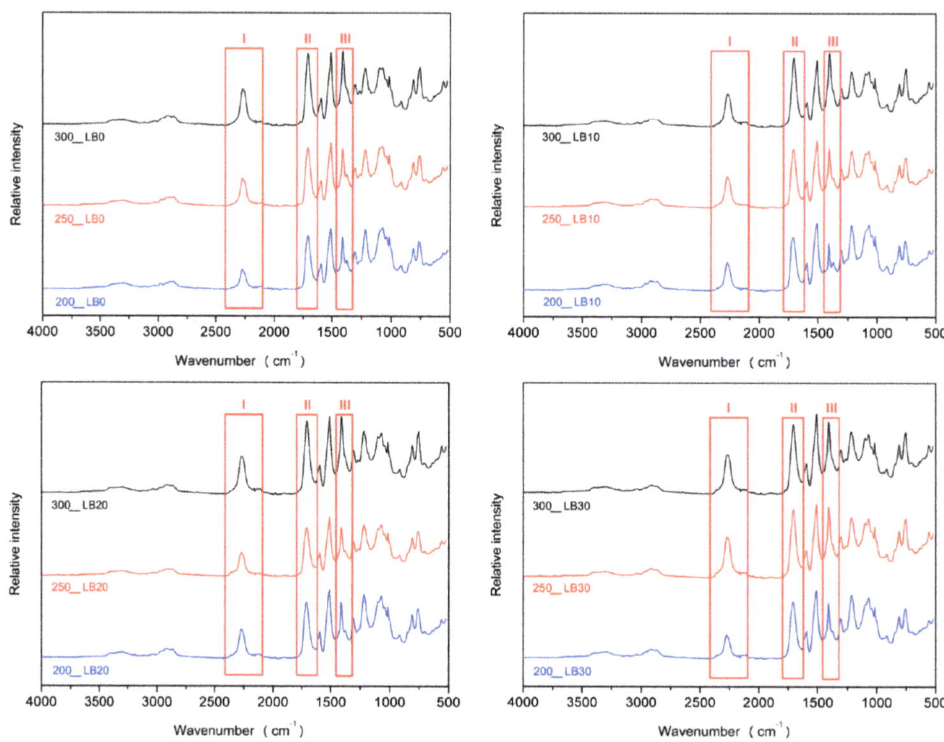

Figure 6. FTIR spectra of foams depending on I_{ISO}.

3.5.2. Principal Component Analysis

To determine the content of polyisocyanurate and polyurethane components in the foams' structure, an analysis of the main factors (PCA) was performed using the Malinowski spectral isolation algorithm. The use of PCA allowed reducing the number of variables, the interpretation of the relationships between components, and the graphical presentation of the configuration of compared variables. The analysis of the main factors for the FTIR spectra of PUR-PIR foams indicated two main factors affecting them, which is shown in Figure 7.

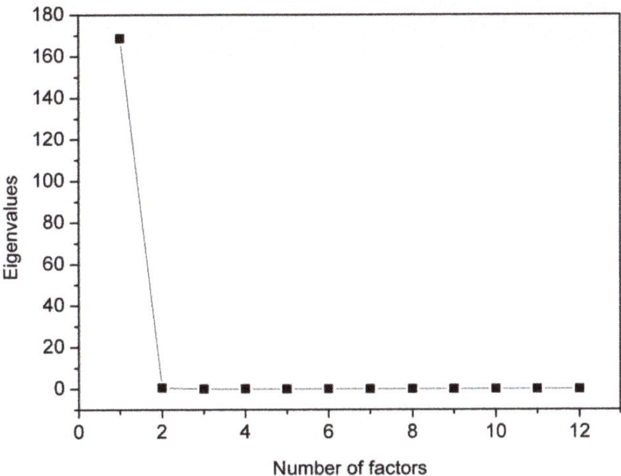

Figure 7. Eigenvalues for 12 factors (12 spectra of PUR-PIR foams).

After determining the number of principal components, the spectra of pure components were isolated using the Malinowski spectra algorithm. It was found that the two components were: the spectrum derived from PUR and from PIR, which are shown in Figure 8.

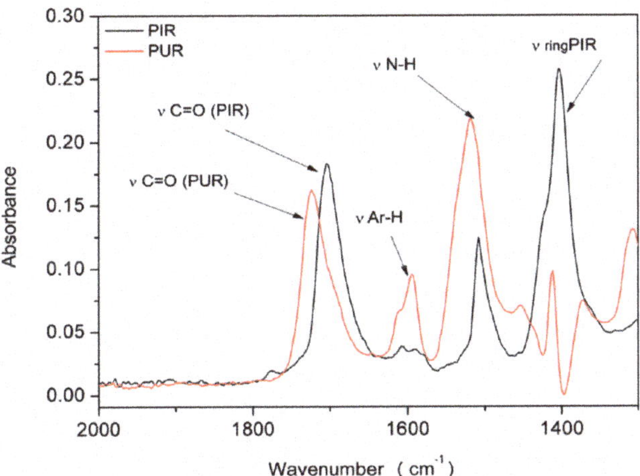

Figure 8. Spectrum of the main factors: PUR and PIR.

In the presented range of wavenumbers of 1300–2000 cm^{-1}, one could notice the separation of the absorbance band for the C=O bond, for individual components. Other researchers also observed the shift of the band associated with PIR towards lower wavenumbers. Xu et al. [57] found out that, in the PIR foams, the carbonyl group bonded to the isocyanurate ring showed a coupling effect, which resulted in a shift in the absorbance peak towards the lower values. The other bands presented in Figure 8 are characteristic of individual components. Absorbance bands in the range of 1595 cm^{-1} in polyurethane confirmed the presence of aromatic rings derived from isocyanate [58]. The lack of this absorbance band in the case of PIR foam with the simultaneous appearance of the band at 1403cm^{-1} confirmed the formation of isocyanate group trimerization products [57].

To determine the percentage content of particular components in PUR-PIR foams, relative concentrations were calculated according to the following formula (2):

$$C_{w,a} = C_a / (C_a + C_b) \tag{2}$$

where: $C_{w,a}$ is the relative concentration of Component a in the mixture; C_a is the concentration of Component a; and C_b is the concentration of Component b. In Figure 9 are presented the relative concentrations of PUR and PIR for the analyzed foams. Samples 1–4 were foams obtained with I_{ISO} equal to 200 with biopolyol content from 0 to 30 wt%, while Samples 5–8 and 9–12 were obtained with I_{ISO} of 250 and 300, respectively.

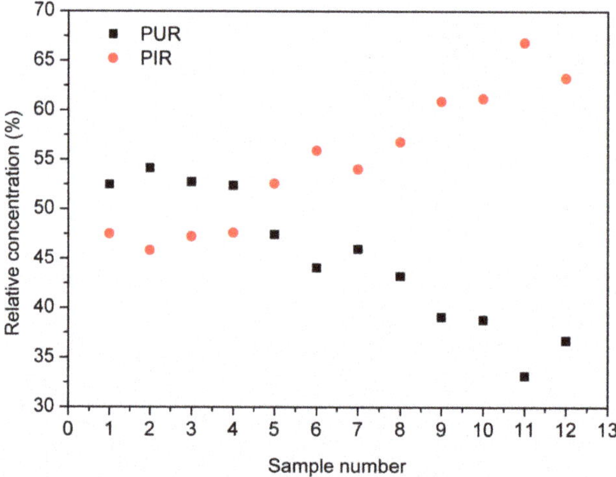

Figure 9. Relative concentrations of major factors.

The data presented in Figure 9 indicated that the increase of the isocyanate index resulted in the increase of relative concentration of polyisocyanurate structures in foams. It could be seen that the concentrations of PUR and PIR parts of foams were strictly related to each other. Their values were strongly dependent on the isocyanate index because only when isocyanates were used in excess, a significant amount of polyisocyanurate rings could be generated. Moreover, the incorporation of biopolyol resulted in an increase of PIR content, which could partially contribute to the decrease of sol fraction content and stiffening of the material, which resulted in the rise of T_g. Although the presented results were probably different from the real ones, because, in the case of such materials like PUR-PIR foams, FTIR should not be considered as a quantitative technique, they provided exciting insight into the analysis of the obtained properties of the foams.

4. Conclusions

The paper presented the method of the preparation and characterization of rigid PUR-PIR foams produced from biopolyol obtained by liquefaction of marine biomass. The foams were obtained with varying contents of biopolyol (0, 10, 20, and 30 wt%) and the isocyanate index (200, 250, and 300). Both factors affected the foaming process. A decrease in the rise time and an increase of the maximum temperature during synthesis were observed for the increasing content of biopolyol independently of I_{ISO}. This was due to the greater reactivity of the systems containing biopolyol. Furthermore, a decrease in the maximum temperature was observed with an increase in I_{ISO}, due to the lower thermal effect of isocyanate trimerization compared to the enthalpy of polyurethane formation. All foams were characterized by apparent density in the range of 49–54 kg/m^3. Computer microtomography was applied to analyze the microstructure of foams containing 0 and 30 wt% of biopolyol. The average values of the cell diameter and volume of all foams were in the range of 95–131 μm and 5.4–11.0 mm$^3 \cdot 10^{-3}$, respectively. Based on the obtained histograms, it was found that the foams containing biopolyol were characterized by a higher content of cells smaller than 60 μm. All foams showed relatively low thermal conductivity λ in the range 24.69–26.98 mW/m·K. Along with the increase in biopolyol content, a decrease in λ by 2.29, 1.24, and 1.26 mW/m·K was observed for $I_{ISO} =$ 200, 250, and 300, respectively. The addition of biopolyol caused enhanced cross-linking of PUR-PIR foams, which was evidenced by a decrease in the amount of sol fraction, an increase in the glass transition temperature, and an increase in compressive strength. To determine the effect of cellular structure on mechanical strength, rigid foams were compressed in two directions, perpendicular and parallel to the rise of foam. Based on the thermogravimetric analysis, it was found that as the content of biopolyol in the formulations increased, their thermal stability increased slightly. FTIR analysis confirmed the occurrence of the characteristic absorption bands for rigid PUR-PIR foams (including urethane groups, isocyanate rings, methylene, ester, and ether groups). Furthermore, along with the increase in I_{ISO}, an increase in the absorption bands derived from N=C=O, C=O, and isocyanurate rings was observed, which confirmed the formation of PIR structures. To determine the share of PUR and PIR structures in foams, a chemometric analysis based on principal component analysis was performed using the Malinowski spectral isolation algorithm. The PCA confirmed that as the I_{ISO} increased, the proportion of relative PIR concentration increased. For $I_{ISO} = 200$, it was about 48%, and for $I_{ISO} = 300$, it was about 65%.

Author Contributions: Conceptualization, P.K. and Ł.P.; methodology, P.K. and J.S.; software, P.K.; validation, J.T.H., J.S., and A.H.; formal analysis, P.K. and Ł.P.; investigation, P.K. and Ł.P.; resources, J.T.H.; writing, original draft preparation, P.K. and A.H.; writing, review and editing, Ł.P. and J.T.H.; visualization, P.K.; supervision, J.T.H.; project administration, Ł.P. All authors read and agreed to the published version of the manuscript.

Funding: This research received no external funding.

Conflicts of Interest: The authors declare no conflict of interest.

References

1. Plaza, M.; Cifuentes, A.; Ibanez, E. In the search of new functional food ingrediens from algae. *Trends Food Sci. Technol.* **2008**, *19*, 31–39. [CrossRef]
2. Wells, M.L.; Potin, P.; Craigie, J.S.; Raven, J.A.; Merchant, S.S.; Helliwell, K.; Smith, A.G.; Camire, M.E.; Brawley, S.H. Algae as nutritional and functional food sources: Revisiting our understanding. *Environ. Boil. Fishes* **2016**, *29*, 949–982. [CrossRef] [PubMed]
3. Priyadarshani, I.; Rath, B. Commercial and industrial applications of micro algae—A review. *J. Algal Biomass Util.* **2012**, *3*, 89–100.
4. Ariede, M.B.; Candido, T.M.; Jacome, A.L.M.; Velasco, M.V.R.; De Carvalho, J.C.M.; Baby, A.R. Cosmetic attributes of algae—A review. *Algal Res.* **2017**, *25*, 483–487. [CrossRef]
5. Ishikawa, C.; Tafuku, S.; Kadekaru, T.; Sawada, S.; Tomita, M.; Okudaira, T.; Nakazato, T.; Toda, T.; Uchihara, J.N.; Taira, N.; et al. Anti-adult T-cell leukemia effects of brown algae fucoxanthin and its deacetylated product, fucoxanthinol. *Int. J. Cancer* **2008**, *123*, 2702–2712. [CrossRef]
6. Jung, H.A.; Jin, S.E.; Ahn, B.R.; Lee, C.M.; Choi, J.S. Anti-inflammatory activity of edible brown alga Eisenia bicyclis and its constituents fucosterol and phlorotannins in LPS-stimulated RAW264.7 macrophages. *Food Chem. Toxicol.* **2013**, *59*, 199–206. [CrossRef]

7. Park, M.-K.; Jung, U.; Roh, C. Fucoidan from Marine Brown Algae Inhibits Lipid Accumulation. *Mar. Drugs* **2011**, *9*, 1359–1367. [CrossRef]

8. Li, Y.; Lee, S.-H.; Le, Q.-T.; Kim, M.-M.; Kim, S.-K. Anti-allergic Effects of Phlorotannins on Histamine Release via Binding Inhibition between IgE and FcεRI. *J. Agric. Food Chem.* **2008**, *56*, 12073–12080. [CrossRef]

9. Mata, T.; Martins, A.; Caetano, N. Microalgae for biodiesel production and other applications: A review. *Renew. Sustain. Energy Rev.* **2010**, *14*, 217–232. [CrossRef]

10. Rowley, J.A.; Madlambayan, G.; Mooney, D.J. Alginate hydrogels as synthetic extracellular matrix materials. *Biomaterials* **1999**, *20*, 45–53. [CrossRef]

11. Hassan, M.M.; Mueller, M.; Wagners, M.H. Exploratory study on seaweed as novel filler in polypropylene composite. *J. Appl. Polym. Sci.* **2008**, *109*, 1242–1247. [CrossRef]

12. Na Sim, I.; Han, S.O.; Seo, Y.B. Dynamic mechanical and thermal properties of red algae fiber reinforced poly(lactic acid) biocomposites. *Macromol. Res.* **2010**, *18*, 489–495. [CrossRef]

13. Luan, L.; Wu, W.; Wagner, M.H. Rheological behavior of lubricating systems in polypropylene/seaweed composites. *J. Appl. Polym. Sci.* **2011**, *121*, 2143–2148. [CrossRef]

14. Jang, Y.H.; Han, S.O.; Na Sim, I.; Kim, H.-I. Pretreatment effects of seaweed on the thermal and mechanical properties of seaweed/polypropylene biocomposites. *Compos. Part A Appl. Sci. Manuf.* **2013**, *47*, 83–90. [CrossRef]

15. Oh, S.-T.; Kim, W.-R.; Kim, S.-H.; Chung, Y.-C.; Park, J. The preparation of polyurethane foam combined with pH-sensitive alginate/bentonite hydrogel for wound dressings. *Fibers Polym.* **2011**, *12*, 159–165. [CrossRef]

16. Yun, J.-K.; Yoo, H.-J.; Kim, H.-D. Preparation and properties of waterborne polyurethane-urea/sodium alginate blends for high water vapor permeable coating materials. *J. Appl. Polym. Sci.* **2007**, *105*, 1168–1176. [CrossRef]

17. Yuvarani, I.; Senthilkumar, S.; Venkatesan, J.; Kim, S.-K.; Al-Kheraif, A.A.; Anil, S.; Sudha, P.N. Chitosan Modified Alginate-Polyurethane Scaffold for Skeletal Muscle Tissue Engineering. *J. Biomater. Tissue Eng.* **2015**, *5*, 665–672. [CrossRef]

18. Chen, H.-B.; Shen, P.; Chen, M.-J.; Zhao, H.-B.; Schiraldi, D.A. Highly Efficient Flame Retardant Polyurethane Foam with Alginate/Clay Aerogel Coating. *ACS Appl. Mater. Interfaces* **2016**, *8*, 32557–32564. [CrossRef]

19. Patil, C.K.; Jirimali, H.D.; Paradeshi, J.S.; Chaudhari, B.; Alagi, P.; Hong, S.C.; Gite, V.V. Synthesis of biobased polyols using algae oil for multifunctional polyurethane coatings. *Green Mater.* **2018**, *6*, 165–177. [CrossRef]

20. Kadam, A.; Pawar, M.; Thamke, V.; Yemul, O. Polyester amide based polyurethane coatings from algae oil and their larvicidal, anti-ant properties. *Prog. Org. Coat.* **2017**, *107*, 43–47. [CrossRef]

21. Petrovic, Z.S.; Wan, X.; Bilić, O.; Zlatanić, A.; Hong, J.; Javni, I.; Ionescu, M.; Milic, J.; DeGruson, D. Polyols and Polyurethanes from Crude Algal Oil. *J. Am. Oil Chem. Soc.* **2013**, *90*, 1073–1078. [CrossRef]

22. Pawar, M.S.; Kadam, A.S.; Dawane, B.S.; Yemul, O.S. Synthesis and characterization of rigid polyurethane foams from algae oil using biobased chain extenders. *Polym. Bull.* **2015**, *73*, 727–741. [CrossRef]

23. Kosmela, P.; Gosz, K.; Kazimierski, P.; Hejna, A.; Haponiuk, J.T.; Skarżyński, Ł. Chemical structures, rheological and physical properties of biopolyols prepared via solvothermal liquefaction of Enteromorpha and Zostera marina biomass. *Cellulose* **2019**, *26*, 5893–5912. [CrossRef]

24. Hejna, A.; Kosmela, P.; Kirpluks, M.; Cabulis, U.; Klein, M.; Haponiuk, J.; Skarżyński, Ł. Structure, Mechanical, Thermal and Fire Behavior Assessments of Environmentally Friendly Crude Glycerol-Based Rigid Polyisocyanurate Foams. *J. Polym. Environ.* **2017**, *26*, 1854–1868. [CrossRef]

25. Hejna, A.; Kirpluks, M.; Kosmela, P.; Cabulis, U.; Haponiuk, J.; Skarżyński, Ł. The influence of crude glycerol and castor oil-based polyol on the structure and performance of rigid polyurethane-polyisocyanurate foams. *Ind. Crop. Prod.* **2017**, *95*, 113–125. [CrossRef]

26. Hejna, A.; Haponiuk, J.; Skarżyński, Ł.; Klein, M.; Formela, K. Performance properties of rigid polyurethane-polyisocyanurate/brewers' spent grain foamed composites as function of isocyanate index. *e-Polymers* **2017**, *17*, 427–437. [CrossRef]

27. Cellular Plastics and Rubbers. *Determination of Apparent (Bulk) Density*; PN-EN ISO 845-2000; Polish Committee for Standardization: Warsaw, Poland, 2000.

28. *Rigid Cellular Plastics: Determination of Compression Properties*; EN ISO 844:2007; ISO International Standards: Geneva, Switzerland, 2007.

29. *Standard Test Method for Steady-State Thermal Transmission Properties by Means of the Heat Flow Meter Apparatus*; ASTM C518; American Society of Testing and Materials C518: Montgomery, PA, USA, 2010.

30. Skarżyński, Ł.; Tejchman, J. Experimental Investigations of Fracture Process in Concrete by Means of X-ray Micro-computed Tomography. *Strain* **2015**, *52*, 26–45. [CrossRef]

31. Suchorzewski, J.; Tejchman, J.; Nitka, M. Experimental and numerical investigations of concrete behaviour at meso-level during quasi-static splitting tension. *Theor. Appl. Fract. Mech.* **2018**, *96*, 720–739. [CrossRef]

32. Kosmela, P.; Hejna, A.; Formela, K.; Haponiuk, J.; Skarżyński, Ł. The Study on Application of Biopolyols Obtained by Cellulose Biomass Liquefaction Performed with Crude Glycerol for the Synthesis of Rigid Polyurethane Foams. *J. Polym. Environ.* **2017**, *26*, 2546–2554. [CrossRef]

33. Bykova, T.A.; Lebedev, B.V.; Kiparisova, E.G.; Tarasov, E.N.; Frenkel, T.M.; Pankratov, V.A.; Vinogradova, S.V.; Korshank, V.V. Thermodynamics of phenyl isocyanate, the process of its cyclotrimerization, and the triphenyl isocyanurate that is formed, in the 0–330°K interval. *Zhurnal Obs. Khimii* **1985**, *55*, 2303–2308.

34. Zhang, X.; Jeremic, D.; Kim, Y.; Street, J.; Shmulsky, R. Effects of Surface Functionalization of Lignin on Synthesis and Properties of Rigid Bio-Based Polyurethanes Foams. *Polymers* **2018**, *10*, 706. [CrossRef]

35. Mondal, P.; Khakhar, D. Rigid polyurethane–clay nanocomposite foams: Preparation and properties. *J. Appl. Polym. Sci.* **2006**, *103*, 2802–2809. [CrossRef]

36. Fan, H.; Tekeei, A.; Suppes, G.J.; Hsieh, F.-H. Rigid polyurethane foams made from high viscosity soy-polyols. *J. Appl. Polym. Sci.* **2012**, *127*, 1623–1629. [CrossRef]

37. Ciecierska, E.; Jurczyk-Kowalska, M.; Bazarnik, P.; Gloc, M.; Kulesza, M.; Kowalski, M.; Krauze, S.; Lewandowska, M. Flammability, mechanical properties and structure of rigid polyurethane foams with different types of carbon reinforcing materials. *Compos. Struct.* **2016**, *140*, 67–76. [CrossRef]

38. Szycher, M. *Szycher's Handbook of Polyurethanes*, 1st ed.; CRC Press: Boca Raton, FL, USA, 1999.

39. Glicksman, L.R. Heat transfer in foams. In *Low Density Cellular Plastics*; Springer: Berlin/Heidelberg, Germany, 1994; pp. 104–152.

40. Randall, D.; Lee, S. *The Polyurethanes Boo*; John Wiley & Sons: New York, NY, USA, 2002.

41. Kim, S.H.; Kim, B.K.; Lim, H. Effect of isocyanate index on the properties of rigid polyurethane foams blown by HFC 365mfc. *Macromol. Res.* **2008**, *16*, 467–472. [CrossRef]

42. Modesti, M.; Lorenzetti, A. An experimental method for evaluating isocyanate conversion and trimer formation in polyisocyanate–polyurethane foams. *Eur. Polym. J.* **2001**, *37*, 949–954. [CrossRef]

43. Kurańska, M.; Barczewski, M.; Uram, K.; Lewandowski, K.; Prociak, A.; Michałowski, S. Basalt waste management in the production of highly effective porous polyurethane composites for thermal insulating applications. *Polym. Test.* **2019**, *76*, 90–100. [CrossRef]

44. Barczewski, M.; Kurańska, M.; Sałasińska, K.; Michałowski, S.; Prociak, A.; Uram, K.; Lewandowski, K. Rigid polyurethane foams modified with thermoset polyester-glass fiber composite waste. *Polym. Test.* **2020**, *81*. [CrossRef]

45. Guo, H.; Gao, Q.; Ouyang, C.; Zheng, K.; Xu, W. Research on properties of rigid polyurethane foam with heteroaromatic and brominated benzyl polyols. *J. Appl. Polym. Sci.* **2015**, *132*. [CrossRef]

46. Shams, A.; Stark, A.; Hoogen, F.; Hegger, J.; Schneider, H. Innovative sandwich structures made of high performance concrete and foamed polyurethane. *Compos. Struct.* **2015**, *121*, 271–279. [CrossRef]

47. Javni, I.; Zhang, W.; Petrovic, Z.S. Soybean-Oil-Based Polyisocyanurate Rigid Foams. *J. Polym. Environ.* **2004**, *12*, 123–129. [CrossRef]

48. Vale, M.; Mateus, M.M.; Dos Santos, R.G.; De Castro, C.A.N.; De Schrijver, A.; Bordado, J.C.; Marques, A.C. Replacement of petroleum-derived diols by sustainable biopolyols in one component polyurethane foams. *J. Clean. Prod.* **2019**, *212*, 1036–1043. [CrossRef]

49. Somani, K.P.; Kansara, S.S.; Patel, N.K.; Rakshit, A.K. Castor oil based polyurethane adhesives for wood-to-wood bonding. *Int. J. Adhes. Adhes.* **2003**, *23*, 269–275. [CrossRef]

50. Zhang, L.; Zhang, M.; Zhou, Y.; Hu, L. The study of mechanical behavior and flame retardancy of castor oil phosphate-based rigid polyurethane foam composites containing expanded graphite and triethyl phosphate. *Polym. Degrad. Stab.* **2013**, *98*, 2784–2794. [CrossRef]

51. Sormana, J.-L.; Meredith, J.C. High-Throughput Discovery of Structure−Mechanical Property Relationships for Segmented Poly(urethane−urea)s. *Macromolecules* **2004**, *37*, 2186–2195. [CrossRef]

52. Fournier, D.; Du Prez, F. "Click" Chemistry as a Promising Tool for Side-Chain Functionalization of Polyurethanes. *Macromolecules* **2008**, *41*, 4622–4630. [CrossRef]

53. Samborska-Skowron, R.; Balas, A. Jakościowa identyfikacja pierścieni izocyjanurowych w elastomerach uretanowo-izocyjanurowych i w ich hydrolizatach. *Polimery* **2003**, *48*, 371–374. [CrossRef]

54. Jiao, L.; Xiao, H.; Wang, Q.; Sun, J. Thermal degradation characteristics of rigid polyurethane foam and the volatile products analysis with TG-FTIR-MS. *Polym. Degrad. Stab.* **2013**, *98*, 2687–2696. [CrossRef]
55. Pretsch, T.; Jakob, I.; Müller, W. Hydrolytic degradation and functional stability of a segmented shape memory poly(ester urethane). *Polym. Degrad. Stab.* **2009**, *94*, 61–73. [CrossRef]
56. Romero, R.R.; Grigsby, R.A.; Rister, E.L.; Pratt, J.K.; Ridgway, D. A Study of the Reaction Kinetics of Polyisocyanurate Foam Formulations using Real-time FTIR. *J. Cell. Plast.* **2005**, *41*, 339–359. [CrossRef]
57. Xu, Q.; Hong, T.; Zhou, Z.; Gao, J.; Xue, L. The effect of the trimerization catalyst on the thermal stability and the fire performance of the polyisocyanurate-polyurethane foam. *Fire Mater.* **2017**, *42*, 119–127. [CrossRef]
58. Yarahmadi, N.; Vega, A.; Jakubowicz, I. Accelerated ageing and degradation characteristics of rigid polyurethane foam. *Polym. Degrad. Stab.* **2017**, *138*, 192–200. [CrossRef]

Article

The Impact of Ground Tire Rubber Oxidation with H_2O_2 and $KMnO_4$ on the Structure and Performance of Flexible Polyurethane/Ground Tire Rubber Composite Foams

Aleksander Hejna *, Adam Olszewski, Łukasz Zedler, Paulina Kosmela and Krzysztof Formela

Department of Polymer Technology, Gdańsk University of Technology, Gabriela Narutowicza 11/12, 80-233 Gdańsk, Poland; adam.olszewski1995@gmail.com (A.O.); lukasz.zedler@pg.edu.pl (Ł.Z.); paulina.kosmela@pg.edu.pl (P.K.); krzform1@pg.edu.pl (K.F.)
* Correspondence: aleksander.hejna@pg.gda.pl

Abstract: The use of waste tires is a very critical issue, considering their environmental and economic implications. One of the simplest and the least harmful methods is conversion of tires into ground tire rubber (GTR), which can be introduced into different polymer matrices as a filler. However, these applications often require proper modifications to provide compatibility with the polymer matrix. In this study, we examined the impact of GTR oxidation with hydrogen peroxide and potassium permanganate on the processing and properties of flexible polyurethane/GTR composite foams. Applied treatments caused oxidation and introduction of hydroxyl groups onto the surface of rubber particles, expressed by the broad range of their hydroxyl numbers. It resulted in noticeable differences in the processing of the polyurethane system and affected the structure of flexible composite foams. Treatment with H_2O_2 resulted in a 31% rise of apparent density, while the catalytic activity of potassium ions enhanced foaming of system decreased density by 25% and increased the open cell content. Better mechanical performance was noted for H_2O_2 modifications (even by 100% higher normalized compressive strength), because of the voids in cell walls and incompletely developed structure during polymerization, accelerated by $KMnO_4$ treatment. This paper shows that modification of ground tire rubber is a very promising approach, and when properly performed may be applied to engineer the structure and performance of polyurethane composite foams.

Keywords: polyurethane foam; ground tire rubber; rubber modification; compatibility; recycling

Citation: Hejna, A.; Olszewski, A.; Zedler, Ł.; Kosmela, P.; Formela, K. The Impact of Ground Tire Rubber Oxidation with H_2O_2 and $KMnO_4$ on the Structure and Performance of Flexible Polyurethane/Ground Tire Rubber Composite Foams. *Materials* **2021**, *14*, 499. https://doi.org/10.3390/ma14030499

Academic Editor: Alfonso Maffezzoli
Received: 15 December 2020
Accepted: 18 January 2021
Published: 21 January 2021

Publisher's Note: MDPI stays neutral with regard to jurisdictional claims in published maps and institutional affiliations.

1. Introduction

Polyurethane (PU) foams are a versatile group of materials commonly applied in various industry branches due to their very broad spectrum of potential properties [1]. Their structure and properties are directly influenced by their chemical composition and the ratio of the two most important components applied during polymerization—Polyols and isocyanates. In the simplest terms, polyurethane foams may be divided into rigid and flexible foams [2]. They are used in different applications, but both groups are prone to potential innovations. Just as in the case of other polymer materials, PU foams are often applied as matrices for polymer composites [3]. Like other materials, one of the main trends associated with foamed PU composites is the search for new fillers, preferably from renewable resources or byproducts of other processes and products [4–6]. Such a phenomenon is commonly observed and is driven by economic and ecological factors. The introduction of such materials could noticeably reduce the use of conventional, petroleum-based raw materials required to manufacture polyurethanes [7]. Among the potential filler candidates for polyurethane foams are the following: polyurethane foam scraps [8], waste lignocellulose fillers [9,10], textiles [11], eggshell waste [12] and rubber wastes [13]. The first solution is present on the market and used to manufacture the underlays for floors or carpet linings [14]. These materials are obtained by the re-foaming of the flexible polyurethane foam waste. Due to the cellular structure of matrix (new foam) and filler (foam scraps), they act as an excellent insulating material and are characterized by a thermal conductivity coefficient in the range of 0.036–0.041 W/(m/K), which is lower than conventional expanded polystyrene (~0.044 W/(m/K)) or mineral wool (~0.055 W/(m/K)) [15]. Moreover, they may also act as acoustic insulation with a sound reduction improvement of 19–44 dB, depending on the thickness.

The cellular structure of polyurethane foam scraps is a great advantage compared to the lignocellulose fillers, textiles or eggshell waste, because it is not affected by the filler particle size [9]. In case of solid fillers, their size and surface development noticeably affect the foaming of the polyurethane system, cellular structure and performance of resulting composites. Typical applications of PU foams such as damping, insulation or sound absorption require a well-developed cellular structure and are very sensitive towards changes [3]. Therefore, filler properties should be properly adjusted, which may require additional operations, especially considering that these materials themselves are rigid and do not show excellent damping or insulation performance. This could be the reason such materials are not industrially produced and applied as insulation materials.

The last material, rubber waste, is an auspicious one, due to the excellent mechanical properties of many primary rubber materials, e.g., car tires. They are commonly used in mechanical recycling resulting in the production of ground tire rubber (GTR). This material can be efficiently introduced into various polymer matrices, including polyurethane foams [16]. Contrary to the lignocellulose materials or textiles, GTR seems to be a more promising material due to its structure and properties. Waste rubber is a viscoelastic material, which when applied as a filler may enhance the damping [17] and sound absorption [18,19] performance of polymeric materials. Therefore, in foamed polyurethane composites it may act similarly to the waste polyurethane scraps, despite the lack of cellular structure.

The main factor limiting the application of ground tire rubber in polymer composites is the insufficient compatibility with polymer matrices. As a result, interfacial interactions between GTR applied as a filler and the continuous polymer phase is too weak for efficient stress transfer. The strength of materials is reduced, limiting one of rubber's main advantages—an excellent mechanical performance. The proper adhesion between phases is particularly important in the case of foamed composites, e.g., based on polyurethanes (PU), whose mechanical performance is strictly associated with the apparent density, which is proportional to the share of solid material [20]. Therefore, it is essential to enhance the interfacial interactions between the matrix and filler. As mentioned above, polyurethanes are obtained by the reactions between polyols and isocyanates. Therefore, to enhance the compatibility of GTR with PU, it is beneficial to introduce hydroxyl or isocyanate functional groups onto its surface. The more feasible approach is related to the hydroxyl groups, which can be introduced during partial devulcanization or oxidation of the GTR surface [21]. Different approaches to GTR surface treatment resulting in its activation and incorporation of functional groups have been reported in the literature, which was aimed at the enhancement of adhesion with polymer matrices [22].

In the presented research work, we aimed to investigate GTR treatment's impact with hydrogen peroxide and potassium permanganate on its structure and properties. We proposed the titration-based method to evaluate modification effectiveness for the potential applications of modified GTR in manufacturing polyurethane-based composites. Moreover, modified GTR samples were introduced into a flexible foamed PU matrix. The influence of GTR modifications on the processing of polyurethane systems (kinetic profile of foaming, processing times and temperatures), cellular structure (scanning electron microscopy, helium pycnometry), mechanical (static compression tests), and thermal (thermogravimetric analysis) properties were determined.

2. Materials and Methods

2.1. Materials

Ground tire rubber (GTR) obtained by ambient grinding of used tires (a combination of passenger car and truck tires in 50:50 mass ratio), whose average particle size is approximately 0.6 mm, was produced and provided by Recykl S.A. (Srem, Poland).

The 30% solution of hydrogen peroxide and crystals of potassium permanganate were acquired from Sigma Aldrich (Poznan, Poland). During the evaluation of the chemical structure of modified GTR the following chemicals were applied: acetone, dibutylamine, chlorobenzene, hydrochloric acid, technical grade toluene diisocyanate (TDI) and 3′,3″,5′,5″-tetrabromophenolsulfonphthalein. All chemicals were acquired from Sigma Aldrich (Poznan, Poland) and were used as received.

Polyurethanes were synthesized from a commercially available polyurethane system consisting of SPECFLEX® NF 706 polyol and SPECFLEX® NE 434 isocyanate, acquired from M. B. Market Ltd. (Czestochowa, Poland). The densities of used components at 25 °C were equal to 1.03 and 1.21 g/cm^3, respectively, while their viscosity values at 25 °C equal 1340 and 66 mPa/s. According to the manufacturer, an applied polyurethane system is recommended to produce highly flexible, formed polyurethane foams.

2.2. Modifications of GTR

GTR was modified with a 30% solution of hydrogen peroxide and a 15% solution of potassium permanganate. The solution of potassium permanganate was prepared by dissolving its crystals in distilled water. Particles of ground tire rubber and the proper solution in different weight ratios: 1:2, 1:1 and 2:1, were mixed for 5 min at room temperature using a mechanical stirrer. They were left in solutions at room temperature for 72 h, then strained and dried at 70 °C for 8 h. For comparison, GTR dried at 70 °C for 8 h was used as reference.

2.3. Preparation of Polyurethane/GTR Composite Foams

Polyurethane/GTR composite foams were prepared on a laboratory scale by a single-step method. Predetermined amounts of polyol and isocyanate were mixed at a 100:70 mass ratio for 5 s at 1800 rpm. In the case of modified foams, incorporated fillers were previously mixed with polyol components for 1 min at 1800 rpm. Total mass of foam was set at 150 g. The resulting mixture was left for a free rise. After, the samples were conditioned at room temperature for 24 h. Table 1 contains the details of foam formulations.

Table 1. Formulations of prepared composite foams.

Component	Foam Symbol							
	P0	P1	P2	P3	P4	P5	P6	P7
	Component Content, wt.%							
Polyol	58	47	47	47	47	47	47	47
Isocyanate	42	33	33	33	33	33	33	33
GTR	-	20	-	-	-	-	-	-
GTR:H$_2$O$_2$ 2:1	-	-	20	-	-	-	-	-
GTR:H$_2$O$_2$ 1:1	-	-	-	20	-	-	-	-
GTR:H$_2$O$_2$ 1:2	-	-	-	-	20	-	-	-
GTR:KMnO$_4$ 2:1	-	-	-	-	-	20	-	-
GTR:KMnO$_4$ 1:1	-	-	-	-	-	-	20	-
GTR:KMnO$_4$ 1:2	-	-	-	-	-	-	-	20

2.4. Measurements

Changes in the chemical structure of GTR were evaluated using a modified method for the determination of free isocyanate group content by titration with dibutylamine, according to ASTM D-2572 [23]. The 0.5 g samples of GTR were put in a glass flask with 0.5 g of toluene diisocyanate and 20 cm^3 of acetone. Mixtures were thoroughly mixed, sealed and stored at room temperature for 24 h. Then, proper amounts of dibutylamine solution in chlorobenzene and 3′,3″,5′,5″-tetrabromophenolsulfonphthalein were added. Then, mixtures were titrated with 0.1 M hydrochloric acid until the color changed to yellow. Obtained results were compared with the free isocyanate content of neat toluene diisocyanate to determine the number of functional groups at the rubber surface able to react with isocyanates. Such evaluation is essential for the potential application of modified GTR as a filler for polyurethane materials.

The free isocyanate content of the GTR/TDI mixture ($\%_{NCO}$) was calculated according to the following Equation (1):

$$\%_{NCO} = (4.202 \times (V_B - V_S) \times N_{HCl})/m_{TDI} \tag{1}$$

where: V_B—The volume of HCl required for titration of the blank sample, ml; V_S—The volume of HCl required for titration of analyzed sample, ml; N_{HCl}—Molarity of HCl, M; m_{TDI}—The mass of TDI placed in the flask, g.

Based on these values, the assumed hydroxyl numbers (L_{OH}) of GTR were calculated. During calculations, it was assumed that all of the consumed isocyanate groups reacted with the GTR particles. Another assumption was that all of the functional groups present on the surface of GTR were hydroxyls. Considering these assumptions, the number of hydroxyl groups, which took part in reactions was calculated following the Equation (2):

$$X_{OH} = X_{NCO} = ((\%_{NCO\text{-}TDI} - \%_{NCO}) \times m_{TDI} \times 2)/(M_{TDI} \times 100) \tag{2}$$

where: $\%_{NCO\text{-}TDI}$—Free isocyanate content in TDI, equal to 42.7%; M_{TDI}—The molar mass of TDI, equal to 174.2 g/mol.

Then, the hydroxyl number of GTR was calculated from the Formula (3):

$$L_{OH} = 56{,}100 \times X_{NCO}/m_{GTR} \tag{3}$$

where: m_{GTR}—The mass of GTR placed in the flask, g.

The thermogravimetric (TGA) analysis of GTR and composites was performed using the TG 209 F3 apparatus from Netzsch (Selb, Germany). Samples of composites weighing approx. 10 mg were placed in a ceramic dish. The study was conducted in an inert gas atmosphere—Nitrogen in the range from 30 to 900 °C with a temperature increase rate of 10 °C/min.

For all the samples, the following processing times were determined: rise time (time of volumetric expansion) and the tack-free time (from the end of volumetric expansion to the point when the surface stopped being tacky to the touch). Moreover, during polymerization, the temperature of the foam surface was measured with infrared thermal imaging camera model Testo 872 (Testo SE & Co. KGaA, Lenzkirch, Germany).

After conditioning, foamed polyurethane composites were cut into samples whose properties were later determined following the standard procedures.

The samples' morphology was evaluated by using a Hitachi model S3400 (Tokyo, Japan) scanning electron microscopy.

The samples' apparent density was calculated following PN-EN ISO 845 [24], as a ratio of the sample weight to the sample volume (g/cm^3). The cube-shaped samples were measured with a slide caliper with an accuracy of 0.1 mm and weighed using an electronic analytical balance with an accuracy of 0.0001 g.

The content of open cells in foamed PU/GTR composites was determined using Ultrapyc 5000 Foam gas pycnometer from Anton Paar (Warszawa, Poland). The following measurement settings were applied: gas—Helium; target pressure—3.0 psi; foam mode—On; measurement type—Uncorrected; flow direction—Sample first; temperature control—On; target temperature—20.0 °C; flow mode—Monolith; cell size—Small, 10 cm^3; preparation mode—Flow; time of the gas flow—0.5 min.

The compressive strength of studied samples was estimated following ISO 604 [25]. The cylindric samples with dimensions of 20 mm × 20 mm (height and diameter) were measured with a slide caliper with an accuracy of 0.1 mm. The compression test was performed on a Zwick/Roell Z020 tensile tester (Ulm, Germany) at a constant speed of 15%/min until reaching 70% deformation.

3. Results and Discussion

3.1. Changes in GTR Structure

In Table 2, the free isocyanate contents ($\%_{NCO}$) of GTR/TDI mixtures prepared according to developed methodology and the decrease of their value comparing to neat TDI (Δ_{NCO}) are shown. Before the modification, GTR particles also contained the functional groups able to react with isocyanates on its surface. Considering the data presented by Vilar [26], at room temperature and without the catalyst, isocyanates are reacting the most rapidly with amines (relative reaction rates from 200 for aromatic amines to even 100,000 for primary aliphatic ones), which are hardly present in GTR. Lower reaction rates are noted for primary hydroxyls and water (100), followed by carboxylic acids (40), secondary hydroxyls (30) and ureas (15) [26]. Therefore, the presence of these groups determines the reactivity of ground tire rubber particles with isocyanates. The potential reactions with the free isocyanate groups are presented in Figure 1.

Table 2. Free isocyanate contents of GTR/TDI mixtures, amount of isocyanate groups of TDI consumed by GTR and calculated hydroxyl numbers of GTR.

Sample	$\%_{NCO}$, %	Δ_{NCO}, %	L_{OH} mg KOH/g
GTR	33.3 ± 1.1	9.4 ± 1.1	61.7 ± 3.0
GTR:H_2O_2 2:1	37.0 ± 0.3	5.7 ± 0.3	36.4 ± 1.6
GTR:H_2O_2 1:1	37.3 ± 1.1	5.4 ± 1.1	34.5 ± 2.4
GTR:H_2O_2 1:2	37.9 ± 1.1	4.8 ± 1.1	32.1 ± 2.3
GTR:$KMnO_4$ 2:1	12.1 ± 0.5	30.6 ± 0.5	205.9 ± 9.9
GTR:$KMnO_4$ 1:1	7.4 ± 0.5	35.3 ± 0.5	226.3 ± 6.2
GTR:$KMnO_4$ 1:2	3.7 ± 0.2	39.0 ± 0.2	248.9 ± 3.3

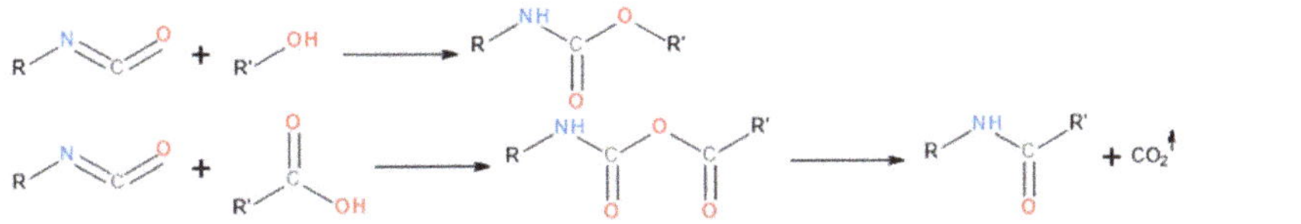

Figure 1. Possible reactions of isocyanates with functional groups present on the surface of modified GTR.

The TDI's initial free isocyanate content mixed with unmodified GTR equaled 33.3%, indicating its drop by 9.4%. As a result, the hydroxyl number of neat GTR was determined as 61.7 mg KOH/g. The presence of the functional groups on the surface of unmodified GTR was associated with the shredding of tires, which is performed under air atmosphere, which together with the high shear forces enables oxidation of rubber particles [27].

Performed modifications of GTR caused noticeable changes in the chemical structure of their surface. It can be seen that the introduction of hydrogen peroxide resulted in the rise of the free isocyanate content, pointing to the reduced reactivity of GTR with TDI. Previous reports indicated rubber surface activation by creating carboxylic sites [28]. Hydrogen peroxide causes the generation of carbonium ions on the surface, which are converted into carboxylic sites. Potential reactions are presented in Figure 2. According to Shatanawi et al. [25], such an effect may be used to enhance the interactions between modified rubber and asphalts. A similar phenomenon was noted by Yehia et al. [29] for natural rubber vulcanizates. Moreover, they confirmed the generation of carboxyl groups by FTIR analysis.

In the presented case, the hydrogen peroxide treatment probably caused oxidation of the hydroxyls present on the surface of GTR particles, which resulted in the generation of carboxyl groups [30]. As mentioned above, carboxyls show lower reactivity with isocyanate groups. Hence, the hydroxyl number determined by the applied methodology was reduced. Although presented values should not be treated as the quantitative indicator or hydroxyl group content, they could be helpful for the adjustment of polyurethane foam recipes because they represent the content of groups reacting with isocyanate—One of the main components of polyurethanes.

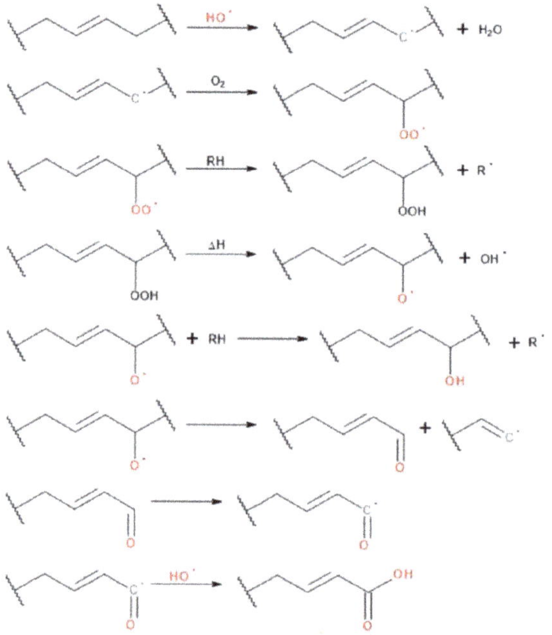

Figure 2. Schematic reactions occurring during oxidation of rubber.

On the other hand, potassium permanganate application caused a significant drop in GTR/TDI mixtures' isocyanate content. Such an effect is associated with the chemical reactions occurring during the modification. As commonly known, $KMnO_4$ is a strong oxidizing agent, independently of the media and pH value. In neutral solution, which was applied for modification it gets reduced to the brown manganese dioxide and four OH^- groups are released [31]. Moreover, when $KMnO_4$ is attacking the alkene double bond, which could be present in GTR, two hydroxyl groups are generated [32]. As a result, modification with $KMnO_4$ resulted in very high hydroxyl numbers. Therefore, compared to the H_2O_2, potassium permanganate could be considered a more effective activator of rubber particles' surface aimed at the preparation of polyurethane materials.

Potassium permanganate is a highly reactive oxidizer. Even at ambient temperature, a violent reaction occurs with the release of MnO_2 and molecular oxygen. The aforementioned compound application as a GTR surface modifier was studied before by Sonnier et al. [33], who confirmed the GTR surface oxidation phenomenon by a 2% solution of $KMnO_4$ resulting in the creation of carbonyl groups. In the present study, GTR was treated with a 15% solution of the compound at three different ratios to conduct the process in a highly aggressive reaction environment. The treatment was also done using a 30% solution of H_2O_2, which also oxidizes the surface of GTR [29].

The impact of H_2O_2 and $KMnO_4$ treatment on the morphology of GTR is presented in Figure 3. Compared to unmodified GTR, the surface of H_2O_2 is more developed [34], however, it does not change significantly with the higher content of the applied modifier. It is in line with hydroxyl numbers determined for GTR:H_2O_2 2:1, GTR:H_2O_2 1:1 and

GTR:H_2O_2 1:2. Increased roughness is also noticeable for $KMnO_4$ modified samples and it is more developed with an increasing amount of the oxidizer. This phenomenon is correlated with the oxidation of the GTR surface and the formation of MnO_2, which was not removed at the end of the modification. The more of the substrate is used, the more products are being formed, influencing the morphology of GTR. More developed specific surface area and potentially formed groups obtained via H_2O_2 and $KMnO_4$ oxidation may improve compatibility between PU matrix and GTR filler, thereby changing the mechanical properties of PU/GTR materials.

Thermogravimetric analysis (TGA) is a convenient method to analyze the changes in the chemical structure of reclaimed/modified rubber [35,36]. To evaluate the applied chemical treatment on GTR, the thermal stability, characteristic peaks and the amount of final residue of the neat and modified GTRs were measured and analyzed. The obtained results are presented in Table 3 and Figure 4.

Two characteristic peaks related to the two main components of GTR were noticed at approx. 362.8–373.6 °C (T_{max1}) and at approx. 424.0–438.7 (T_{max2}). These respectively correspond to thermal degradation of natural rubber and styrene-butadiene rubber [37], which proves that used material is taken from waste tires. All samples were characterized by T_{max1} and T_{max2} close to values for the reference sample, whereas the largest deviation from the neat GTR had sample GTR:$KMnO_4$ 1:2 (T_{max1}—6.1 °C and T_{max2}—14.7 °C). While the deviation can be treated as a measurement error (~1.6 and 3.3%, respectively), it still varies significantly from the rest of the studied samples, especially the T_{max2} value. The shift of the temperature towards a higher value may be due to the start of thermal decomposition of MnO_2 (483.0 °C) obtained during the oxidation reaction.

The difference in residue between the samples is negligible, although it is up to 7.5% by weight (GTR:$KMnO_4$ 2:1) compared to the reference sample. This is because the analyzed sample is extremely small (approx. 10 mg), and GTR is a mix of car/truck tires that differs in composition. Those two factors influence the residue values.

The surface modifications change the thermal behavior of the studied samples. $T_{-2\%}$ values shift towards lower temperatures (215.4, 202.3, 204.1, 227.9, 213.0 and 204.0 °C for GTR:H_2O_2 2:1, GTR: H_2O_2 1:1, GTR: H_2O_2 1:2, GTR:$KMnO_4$ 2:1, GTR:$KMnO_4$ 1:1 and GTR:$KMnO_4$ 1:2, respectively) compared to the reference sample (233.0 °C for neat GTR), which resulted from the oxidation of the waste rubber. Moreover, as a result of the reactions, changes in the structure could occur and affect the volatilization of GTR unreacted components [38] and accelerate the process. The $T_{-5\%}$ values of $KMnO_4$ modified GTR is similar to the reference sample (286.7, 293.6, 287.4 and 285.3 °C for neat GTR, GTR:$KMnO_4$ 2:1, GTR:$KMnO_4$ 1:1 and GTR:$KMnO_4$ 1:2, respectively), while for GTR oxidized with H_2O_2 values drop (273.5, 266.9 and 265.1 °C for GTR:H_2O_2 2:1, GTR: H_2O_2 1:1 and GTR: H_2O_2 1:2, respectively). H_2O_2, as well as $KMnO_4$, oxidize the structure shifting thermal decomposition towards lower temperatures. However, due to the $KMnO_4$ oxidation, MnO_2 is being formed and not removed from the GTR. The presence of the component, which is characterized by high decomposition temperature (483.0 °C) influences the thermal stability as the temperature rises, which is also in accordance with $T_{-10\%}$ and $T_{-50\%}$ values. Significantly lower degradation temperatures of H_2O_2 modified materials may also indicate higher oxidation efficiency of GTR compared to $KMnO_4$.

Table 3. The results of thermogravimetric analysis of neat and modified GTR.

GTR Type	$T_{-2\%}$, °C	$T_{-5\%}$, °C	$T_{-10\%}$, °C	$T_{-50\%}$, °C	Residue$_{890\,°C}$, %	T_{max1}, °C	T_{max2}, °C
GTR	233.0	286.7	331.1	435.0	35.86	369.0	424.0
GTR:H_2O_2 2:1	215.4	273.5	325.3	434.3	34.68	369.3	426.1
GTR: H_2O_2 1:1	202.3	266.9	321.7	438.7	37.57	368.8	429.6
GTR: H_2O_2 1:2	204.1	265.1	320.7	435.9	35.39	370.8	425.6
GTR:$KMnO_4$ 2:1	227.9	293.6	338.9	447.3	38.56	369.5	432.0
GTR:$KMnO_4$ 1:1	213.0	287.4	337.2	449.0	37.88	373.6	429.1
GTR:$KMnO_4$ 1:2	204.0	285.3	336.6	457.8	37.85	369.2	438.7

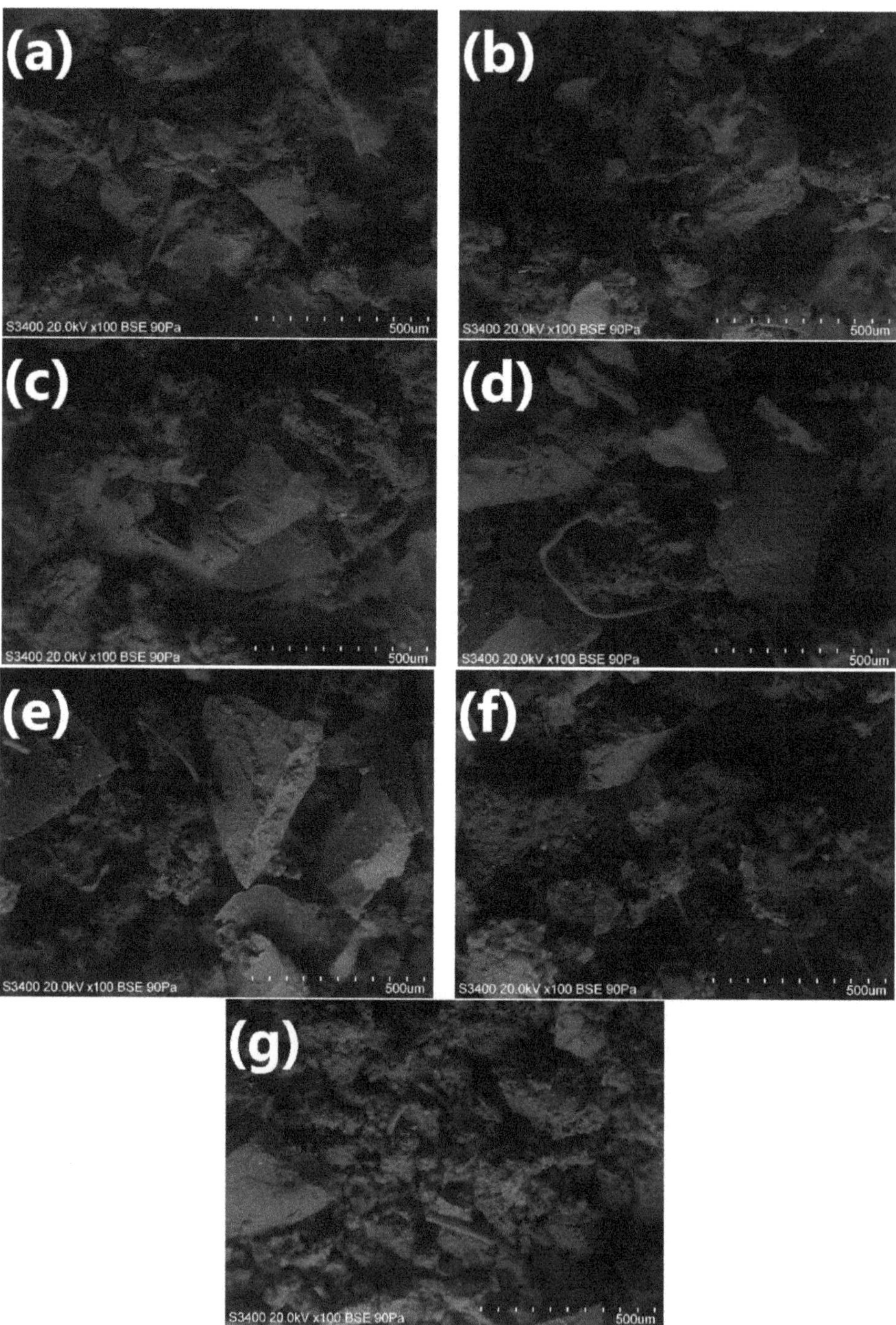

Figure 3. SEM images of studied samples: (**a**) neat GTR; (**b**) GTR:H$_2$O$_2$ 2:1; (**c**) GTR:H$_2$O$_2$ 1:1; (**d**) GTR:H$_2$O$_2$ 1:2; (**e**) GTR:KMnO$_4$ 2:1; (**f**) GTR:KMnO$_4$ 1:1 and (**g**) GTR:KMnO$_4$ 1:2 (magnification ×100).

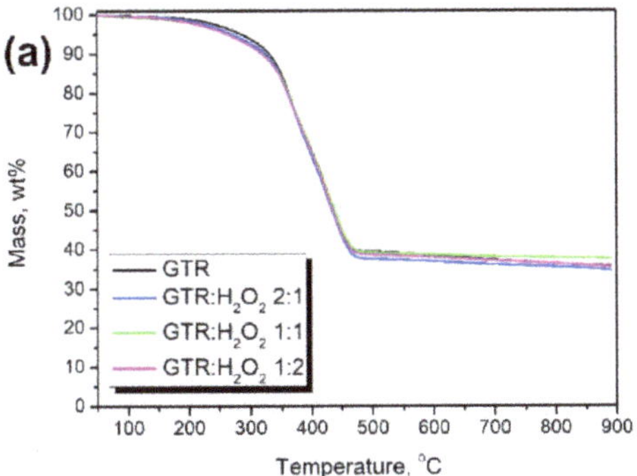
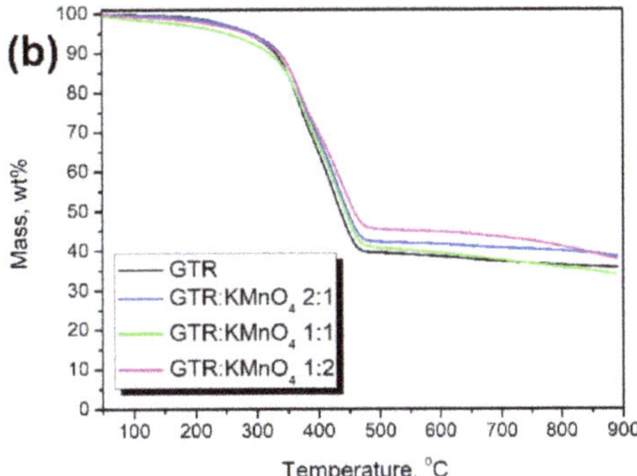

Figure 4. Results of thermogravimetric analysis of applied GTR fillers modified with (**a**) H_2O_2 and (**b**) $KMnO_4$.

3.2. Kinetics of Foaming

In Figure 5, there are presented values of rise time and tack-free time, depending on the foam formulation. The introduction of GTR into the polyol mixture resulted mainly in the elongation of rise time. Such an effect was related to increased viscosity of the polyol mixture due to the incorporation of solid rubber particles, which was also noted in our previous works [39–41]. Nevertheless, in these works, we investigated the rigid polyurethane foams, so different material behavior was noted. Higher isocyanate excess was applied, so the matrix was "stronger" during volumetric expansion and less sensitive to increased viscosity, even with isocyanate groups' partial attraction by functional groups of GTR. For flexible foams, when isocyanate:hydroxyl ratio in the system is lower, the attraction of isocyanate groups by GTR causes the "weakening" of the polyurethane matrix, and hence the elongation of rise time.

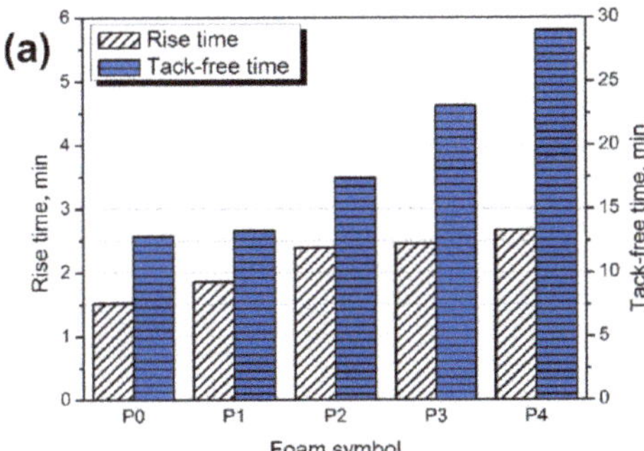
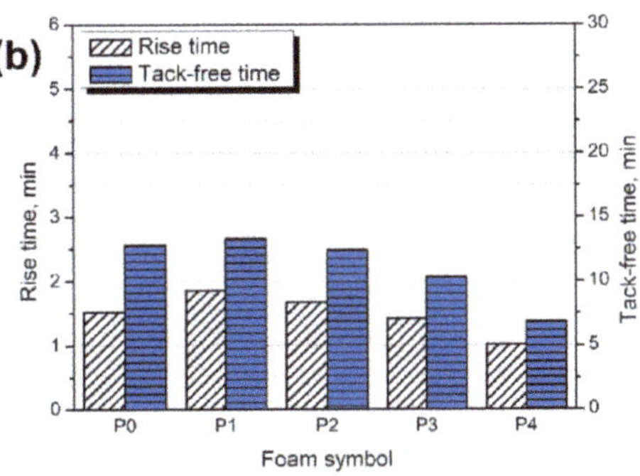

Figure 5. Processing times of reference foam, foam containing neat GTR (**a**) GTR modified with H_2O_2, and (**b**) GTR modified with $KMnO_4$.

Modification of GTR with hydrogen peroxide resulted in elongation of the processing times. Such an effect could be associated with the noticeable increase of the polyol mixture's viscosity related to the incorporation of GTR. A similar phenomenon was noted in our previous works [39,40]. As shown in the SEM images of modified GTR (Figure 3), treatment with hydrogen peroxide caused surface development, which resulted in the enhanced interactions with the polyol mixture.

For modification with $KMnO_4$, processing times were noticeably shortened, which was related to enhanced catalytic activity due to potassium permanganate's basic character [26]. Basic catalysts are commonly applied in polyurethane synthesis and are not selective. They show strong catalytic activity in reactions with hydroxyl groups and water, but also with ureas and urethanes [26]. Therefore, they are accelerating all polyurethane foam synthesis steps, which can be seen in Figure 5, indicating a noticeable reduction of the rise and tack-free times. In the presented case, the catalytic effect was so strong that it overcame the viscosity increase associated with the significant development of the GTR surface after modification with potassium permanganate (see Figure 3).

In Figure 6, the presented graphs show the temperature build-up on the foam surface measured by the thermographic camera. Exemplary photographs are shown in Figure 7. The introduction of GTR into polyurethane foam reduced the maximum foam temperature during polymerization. It was associated with the interactions between GTR and isocyanates and confirms the results of previous works [39,40]. Moreover, the maximum temperature is reached almost 30 s later than for unfilled polyurethane foam, which confirms the processing times' elongation due to the polyol mixture's increased viscosity.

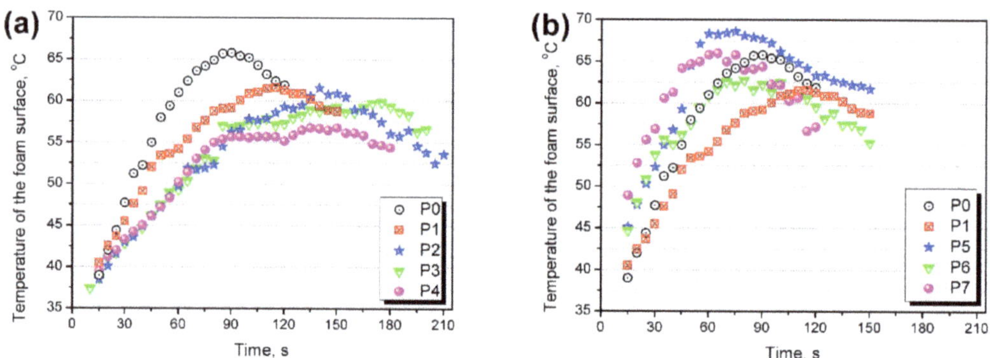

Figure 6. Temperature build-up on the surface of foams containing (**a**) GTR modified with H_2O_2 and (**b**) GTR modified with $KMnO_4$.

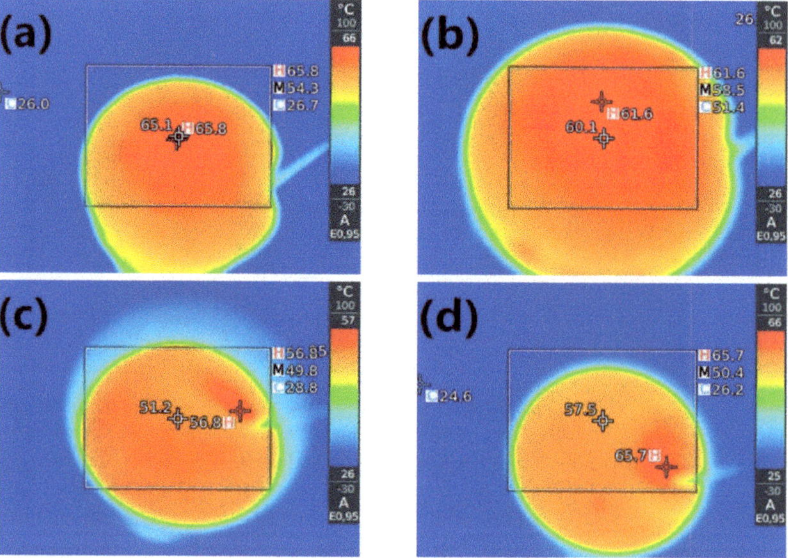

Figure 7. Photographs from thermographic camera showing the maximum temperatures during foaming of (**a**) P0, (**b**) P1, (**c**) P4, and (**d**) P7 samples.

Modification of GTR by the hydrogen peroxide resulted in a slight reduction of the maximum foam surface temperature. It was caused by the increased viscosity of the polyol mixture related to the more developed surface area of modified rubber particles (see Figure 3). Due to the retardation of polymerization, heat build-up inside the foam was lower, and due to the heat convection to the environment, lower surface temperatures were noted.

In the case of $KMnO_4$ modification, the catalytic activity fastened the temperature rise compared to P0 and P1 samples and increased the maximum temperature observed at the rising foam surface. Due to the acceleration of chemical reactions, the heat generated during polymerization could not dissipate and was accumulated inside the foam, increasing its temperature.

3.3. Structure and Properties of PU/GTR Composite Foams

In Figure 8, there are presented values of the apparent density and open cell content, parameters of foams' cellular structure directly influenced by the course of the polymerization process. Incorporation of GTR into the foamed polyurethane matrix resulted in a noticeable increase in apparent density, which was associated with changes in the viscosity of the polyol mixture and differences in density between matrix and filler. The apparent density of foam P1 was noticeably affected when GTR particles applied as a filler were subjected to oxidation with hydrogen peroxide and potassium permanganate. However, significantly different behavior was noted for both modifiers.

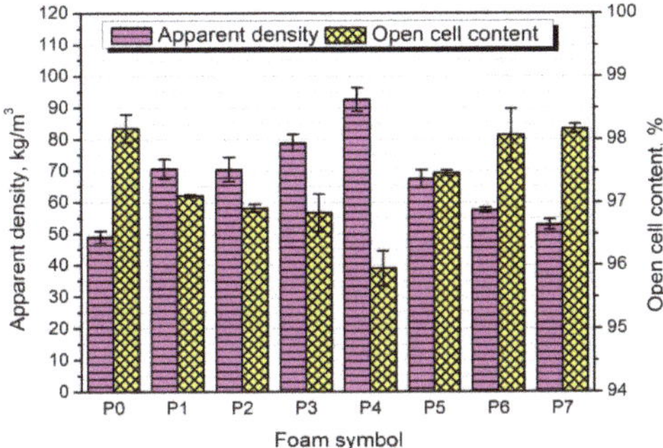

Figure 8. Values of the apparent density and open cell content for prepared PU/GTR composite foams.

The pretreatment of GTR particles with the hydrogen peroxide caused an increase in apparent density proportional to the applied modifier content. Such an effect is associated with the elongation of the processing times. As mentioned above, peroxide treatment of GTR resulted in the significant development of particles' surface area, causing the enhanced viscosity of the reacting mixture and reducing the level of foams' volumetric expansion. On the other hand, the catalytic effect of applied $KMnO_4$ treatment resulted in the acceleration of all reactions occurring inside the foam, including the generation of carbon dioxide. Moreover, due to the higher temperature of the reacting mixture, its viscosity was lower, and gas could evaporate faster, increasing the volumetric expansion. Therefore, the apparent density was significantly reduced, and foam P7 almost reached the level of unfilled foam (52.9 and 49.0 kg/m^3, respectively).

The content of open cells inside foam was also related to the type of applied filler. Its value equaled 98.2% for unfilled foam, typical for flexible, open-cell foams [42]. Viscosity changes and a higher density of GTR, which reduced the foam's volumetric expansion, also caused a slight drop of open cell content to 97.1%. Similar to the apparent density, modification of rubber particles with H_2O_2 caused changes related to the enhanced level of interfacial interactions. Increased viscosity of the polyol mixture caused trapping of carbon dioxide generated during the foaming of material. The adverse effect was noted for potassium permanganate treatment. Due to accelerated foam rise and the lower level of the interactions between GTR and polyurethane matrix, generated gas could rapidly foam the reacting mixture, which was simultaneously solidifying, and then escape from foam. The higher temperature of foam compared to hydrogen peroxide modification also facilitated the gas movement.

Figure 9 presents the images of foams' cellular structure obtained with the scanning electron microscope. It can be seen that the reference foam, without rubber, shows relatively regular cells with noticeable holes in cell walls resulting in high open cell content. Such an effect is typical for the flexible polyurethane foams, which are often used in applications that require a high content of open cells [42]. The introduction of unmodified GTR caused some irregularities in the cell size and shape, which was also observed in our previous work and by other researchers [43,44]. Nevertheless, the disruption of the cellular structure was not very significant. For foams P2 and P4 containing the GTR modified with the hydrogen peroxide, the changes were more visible. A noticeable rise in cell size was observed, and for foam P4, the regularity of the structure was disrupted. It was probably associated with enhancing the polyol mixture's viscosity caused by the higher specific weight of GTR and enhanced interactions with polyurethane components. In the SEM image of sample P4, it can be seen that the cellular structure was even collapsed, which contributed to the increase in the foam's apparent density. For foams containing $KMnO_4$ modified GTR, changes were also evident. Due to the catalytic activity of potassium ions, the cells were bigger and with a higher portion of holes. Such an effect was related to the accelerated gas evaporation during foaming and increased the open cell content in foams (see Figure 8).

In Figure 10 there are presented values of the compressive strength of prepared polyurethane/ground tire rubber composites at various levels of deformation. It can be seen that the incorporation of GTR caused a noticeable increase in the foam's strength. The compressive performance enhancement was even more substantial when GTR particles were pretreated with the hydrogen peroxide solution. Such an effect is associated with improved interfacial interactions related to rubber particles' surface development (see Figure 3). On the other side, modification with $KMnO_4$ caused a drastic drop of foams' compressive strength, even below the values noted for neat, unfilled foam. Such an effect was related to the weakening of the PU cellular structure, caused by the significant increase of holes in cell walls (see Figure 9). As a result, the foams' structure was able to withstand only low forces without breaking.

The compressive performance of cellular materials is strictly associated with their apparent density. A higher density of foam implicates a more compact structure, which can withstand higher forces. Therefore, for the real comparison of different cellular materials, the impact of apparent density should be eliminated. In Figure 11 there are presented normalized values of foams' compressive strength. The incorporation of neat GTR results in a "true" reinforcement effect compared to the unfilled material. This effect was enhanced by the modification with H_2O_2 and its increased loading. For $KMnO_4$, values are significantly lower, even despite the decreased apparent density, indicating that GTR did not cause a reinforcing effect after this modification. Such an effect could be associated with the changes in the isocyanate:hydroxyl ratio, due to the high hydroxyl numbers of $KMnO_4$ modified GTR.

In Table 4, there are presented the results of the thermogravimetric analysis of prepared foams. Moreover, Figure 12 shows the course of degradation, and differential thermogravimetric curves, indicating the rate of decomposition at particular temperatures. They are presented only for the 200–550 °C temperature range, when the actual thermal degradation occurred. It can be seen that the reference foam is characterized by thermal stability typical for flexible polyurethane foams [45]. The onset of the thermal degradation process, measured as a temperature associated with the 2 wt.% mass loss, was determined as 264.2 °C. It is significantly above the temperature range required for the typical applications of the flexible polyurethane foams, which is usually in the range of 160–180 °C [26]. Reference foam is characterized by the almost single-step decomposition process, with the maximum rate of decomposition at 385.5 °C, which is typical for the polyurethane soft segment decomposition [46]. Some portion of the material may degrade at slightly lower temperatures, which corresponds to the decomposition of urethane bonds. The introduction of neat and modified GTR reduced the thermal stability of prepared foams. Such an effect may be related to the lower stability of introduced rubber particles than the polyurethane matrix, due to its oxidation (see Table 3). Similar effects were observed in our previous works, when partially devulcanized rubber was introduced into PU foams [39,40]. The effect was more pronounced for GTR modified with potassium permanganate, which is associated with the enhanced catalytic activity and weaker polyurethane matrix. Moreover, due to the high hydroxyl numbers of modified rubbers, the matrix could not be fully developed, and unbound macromolecular chains of polyol could reduce the decomposition temperature.

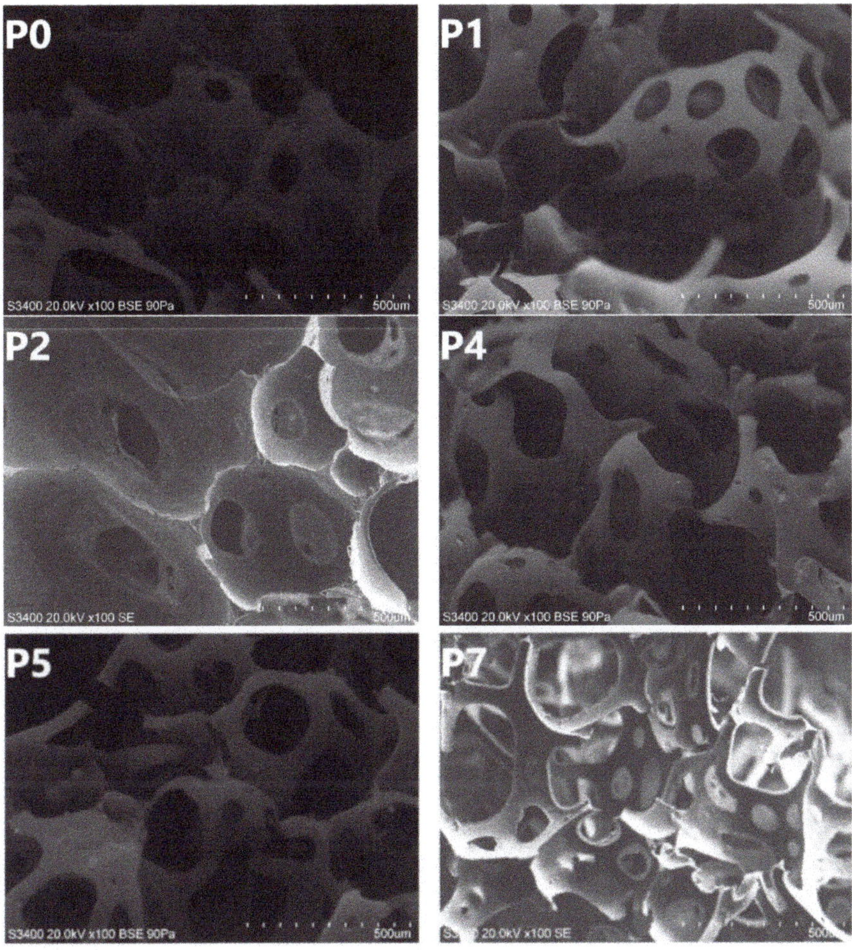

Figure 9. SEM images of prepared foams P0, P1, P2, P4, P5, and P7 (according to Table 1).

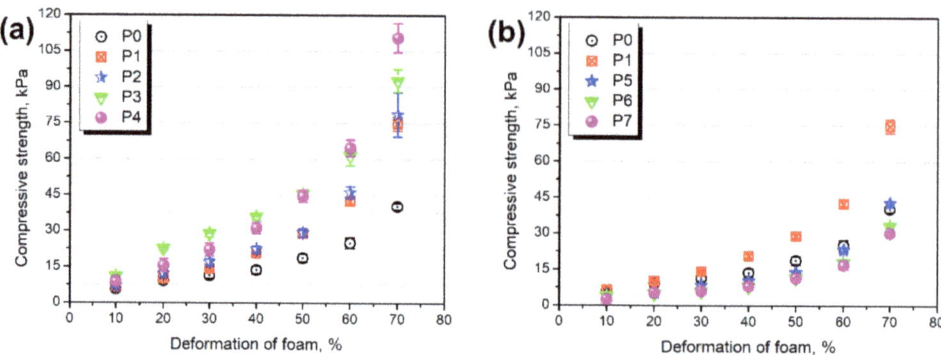

Figure 10. Compressive strength of foams containing (**a**) GTR modified with H_2O_2 and (**b**) GTR modified with $KMnO_4$.

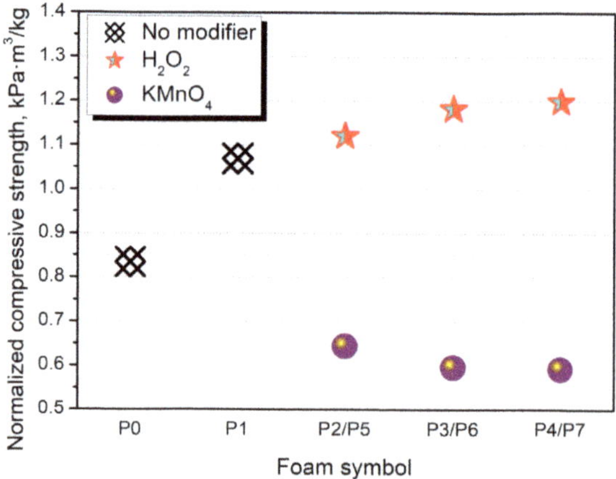

Figure 11. Normalized compressive strength of PU/GTR composite foams.

Table 4. Results of thermogravimetric analysis for prepared foams.

Foam Symbol	$T_{-2\%}$, °C	$T_{-5\%}$, °C	$T_{-10\%}$, °C	$T_{-50\%}$, °C	$Residue_{890\,°C}$, %	T_{max1}, °C	T_{max2}, °C	T_{max3}, °C
P0	264.2	290.1	313.0	380.7	12.98	-	-	385.5
P1	247.6	279.0	306.6	390.3	15.32	-	324.2	393.9
P2	249.2	284.2	313.2	392.9	15.33	297.8	-	393.8
P3	257.8	287.6	315.6	394.5	15.29	287.7	318.9	395.0
P4	248.7	282.2	313.6	394.6	17.24	278.7	322.1	395.7
P5	229.8	263.5	297.3	393.0	16.42	269.2	-	391.8
P6	221.8	251.9	287.9	392.3	18.32	251.5	-	391.7
P7	203.9	234.6	277.3	393.4	16.34	230.0	309.6	395.8

The introduction of rubber particles also resulted in changes in the course of decomposition, which was no longer a one-step process. The value of T_{max3}, which is related to the main decomposition step, was shifted towards higher temperatures because of the presence of rubber and the excellent stability of the untreated, crosslinked parts, mostly styrene-butadiene rubber present in tires (see Table 3). It was marked with an arrow in Figure 10. Also, it can be seen that the values of T_{max1} and T_{max2} correspond to the decomposition of GTR—probably the oxidized part and the urethane and amide bonds generated during interactions of modified rubber with isocyanate groups. These bonds show lower thermal stability comparing to the soft segments originated from polyols' macromolecules [46]. Also, for $KMnO_4$ treatment, the incomplete development of matrix and unbound fragments of polyol chains could reduce material stability. For samples P5–P7, the additional degradation steps at lower temperatures are particularly noticeable.

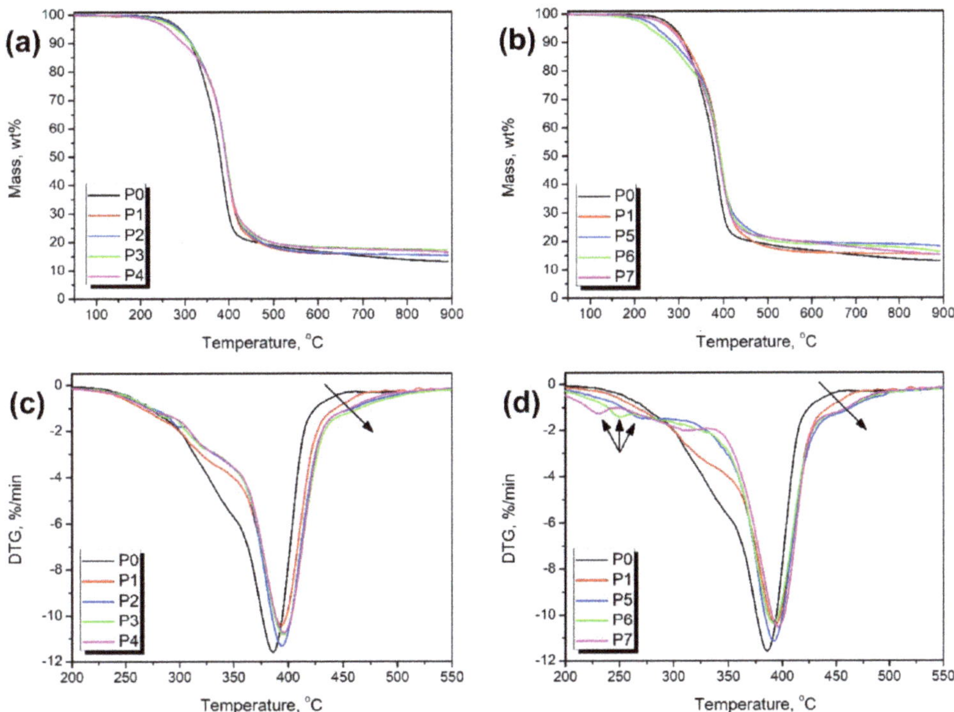

Figure 12. Plots of (**a,b**) mass loss and (**c,d**) differential thermogravimetric curves of foams containing (**a,c**) GTR modified with H_2O_2 and (**b,d**) GTR modified with $KMnO_4$.

4. Conclusions

The presented research paper aimed to investigate the impact of GTR treatment with hydrogen peroxide and potassium permanganate on its microstructure, chemical structure of surface and thermal stability. Applied modifications resulted in the development of GTR surface due to partial oxidation, which could be observed in the scanning electron microscope images. Moreover, application of $KMnO_4$ significantly increased the hydroxyl number of modified GTR samples, associated with the incorporation of hydroxyl groups onto the surface. Partial oxidation of the surface of GTR slightly reduced the thermal stability of modified GTR samples. Nevertheless, the onset of degradation still exceeded the value of 200 °C, which guarantees the safe processing window for manufacturing of foamed polyurethane/GTR composites, without the further decomposition of rubber particles.

Modified GTR samples were introduced into a flexible foamed polyurethane matrix. The impact on the processing, cellular structure, mechanical and thermal performance was investigated. The introduction of neat GTR and GTR modified with hydrogen peroxide caused the elongation of processing times and reduced processing temperatures. Such an effect was mostly due to the increased viscosity of the polyol mixture, caused by the introduction of solid particles. The treatment with $KMnO_4$ caused the opposite effect, due to the catalytic activity of potassium ions. Changes in processing were mirrored in the cellular structure of foams. For neat and H_2O_2-modified GTR, the typical rise of the apparent density and drop of open cell content was observed. At the same time, $KMnO_4$

treatment catalyzed the polymerization, resulting in enhanced gas generation, reduction of the apparent density and increase of open cell content. The compressive properties were directly affected by the quality of cellular structure and apparent density of foams. Therefore, superior mechanical performance was observed for H_2O_2 modifications, because of the incompletely developed structure during polymerization accelerated by $KMnO_4$ treatment. Considering thermal properties, changes in the chemical structure of modified GTR were reflected in changes of foamed polyurethane/GTR composites. Generally, the presented results indicate that the chemical modification of ground tire rubber should be considered an auspicious method for compatibilization of polyurethane/GTR composites. Through proper treatment of GTR, composites with the desired properties could be obtained. Future trends in this area should include modifications of GTR using continuous methods, enhancing the economic and ecological aspects of the process, as well as more in-depth evaluation of the changes during processing of polyurethane systems.

Author Contributions: Conceptualization, A.H., Ł.Z. and P.K.; methodology, A.O., Ł.Z. and P.K.; software, A.H.; validation, A.H., P.K. and K.F.; formal analysis, A.H.; investigation, A.H., A.O., Ł.Z. and P.K.; resources, A.H. and K.F.; data curation, A.H. and Ł.Z.; writing—Original draft preparation, A.H. and Ł.Z.; writing—Review and editing, P.K. and K.F; visualization, A.H.; supervision, A.H. and K.F.; project administration, A.H.; funding acquisition, A.H. All authors have read and agreed to the published version of the manuscript.

Funding: This work was supported by The National Centre for Research and Development (NCBR, Poland) in the frame of LIDER/3/0013/L-10/18/NCBR/2019 project—Development of technology for the manufacturing of foamed polyurethane-rubber composites for the use as damping materials.

Institutional Review Board Statement: Not applicable.

Informed Consent Statement: Not applicable.

Data Availability Statement: Data is contained within the article. The data presented in this study are available in The Impact of Ground Tire Rubber Oxidation with H_2O_2 and $KMnO_4$ on the Structure and Performance of Flexible Polyurethane/Ground Tire Rubber Composite Foams.

Conflicts of Interest: The authors declare no conflict of interest.

References

1. Papiński, J.; Żabski, L. Zrozumieć poliuretany. *Mater. Bud.* **2011**, *1*, 57–58.
2. Swinarew, B. Poliuretany—Nowoczesne wszechstronne materiały. Część II—Pianki poliuretanowe. *Przetw. Tworzyw* **2015**, *5*, 428–434.
3. Kurańska, M.; Prociak, A. Właściwości termoizolacyjne i mechaniczne spienionych kompozytów poliuretanowych z włóknami konopnymi. *Chemik* **2011**, *65*, 1055–1058.
4. Auguścik-Królikowska, M.; Ryszkowska, J.; Szczepkowski, L.; Kwiatkowski, D.; Kołbuk-Konieczny, D.; Szymańska, J. Viscoelastic polyurethane foams with the addition of mint. *Polimery* **2020**, *65*, 196–207. [CrossRef]
5. Kurańska, M.; Michałowski, S.; Radwańska, J.; Jurecka, M.; Zieleniewska, M.; Szczepkowski, L.; Ryszkowska, J.; Prociak, A. Bio-poliole z oleju rzepakowego jako surowce do kompozytów naturalnych z napełniaczami naturalnymi dla kosmetyki. *Przem. Chem.* **2016**, *95*, 256–262. [CrossRef]
6. Paciorek-Sadowska, J.; Borowicz, M.; Czupryński, B.; Liszkowska, J. Kompozyty sztywnych pianek poliuretanowo-poliizocyjanurowych z korą dębu szypułkowego. *Polimery* **2017**, *62*, 666–672. [CrossRef]
7. Członka, S.; Strąkowska, A.; Kairytė, A. Application of Walnut Shells-Derived Biopolyol in the Synthesis of Rigid Polyurethane Foams. *Materials* **2020**, *13*, 2687. [CrossRef]
8. Barnat, W.; Miedzińska, D.; Niezgoda, T. Pianki poliuretanowe—Właściwości, zastosowania, recykling. *Arch. Gospod. Odpadami Ochr. Sr.* **2011**, *13*, 13–17.
9. Tiuc, A.E.; Nemeş, O.; Vermesan, H.; Toma, A.C. New sound absorbent composite materials based on sawdust and polyurethane foam. *Compos. Part B Eng.* **2019**, *165*, 120–130. [CrossRef]
10. El-Meligy, M.G.; Mohamed, S.H.; Mahani, R.M. Study mechanical, swelling and dielectric properties of prehydrolysed banana fiber—Waste polyurethane foam composites. *Carbohydr. Polym.* **2010**, *80*, 366–372. [CrossRef]
11. Tiuc, A.E.; Vermeşan, H.; Gabor, T.; Vasile, O. Improved Sound Absorption Properties of Polyurethane Foam Mixed with Textile Waste. *Energy Procedia* **2016**, *85*, 559–565. [CrossRef]

12. Zieleniewska, M.; Leszczyński, M.K.; Szczepkowski, L.; Bryśkiewicz, A.; Krzyżowska, M.; Bień, K.; Ryszkowska, J. Development and applicational evaluation of the rigid polyurethane foam composites with egg shell waste. *Polym. Degrad. Stabil.* **2016**, *132*, 78–86. [CrossRef]

13. Cachaço, A.G.; Afonso, M.D.; Pinto, M.L. New applications for foam composites of polyurethane and recycled rubber. *J. Appl. Polym. Sci.* **2013**, *129*, 2873–2881. [CrossRef]

14. Pianomat. Available online: http://www.pianomat.pl/en/ (accessed on 11 January 2021).

15. Tiuc, A.E.; Moga, L. Improvement of acoustic and thermal comfort by turning waste into composite materials. *Rom. J. Acoust. Vib.* **2013**, *10*, 77–82.

16. Hejna, A.; Korol, J.; Przybysz-Romatowska, M.; Zedler, Ł.; Chmielnicki, B.; Formela, K. Waste tire rubber as low-cost and environmentally-friendly modifier in thermoset polymers—A review. *Waste Manag.* **2020**, *108*, 106–118. [CrossRef] [PubMed]

17. Roche, N.; Ichchou, M.N.; Salvia, M.; Chettah, A. Dynamic Damping Properties of Thermoplastic Elastomers Based on EVA and Recycled Ground Tire Rubber. *J. Elastomers Plast.* **2011**, *43*, 317–340. [CrossRef]

18. Ubaidillah, H.; Yahya, I.; Kristiani, R.; Muqowi, E.; Mazlan, S.A. Perfect sound insulation property of reclaimed waste tire rubber. *AIP Conf. Proc.* **2016**, *1717*. [CrossRef]

19. Zhang, X.; Lu, Z.; Tian, D.; Li, H.; Lu, C. Mechanochemical devulcanization of ground tire rubber and its application in acoustic absorbent polyurethane foamed composites. *J. Appl. Polym. Sci.* **2012**, *127*, 4006–4014. [CrossRef]

20. Mosiewicki, M.A.; Dell'Arciprete, G.A.; Aranguren, M.I.; Marcovich, N.E. Polyurethane Foams Obtained from Castor Oil-based Polyol and Filled with Wood Flour. *J. Compos. Mater.* **2009**, *43*, 3057–3072. [CrossRef]

21. Formela, K.; Klein, M.; Colom, X.; Saeb, M.R. Investigating the combined impact of plasticizer and shear force on the efficiency of low temperature reclaiming of ground tire rubber (GTR). *Polym. Degrad. Stab.* **2016**, *125*, 1–11. [CrossRef]

22. Fazli, A.; Rodrigue, D. Recycling Waste Tires into Ground Tire Rubber (GTR)/Rubber Compounds: A Review. *J. Compos. Sci.* **2020**, *4*, 103. [CrossRef]

23. ASTM D2527-83(2017) Standard Specification for Rubber Seals—Splice Strength. Available online: https://www.astm.org/Standards/D2527.htm (accessed on 11 January 2021).

24. PN-EN ISO 845:2010—Wersja Polska—Tworzywa Sztuczne Porowate i Gumy—Oznaczanie Gęstości Pozornej. Available online: https://sklep.pkn.pl/pn-en-iso-845-2010p.html (accessed on 11 January 2021).

25. ISO 604:2002 Plastics—Determination of Compressive Properties. Available online: https://www.iso.org/standard/31261.html (accessed on 11 January 2021).

26. Vilar, W.D. *Química e Tecnologia dos Poliuretanos*, 2nd ed.; Vilar Consultoria Técnica Ltda.: Rio de Janeiro, Brazil, 1998.

27. Zedler, Ł.; Kosmela, P.; Olszewski, A.; Burger, P.; Formela, K.; Hejna, A. Recycling of Waste Rubber by Thermo-Mechanical Treatment in a Twin-Screw Extruder. *Proceedings* **2020**, in press. [CrossRef]

28. Shatanawi, K.M.; Biro, S.; Naser, M.; Amirkhanian, S.N. Improving the rheological properties of crumb rubber modified binder using hydrogen peroxide. *Road Mater. Pavement Des.* **2013**, *14*, 723–734. [CrossRef]

29. Yehia, A.A.; Mull, M.A.; Ismail, M.N.; Hefny, Y.A.; Abdel-Bary, E.M. Effect of chemically modified waste rubber powder as a filler in natural rubber vulcanizates. *J. Appl. Polym. Sci.* **2004**, *93*, 30–36. [CrossRef]

30. Fournier, H.M. Transformation des alcools primaires saturès en acides monobasiques correspondants. *Comptes Rendus Acad. Sci.* **1907**, *144*, 331.

31. Dash, S.; Patel, S.; Mishra, B.K. Oxidation by permanganate: Synthetic and mechanistic aspects. *Tetrahedron* **2009**, *65*, 707–739. [CrossRef]

32. Wiberg, K.B.; Saegebarth, K.A. The Mechanisms of Permanganate Oxidation. IV. Hydroxylation of Olefins and Related Reactions. *J. Am. Chem. Soc.* **1957**, *79*, 2822–2824. [CrossRef]

33. Sonnier, R.; Leroy, E.; Clerc, L.; Bergeret, A.; Lopez-Cuesta, J.M. Polyethylene/ground tyre rubber blends: Influence of particle morphology and oxidation on mechanical properties. *Polym. Test.* **2007**, *26*, 274–281. [CrossRef]

34. Li, B.; Li, H.; Wei, Y.; Zhang, X.; Wei, D.; Li, J. Microscopic Properties of Hydrogen Peroxide Activated Crumb Rubber and Its Influence on the Rheological Properties of Crumb Rubber Modified Asphalt. *Materials* **2019**, *12*, 1434. [CrossRef]

35. Kleps, T.; Piaskiewicz, M.; Parasiewicz, W. The use of thermogravimetry in the study of rubber devulcanization. *J. Therm. Anal. Calorim.* **2000**, *60*, 271–277. [CrossRef]

36. Scuracchio, C.H.; Waki, D.A.; Da Silva, M.L.C.P. Thermal analysis of ground tire rubber devulcanized by microwaves. *J. Therm. Anal. Calorim.* **2007**, *87*, 893–897. [CrossRef]

37. Nadal Gisbert, A.; Crespo Amorós, J.E.; López Martínez, J.; Macias Garcia, A. Study of thermal degradation kinetics of elastomeric powder (ground tire rubber). *Polym. Plast. Technol. Eng.* **2007**, *47*, 36–39. [CrossRef]

38. Zhang, X.; Saha, P.; Cao, L.; Li, H.; Kim, J. Devulcanization of waste rubber powder using thiobisphenols as novel reclaiming agent. *Waste Manag.* **2018**, *78*, 980–991. [CrossRef] [PubMed]

39. Piszczyk, L.; Hejna, A.; Formela, K.; Danowska, M.; Strankowski, M. Rigid Polyurethane Foams Modified with Ground Tire Rubber—Mechanical, Morphological and Thermal Studies. *Cell. Polym.* **2015**, *34*, 45–62. [CrossRef]

40. Piszczyk, Ł.; Hejna, A.; Danowska, M.; Strankowski, M.; Formela, K. Polyurethane/ground tire rubber composite foams based on polyglycerol: Processing, mechanical and thermal properties. *J. Reinf. Plast. Compos.* **2015**, *34*, 708–717. [CrossRef]

41. Formela, K.; Hejna, A.; Zedler, Ł.; Przybysz, M.; Ryl, J.; Saeb, M.R.; Piszczyk, Ł. Structural, thermal and physico-mechanical properties of polyurethane/brewers' spent grain composite foams modified with ground tire rubber. *Ind. Crop. Prod.* **2017**, *108*, 844–852. [CrossRef]
42. Wu, J.W.; Sung, W.F.; Chu, H.S. Thermal conductivity of polyurethane foams. *Int. J. Heat Mass Transf.* **1999**, *43*, 2211–2217. [CrossRef]
43. Piszczyk, Ł.; Hejna, A.; Formela, K.; Danowska, M.; Strankowski, M. Effect of ground tire rubber on structural, mechanical and thermal properties of flexible polyurethane foams. *Iran. Polym. J.* **2014**, *24*, 75–84. [CrossRef]
44. Gayathri, R.; Vasanthakumari, R.; Padmanabhan, C. Sound absorption, thermal and mechanical behavior of polyurethane foam modified with nano silica, nano clay and crumb rubber fillers. *Int. J. Sci. Eng. Res.* **2013**, *4*, 301–308.
45. Hejna, A.; Kopczyńska, M.; Kozłowska, U.; Klein, M.; Kosmela, P.; Piszczyk, Ł. Foamed Polyurethane Composites with Different Types of Ash—Morphological, Mechanical and Thermal Behavior Assessments. *Cell. Polym.* **2016**, *35*, 287–308. [CrossRef]
46. Hejna, A.; Kosmela, P.; Kirpluks, M.; Cabulis, U.; Klein, M.; Haponiuk, J.; Piszczyk, Ł. Structure, Mechanical, Thermal and Fire Behavior Assessments of Environmentally Friendly Crude Glycerol-Based Rigid Polyisocyanurate Foams. *J. Polym. Environ.* **2017**, *26*, 1854–1868. [CrossRef]

 *materials*

Article

Change in Conductive–Radiative Heat Transfer Mechanism Forced by Graphite Microfiller in Expanded Polystyrene Thermal Insulation—Experimental and Simulated Investigations

Aurelia Blazejczyk [1],*, Cezariusz Jastrzebski [2] and Michał Wierzbicki [2]

[1] Department of Mechanics and Building Structures, Institute of Civil Engineering, Faculty of Civil and Environmental Engineering, Warsaw University of Life Sciences—SGGW, Nowoursynowska 159 ST., 02-776 Warsaw, Poland

[2] Faculty of Physics, Warsaw University of Technology, Koszykowa 75 ST., 00-662 Warsaw, Poland; cezariusz.jastrzebski@pw.edu.pl (C.J.); michal.wierzbicki@pw.edu.pl (M.W.)

* Correspondence: aurelia_blazejczyk@sggw.edu.pl or pub.wbis.sggw@gmail.com

Received: 23 April 2020; Accepted: 19 May 2020; Published: 9 June 2020

Abstract: This article introduces an innovative approach to the investigation of the conductive–radiative heat transfer mechanism in expanded polystyrene (EPS) thermal insulation at negligible convection. Closed-cell EPS foam (bulk density 14–17 kg·m^{-3}) in the form of panels (of thickness 0.02–0.18 m) was tested with 1–15 μm graphite microparticles (GMP) at two different industrial concentrations (up to 4.3% of the EPS mass). A heat flow meter (HFM) was found to be precise enough to observe all thermal effects under study: the dependence of the total thermal conductivity on thickness, density, and GMP content, as well as the thermal resistance relative gain. An alternative explanation of the total thermal conductivity "thickness effect" is proposed. The conductive–radiative components of the total thermal conductivity were separated, by comparing measured (with and without Al-foil) and simulated (i.e., calculated based on data reported in the literature) results. This helps to elucidate why a small addition of GMP (below 4.3%) forces such an evident drop in total thermal conductivity, down to 0.03 W·m^{-1}·K^{-1}. As proposed, a physical cause is related to the change in mechanism of the heat transfer by conduction and radiation. The main accomplishment is discovering that the change forced by GMP in the polymer matrix thermal conduction may dominate the radiation change. Hence, the matrix conduction component change is considered to be the major cause of the observed drop in total thermal conductivity of EPS insulation. At the microscopic level of the molecules or chains (e.g., in polymers), significant differences observed in the intensity of Raman spectra and in the glass transition temperature increase on differential scanning calorimetry(DSC) thermograms, when comparing EPS foam with and without GMP, complementarily support the above statement. An additional practical achievement is finding the maximum thickness at which one may reduce the "grey" EPS insulating layer, with respect to "dotted" EPS at a required level of thermal resistance. In the case of the thickest (0.30 m) panels for a passive building, above 18% of thickness reduction is found to be possible.

Keywords: expanded polystyrene; graphite particles; thermal conductivity; thermal resistance; thickness effect; phonon-photon transport; Raman spectroscopy; thermal analysis

1. Introduction

Large-scale application of EPS foams with closed cells as thermal insulation in construction engineering requires the sustainable improvement of the thermophysical features of traditional building materials [1–5]. When introducing various technological changes in chemical composition, besides improving measurement and simulation techniques, it is important to understand the physical consequences caused by such changes. A better understanding may allow for the generation of knowledge, as well as let scientists improve the modelling of thermal processes. Moreover, it may allow engineers to develop novel materials and encourage industry stakeholders by more effectively optimizing the production costs of these materials, along with better thermal insulation performance.

A traditional solution for high quality thermal protection in buildings is by using high-thickness insulating layers made of conventional thermal insulation that, together with other more advanced options, comprises "the best building practice". Advanced thermal insulation materials, such as super insulating materials (SIMs) (e.g., vacuum insulation materials (VIMs) and aerogels), phase change materials (PCMs), gas-filled materials (GFMs), and nanoinsulation materials (NIMs), promise to be the most beneficial for applications in the building sector [6–8]. In addition, modern systems, such as adaptive facades, dynamic facades, and active envelopes, provide architectural alternatives, which are a novel platform for energy efficiency, visual comfort, and daylight distribution, as well as branding and image. However, installing conventional thermal insulation of high thickness in external walls remains an attractive option, especially in harsh climate countries, due to the low market prices of materials and the costs of their installation [9–12]. When designing a building, one of the first steps to consider is reaching the required level of thermal protection, in terms of thermal transmittance U-value, which ranges from ca. 0.1 $W \cdot m^{-2} \cdot K^{-1}$ for high-energy standards up to 3 $W \cdot m^{-2} \cdot K^{-1}$ or more for low-energy standard [13]. The challenge is to obtain such values by selecting the most appropriate and cost-effective materials, which provides the thinnest possible insulation with the highest thermal effectiveness.

Conventional polymeric foams applicable in the building sector include both products of very low bulk density (below 20 $kg \cdot m^{-3}$), such as extruded polystyrene (XPS) or EPS (herein under study), and products of low bulk density (up to 45 $kg \cdot m^{-3}$), such as phenolic foam (PhF), polyurethane (PUR), and polyisocyanurate (PIR) foams, as well as XPS and EPS foams. A popular option is the industrial production of polymer composite, such as EPS foams that include graphite fillers [14–16], due to low cost, lightweight, and moderate dispersibility in polymer matrix [17,18], in particular, graphite microparticles (GMP) [19–25]. These are typically applied in the form of panels for the thermal insulation of buildings. Despite the problem of possible overheating due to solar radiation causing panel deformation [26,27], this choice offers designers relevant insulation thickness reduction. The thickness of "pure" EPS panels in the external walls of passive buildings (U-value = 0.1 $W \cdot m^{-2} \cdot K^{-1}$ for Central European climates) can reach up to $d = 0.30$ m. As it is of interest to reduce the thickness (at a given U-value), one may study the material's thermal insulation performance [28–30]. Some optimization is possible by observing heat transfer via the apparent thermal conductivity λ' and resistance R' as a function of the EPS sample thickness d, with relation to the variable bulk density ρ and concentration of fillers, such as GMP.

Following measurements in accredited laboratories, EPS manufacturers have declared its thermal conductivity coefficient λ_D and thermal resistance R_D values [31,32]. However, the literature has barely discussed the so-called thickness effect, which has the apparent impact of the sample thickness (in its lower range) on the material's thermophysical parameters [33,34]. On the other hand, the thickest samples have rarely been measured and reported [35]. The latter may be due to common technical difficulties when measuring insulation panels above 0.1 m in thickness, which is not suitable for most of the (too narrow) plate instruments (GHP and HFM). Besides the main purpose of thermal transport analysis, this work also tries to fill this gap, explaining these technical issues (i.e., experimental knowledge regarding thin and thick case measurements). Therefore, the extra purpose is demonstrating a methical approach that may answer the following two questions:

- How can we overcome the experimental artefact of the thickness effect, as revealed by conventional polymeric foam panels?
- How can we find the true conductivity λ and resistance R values for the panels which are thicker than the gauge of plate instruments?

In each polymeric insulating material, one must initially assess the thickness effect relevance by testing thin slices of the foam panel. Herein, it is done, concerning EPS of very low bulk density, by using an experimental method in accordance with the recommendations of the European Committee for Standardization CEN, as described in Standards [30,32,35,36].

Besides considering technical issues and providing correct data for engineers, this paper aims to contribute to the general discussion about the thermal transport mechanisms in polymeric insulating materials and better understanding them. Example of such discussions describing mainly EPS insulation can be found in the recognized literature [37–49]. In particular, it has been reported that, when incorporating industrially higher concentrations of GMP into an EPS polymer matrix, the total thermal conductivity can be reduced by opacifying the insulation product for thermal radiation, which is absorbed by the carbon atoms in graphite or scattered by the larger graphite microparticles. Unfortunately, the quantitative analysis of the three major thermal components—radiation, air conduction, and polymer matrix conduction—over a wide thickness range of the insulating EPS layers with different GMP industrial concentrations is missing, in general.

Therefore, the main scientific purpose is analysing heat transfer through EPS foam of very low bulk density and separating it into its radiative and conductive thermal components. To the authors' knowledge, this is the first successful approach of separating thermal components to examine EPS in relation to the identified spectral features of the Raman spectrum. As for convection of insulating gas, it is proved to be negligible in EPS, as well as also other polymeric foams with a closed or open-cell structure of diameter up to 0.004 m (4 mm) [1,27,50–52]. The closed cells in the expanded beads of "pure" EPS at various densities are of size ranging from 0.05 to 0.40 mm [27,38,53]. The scientific aim of the present paper was achieved by comparing experimental and literature data [42]. As a result, we observe the effect of change in conductive–radiative heat transfer mechanism triggered by the GMP at two different industrial concentrations. Its apparent impact on each particular thermal component is then analysed and discussed.

2. Materials and Methods

This section briefly outlines characteristics of the tested EPS products and describes the experimental methods used to measure their physical parameters, including the accuracy/uncertainty estimation in the error analysis, as well as comments on the experimental limitations.

2.1. Tested Products

The tested EPS products were manufactured by Polish enterprises, in the form of panels, for thermal insulation of buildings. The three types of commercial EPS products investigated were (Figure 1):

- "white" EPS A—pure;
- "dotted" EPS B—with a low concentration of GMP (only within the expanded beads of the black dotted isles, scattered randomly in the entire panel); and
- "grey" EPS C—with a high concentration of GMP (evenly distributed throughout the polymer matrix and in the entire panel).

It is known, from the literature [45], that the size of industrially applied GMP ranges from 1 to 15 μm (diameter distribution range).

The GMP mass concentration in EPS C can be estimated, from measured density values, as equal to the relative density change $(\rho_C - \rho_B)/\rho_C$. By neglecting the GMP concentration in the EPS B product, a rough prediction of the graphite content in EPS C is up to 7.3%, as $((14.6 + 0.5) - (14.2 - 0.2))/(14.6 + 0.5) = 0.073$ (see Section 3.1 Table 2; also confront with Section 3.6).

Figure 1. The tested products: A—"white" expanded polystyrene (EPS) (pure), B—"dotted" EPS, and C—"grey" EPS.

Each of the tested EPS panels was made by cutting a single massive block (material batch) by a hot wire to the desired thickness on a manufacturing line. The range (from 0.02 to 0.18 m) and the particular thicknesses of the individual panels can be read in Section 3.1 from Figure 3. Then, the panels were subjected to systematic control of their thermal insulation performance. Unlike B and C, the thermal parameters of A were unstable during the first 20 months of seasoning. During the entire experiment, the panels were conditioned under stable laboratory conditions: air temperature (20 ± 2 °C) and relative air humidity (50 ± 10% RH).

2.2. Experimental Limitations for Thermal Measurement

The first experimental limitation arises from the fact that testing the thermal insulation performance of any homogeneously porous material can be conducted using any plate instrument, but only if the maximum nominal size of any diversity in its structure (i.e., grains or pores—whatever is larger) is smaller than one-tenth of the sample thickness [28,29,36]. Slightly varying with bulk density [27,39,40,44], the order of magnitude of the "pure" EPS grains (expanded beads filled with closed cells) size may reach 10^{-3} m; then, the minimum thickness of the EPS sample should be at least 10 times greater, which sets the minimum thickness no less than 10^{-2} m.

The second limitation arises from determining the thermal parameters of EPS of very low or low bulk density. Regarding the experimental method, this is related to the so-called thickness limit estimation [35], which is further explained in Section 2.4.1 and Supplementary Material part 1.

The third limitation is caused by the so-called permissible sample thickness range (from minimum d_{min} to maximum d_{max}), depending on the specific plate instrument chamber dimensions and the measurement section area. This range is suggested by the Standards [28–30,35].

2.3. Bulk Density Measurement Method

The bulk density ρ was determined experimentally for the tested EPS products by measuring (for each panel thickness d):

- the panel mass, with an accuracy of $\pm 10^{-5}$ kg,
- the panel dimensions (length x and width y), with an accuracy of $\pm 5 \times 10^{-4}$ m; and thickness d, with an accuracy of $\pm 10^{-6}$ m,

  and calculating:

- the panel volume (as a regular rectangular prism), with accuracy no worse than $\pm 10^{-4}$ m^3 and
- the panel density (as the mass to volume ratio) and the bulk density of the material (as an average over the thickness range), with accuracy no less than $\pm 2 \times 10^{-1}$ kg·m^{-3}.

A standard precise laboratory balance (Radwag PS 2100.R1) was used for mass measurement. The length and width were found using a professional EC Class 1 measuring tape. The thickness value was given by an HFM measuring system. The environment was controlled, in terms of air temperature (20 ± 2 °C) and relative air humidity (50 ± 10% RH).

Details of the expanded uncertainty calculations regarding the bulk density measurements are explained in Supplementary Material part 2. The averaged expanded uncertainty $<U(\rho)>$ values of the bulk density $\rho(d)$ measurement were calculated, from all experimental uncertainties $U(\rho)$ within the whole thickness range under study, individually for each type of EPS product. The results are given in Section 3.1.

2.4. Thermal Measurement Method

This section describes a practical approach for the measurement of the thin and thick EPS panels, which takes into consideration the limitations mentioned in Section 2.2, as well as an experimental setup that allows us to obtain correct thermal measurements as output.

2.4.1. The Thickness Effect and Thickness Limit of (Non-)Linearity

The thickness effect refers to the apparent sublinear growth of the apparent thermal conductivity $\lambda'(d)$, up to a certain level achieved at the critical thickness limit d_L, above which the function takes a seemingly constant value (yet still below the upper bound—herein assigned to the thermal transmittance λ'_t). Above d_L, the value of $\lambda'(d)$ approaches the true material conductivity $\lambda(d)$, so one may assign the values of $\lambda(d)$ to the values of $\lambda'(d)$. The data seem to oscillate around $<\lambda>$ (the material-representing averaged value).

The thickness limit, d_L, results from several factors, such as the material property, sample features, the experimental set-up, and so on. The limit could be assigned to the minimal thickness above which the thermal transmittance λ'_t can be determined from the thickness-independent ratio $\Delta d/\Delta R'$ [28,29,33,36]. As λ'_t corresponds to the inverse gradient of the oblique asymptote of $R'(d)$, it can be determined by a linear fit to all data above d_L.

When $d > d_L$, the so-called transfer factor $\mathfrak{I}$ (see Supplementary Material part 1) practically does not depend on the thickness (within experimental inaccuracy tolerance). In this region, $\mathfrak{I}$ does not differ from λ'_t by more than 2% [28,29,36]. In other words, in the case of sufficiently large thermal insulation thickness, the asymptotic values of $\lambda'(d)$ and $\mathfrak{I}(d)$ are equal to the value of λ'_t. Thus, the level of λ'_t is the limiting horizontal asymptote achieved by both functions.

As shown in Supplementary Material part 1, for the EPS A, B, and C products, the respective thickness limits d_{LA}, d_{LB}, and d_{LC} can be obtained by comparing the thickness-dependent $1 - L(d)$ values (which are linearity measures of the λ' or R' curves at a given d) with the experimental tolerance for nonlinearity of the measured data at the optimum level of 0.02. In general, the lower the level of $1 - L$, the larger the value of d_L; thus, as $1 - L$ approaches 0.00, the thickness limit tends to infinity, which means that none of the EPS products could comply with the requirements and determination of their thermal parameters would be impossible. If a less rigorous condition, such as $1 - L \leq 0.03$, is applied, then the resultant d_L would obviously be inside of the curved region. If a more rigorous condition, such as $1 - L \leq 0.01$, is applied, then the thicknesses limits would be unreasonably far from the curved region. Thus, by applying the optimal experimental condition of $1 - L \leq 0.02$, one can finally obtain the reasonable thickness limits (see Section 3.2).

The thickness effect has an impact not only on the measurement of thin panels but also on the thickest panels, exceeding the maximum distance d_{max} between "hot" and "cold" plates in the HFM instrument. In the latter case, the determination of the material thermophysical parameters R and λ could be achieved by the proposed procedure, which includes cutting thinner panels from the thick product block (to a thickness no smaller than d_L), measuring the $R'(d)$ value, correcting data and extrapolating the obtained values of $R(d)$ up to the considered product thickness and, finally, conversion from resistance R to conductivity λ of the product. Nevertheless, before cutting the panels, one has to take into consideration the thickness effect, evaluating its relevance by calculation of d_L. Once the thickness limit is known, the total thermal resistance of the product block can be calculated (even without $R'(d)$ correction), according to the practical formula (1):

$$R_{product}^{EPS} = \frac{\sum_{i=1}^{n} \frac{R_i}{d_i}}{n} \cdot d_{product}^{EPS} \tag{1}$$

where $i = 1, 2, \ldots, n$ and n is the number of panels cut from the block of product under consideration, where the condition $d \geq d_L$ is satisfied for each panel.

2.4.2. HFM Setup

Thermal tests in steady state conditions were carried out by using the HFM FOX 600 plate instrument, made in the USA by the LaserComp Company.

As mentioned in Section 2.2, the Standards recommend limiting thermal measurements to a permissible sample thickness range. These limits depend on the geometry of the experimental setup. The HFM chamber dimensions were 0.600×0.600 m^2, the dimensions of its measuring section area (located at the heating/cooling black plate centre) were 0.300×0.300 m^2 (also assigned to the size of the sample area), and the dimensions of the tested panels were 0.600×0.500 m^2. In this case, the permissible thickness should range from $d_{min} = 0.030$ m to $d_{max} = 0.150$ m [30]. Yet, as has been demonstrated, based on uncertainty analysis, the thermophysical parameters obtained for samples between 0.020 and 0.180 m were reliable. In general, the HFM allows for mounting and precisely measuring samples of thickness from 0.005 m (see Section 3.4) up to 0.200 m (max gauge space).

The HFM was calibrated using the certified Standard Reference Material (SRM–IRMM–440), as recommended by the Institute for Reference Materials and Measurements (IRMM) in Geel. The SRM characteristics are shown in Table 1, for comparison of the measurements completed by IRMM in Geel with the author's results. The tiny value of the correction parameter C, resulted even less than $U(\lambda^C_{SRM})$, reflecting proper experimental setup and testing conditions.

Table 1. Data concerning the Standard Reference Material (SRM) sample (size, bulk density, thermal conductivity, and the correction parameter).

SRM Type	SRM Dimensions (m)			SRM Bulk Density ρ_{SRM} (kg·m^{-3}) Average Value given in the Certificate
	Length x	Width y	Thickness d	
Glass wool IRMM-440	0.500	0.500	0.0347	75.8

SRM thermal conductivity coefficients at $T_m = 10$ °C (W·m^{-1}·K^{-1})		Correction parameter C at the coverage factor k = 2.0 $C = \lambda^C_{SRM} - \lambda^M_{SRM}$ (W·m^{-1}·K^{-1})	Expanded uncertainties (W·m^{-1}·K^{-1})	
λ^C_{SRM} Average value given in the Certificate	λ^M_{SRM} Average value measured by HFM FOX 600		$U(\lambda^C_{SRM})$	$U(\lambda^M_{SRM})$
0.03048	0.03057	-9×10^{-5}	2.8×10^{-4}	3.2×10^{-4}

In order to estimate the thermal radiation component (through the insulation), the surface emissivity values individually for the HFM black plate and 10 μm Al-foil were measured earlier at $T_m = 10$ °C (no sample inside the HFM chamber), according to the Standard [30]. The latter measurement was carried out by placing two symmetrical Al-foil layers attached to the HFM bottom and the top black plate. The resulting values were as follows: 0.873 for the HFM black plate, 0.042 for the rough Al-foil side, and 0.032 for the polished Al-foil side.

2.4.3. Conductivity-Resistance Measurement Method

Standard Method

Measurements of thermal insulation performance of the tested EPS products were carried out in compliance with the recommendations of the Standards [28,29,32,33,35] and the acquired data processing (for the conductivity and resistance correction) was completed according to the relations and procedure described in Supplementary Material part 1.

The environment was controlled, in terms of air temperature (20 ± 2 °C) and relative air humidity (50 ± 10% RH). The HFM was set to the average test temperature T_m = 10 °C. The temperature difference applied to the sample was ΔT = 20 °C. The heating (bottom) black plate temperature was set as 20 °C and the cooling (top) black plate temperature was set as 0 °C, such that the heat flow was directed upwards. A schematic design of the measurement system in standard method is shown in Figure 2a. For each EPS panel tested (see Section 2.1), the following parameters were measured (i.e., the HFM output):

- the sample thickness d, with accuracy Δd of ±10^{-6} m;
- the temperature difference ΔT between upper and lower sample surfaces, with accuracy $\Delta(\Delta T)$ of ±0.1 °C;
- the heat flux density q through the sample, with accuracy Δq of ±10^{-1} W·m^{-2};
- the apparent thermal resistance R', with accuracy $\Delta R'$ of ±10^{-4} m^2·K·W^{-1}; and
- the apparent thermal conductivity coefficient λ', with accuracy $\Delta\lambda'$ of ±10^{-5} W·m^{-1}·K^{-1}.

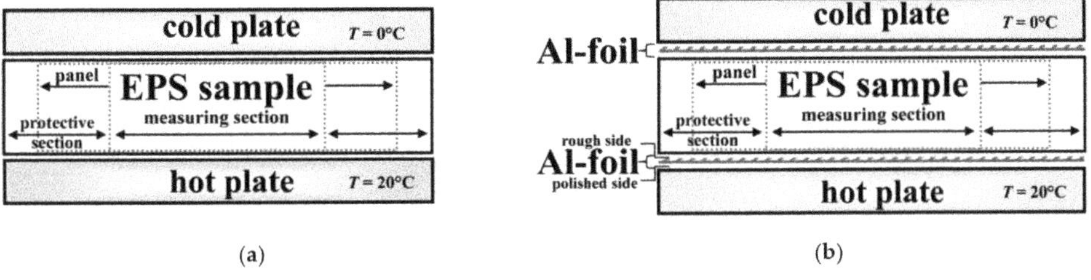

Figure 2. Schematic design of the measurement system in: (**a**) standard and (**b**) nonstandard method.

Details of the expanded uncertainty calculations regarding the thermal conductivity measurements are explained in Supplementary Material part 2. The average expanded uncertainty $<U(\lambda)>$ values of $\lambda(d)$ were calculated from all experimental uncertainties $U(\lambda)$, within the whole thickness range under study, individually for each type of EPS product. The results are given in Section 3.3.

Nonstandard Method

Furthermore, during nonstandard HFM measurement, i.e., the sample placed between two Al-foil layers (applied at the bottom and on the top of the insulation) and HFM plates, as shown in Figure 2b, one may try to assume that the expected level of the total thermal conductivity may be considered as the system response, in which radiation (primary—external, emitted by the HFM "hot" plate and secondary—internal, radiation generated across the insulation) is sufficiently blocked (cut off) by reflection from the bottom and upper Al-foil, respectively. However, from primary continuous radiative–conductive heat flux, only phonons can pass through the Al-foil (either the "hot" or the "cold" one). One must notice that the Al-foil emissivity value (see Section 2.4.2) is no more than 0.04 and that above 96% of the photon flux, either "primary" or "secondary", can be reflected back by the upper Al-foil (in the extreme scenario). Hence, in order to estimate the total thermal conductivity with simulated Al-foil effect, only for the tested EPS B and C products, the difference between the heat flux without Al-foil and the heat flux with Al-foil may correspond to the thermal radiation component, which allows us to calculate its contribution to the total thermal flux individually for each product. This topic is presented further in Section 4.1.2 and Supplementary Material part 3, also explaining the impact of GMP on total thermal conductivity.

2.5. Micro-Raman Measurement

The micro-Raman Spectroscopy (μ-RS) measurements were carried out by using Renishaw's inVia Reflex Spectrometer. The μ-RS tests were performed to investigate comprehensively microstructural changes (e.g., at the molecular level) of the insulation samples. The Raman spectra were collected at room temperature and normal conditions, in backscattering geometry with the 633 nm line of a He–Ne-ion laser and with the 514 nm line of an Ar-ion laser as excitation wavelengths. The results are given in Section 3.5.

2.6. Thermal Analysis Measurement

Thermogravimetric analysis (TGA) and differential scanning calorimetry (DSC) measurements were carried out by using TA Instruments equipment i.e., SDT Q600 and Q2000, respectively.

The conventional DSC tests were performed to measure the amorphous glass transition (T_g) or the crystalline melting (T_{cm}) temperatures of the insulation samples. The DSC thermograms were collected at the temperature range of -100 to 350 °C at a heating rate of 10 °C·min^{-1} and a nitrogen flow rate of 50 mL·min^{-1}. In order to reduce the impact of pressure increase on the measurement results, the samples were put into the nonhermetic aluminium calorimetric containers; appropriate empty aluminium containers were used as reference. The temperature scale was calibrated with the melting point of indium. The estimated error in the determination of T_g or T_{cm} was ± 2 °C.

The TGA tests were performed to measure temperature dependence of the mass loss of the insulation samples. The TGA thermograms were collected at the temperature range of 25–550 °C, at a heating rate of 10 °C·min^{-1}, a nitrogen flow rate of 100 mL·min^{-1}, and in the hermetic aluminium containers. The results are given in Section 3.6.

3. Experimental Results

3.1. Bulk Density and Homogeneity Assessment

Each measured bulk density $\rho(d)$ value, representing an EPS panel of a given d, was found as the average of several single measurements and plotted, with a vertical error bar corresponding to its expanded uncertainty, $U(\rho)$ and a horizontal error bar covered by the symbol used in Figure 3.

Table 2. Data concerning density measurements of the materials expanded polystyrene (EPS) A, B, and C.

EPS Type	Panels Mass Range (kg·10^{-3})	Panels Dimensions (m)			Bulk Density (kg·m^{-3})			Average Expanded Uncertainty $<U(\rho)>$ (kg·m^{-3})
		Length x	Width y	Thickness Range d	$<\rho_A>$	$<\rho_B>$	$<\rho_C>$	
					Average Value with Double Standard Deviation $\pm 2\sigma$			
A "white"	103.14–756.90	0.60	0.50	0.02–0.15	16.9 ± 1.2			2.1 × 10^{-1}
B "dotted"	87.06–666.00	0.60	0.50	0.02–0.18		14.2 ± 0.2		1.8 × 10^{-1}
C "grey"	90.48–567.84	0.60	0.50	0.02–0.13			14.6 ± 0.5	2.2 × 10^{-1}

Next, the averaged bulk density $<\rho>$ was calculated, from all experimental $\rho(d)$ values, individually for each tested EPS product. The resulting $<\rho_A>$, $<\rho_B>$, and $<\rho_C>$ values are reported in Table 2, including average expanded uncertainties $<U(\rho)>$, and presented in Figure 3.

The tested EPS A density was slightly higher, while the EPS B and C products revealed comparable density values.

Homogeneity assessment for each tested EPS product was performed based on $\rho(d)$ measurements, throughout the entire thickness range. As seen from Figure 3 and Table 2, the $\rho_A(d)$ points for the EPS A product reveal the widest spread of bulk density, extremely fluctuating around the average value $<\rho_A>$; furthermore, its corresponding standard deviation was the largest. This indicates the relatively poor homogeneity (possibly due to differences in density between the pre-expanded beads and expanded beads, which were mixed for recycling purposes during the final block foaming process [4,43,54,55]). On the contrary, EPS B was the most homogenous product (no recycling, in this case).

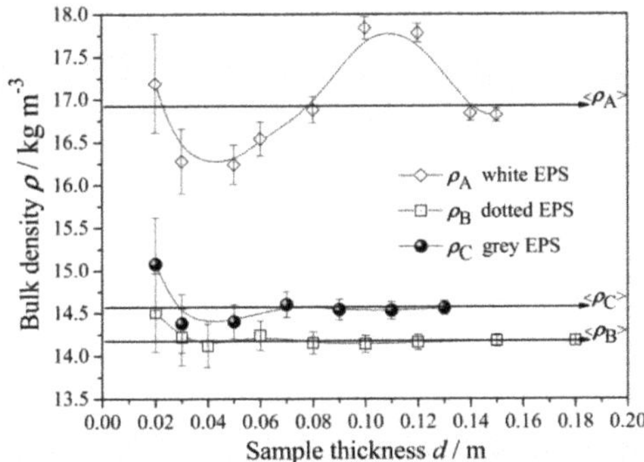

Figure 3. The measured bulk density ρ versus the panel thickness d for the EPS A, B, and C products. The error bars correspond to the expanded uncertainties, $U(\rho)$. The horizontal arrows indicate the levels of the average density $<\rho>$ values (see Table 2).

3.2. Thickness Limits Results

All basic calculations of the thermophysical parameters, including the heat transfer factor $\mathfrak{I}$, the thickness effect function $L(d)$, and the thickness limit d_L, are elaborated and shown in Supplementary Material part 1.

In order to determine the thermal insulation performance of each tested EPS product, first and foremost, one has to check whether the thickness effect is relevant. This analysis, performed separately for each EPS type, was based on testing the values of the function $L(d)$ over the investigated thickness range. Wherever $1 - L(d) > 0.02$ (see Section 2.4.1), at lower thicknesses, the effect was qualified as relevant (thermal conductivity and resistance were apparently nonlinear functions of thickness); wherever $1 - L(d) \leq 0.02$, at higher thicknesses, the thickness effect was qualified as irrelevant (thermal conductivity and resistance were nearly linear functions of thickness).

In the case of EPS A and B, the thickness effect appeared to be relevant up to the estimated limits (d_{LA} and d_{LB}) shown in Section 3.3 (Table 3 and Figure 4a). As can be seen, in the case of EPS C, the thickness effect appeared to be irrelevant (negligibly small), as the estimated limit d_{LC} appeared to be lower than the permissible minimum d_{min} of the HFM.

Table 3. Data concerning the thermal conductivity of the EPS A, B, and C products (thermal transmissivity declared and averaged coefficients, with corresponding thickness limits).

EPS Type	Thickness Limit d_L (m)	Thermal Transmissivity λ_t (W·m^{-1}·K^{-1})	Thermal Conductivity Coefficients at T_m = 10 °C (W·m^{-1}·K^{-1})				Average Expanded Uncertainty $<U(\lambda)>$ (W·m^{-1}·K^{-1})
			λ_D	$<\lambda_A>$	$<\lambda_B>$	$<\lambda_C>$	
				Average Value with Double Standard Deviation $\pm 2\sigma$			
A "white"	0.043 ± 0.010	0.0400	0.040	0.0394 ± 0.0010			1 × 10^{-3}
B "dotted"	0.060 ± 0.005	0.0386	0.040		0.0377 ± 0.0001		1 × 10^{-3}
C "grey"	0.007 ± 0.001	0.0314	0.032			0.0312 ± 0.0005	8 × 10^{-4}

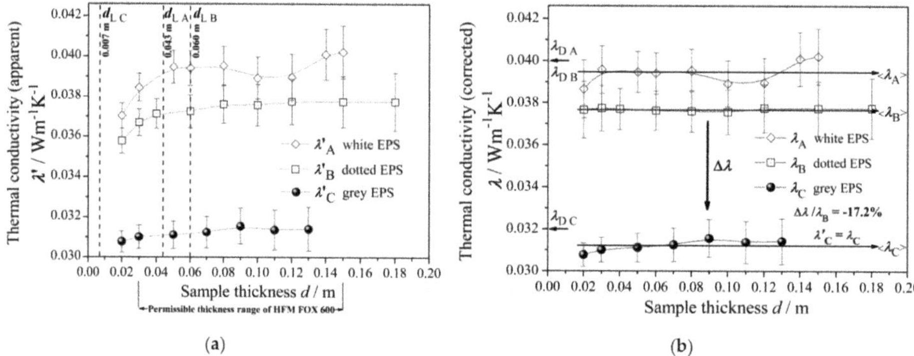

Figure 4. (a) The apparent thermal conductivity coefficient λ' versus the panel thickness d for the EPS A, B, and C products, as measured at T_m = 10 °C. The error bars correspond to the expanded uncertainties $U(\lambda')$. The vertical dashed lines, labelled by d_{LA}, d_{LB}, and d_{LC}, show the corresponding thickness limits; (b) The corrected thermal conductivity coefficient λ at T_m = 10 °C versus the panel thickness d for the EPS A, B, and C. The error bars correspond to the expanded uncertainties $U(\lambda)$. The horizontal right arrows indicate the average conductivity $\langle\lambda\rangle$ values from Table 3. The left arrow on the level 0.04 indicates λ_{DA} and λ_{DB} (as declared for the A and B products), while λ_{DC} appears at the 0.032 level. The vertical down arrow $\Delta\lambda$ shows the effect of graphite microparticles (GMP) on thermal conduction. Referring EPS C to EPS B, the relative drop of conductivity achieved was 17.2%.

3.3. The Thermal Conductivity Results

The declared thermal conductivity coefficient, λ_D, taken from technical data sheet (TDS) of each tested EPS product, was compared with the average corrected thermal conductivity $\langle\lambda\rangle$ (see Table 3 and Figure 4b). The $\langle\lambda_A\rangle$, $\langle\lambda_B\rangle$, and $\langle\lambda_C\rangle$ values, together with their uncertainties $\langle U(\lambda)\rangle$, represent the tested EPS products of any thickness, which were all calculated from $\lambda(d)$, respectively.

Each collected thermal conductivity value $\lambda'(d)$, representing an EPS panel of given d, was found as the average of several single measurements and was plotted, with vertical error bar corresponding to its expanded uncertainty, $U(\lambda')$ and horizontal error bar covered by the symbol used in Figure 4a.

Next, due to relevance of the thickness effect revealed by the EPS A and B products, partial corrections of their apparent thermal conductivities $\lambda'(d)$ were performed, according to the Standards [32,33,35] (see Supplementary Material part 1). Figure 4b shows the corrected thermal conductivity $\lambda(d)$ for EPS A, B, and C. The error bars correspond to the $U(\lambda)$ values. For EPS C, the data was not modified, such that $\lambda'(d) = \lambda(d)$. As seen in Figure 4b, in each case of EPS A, B, and C, the $\lambda(d)$ values oscillate around $\langle\lambda\rangle$ and below the λ_D level within the whole thickness range. Each obtained result satisfies the standard inequality:

$$\lambda_D \geq 0.44\sigma + \langle\lambda\rangle \tag{2}$$

which is used to qualify the material as complying with requirements of the Standards [31,32].

In order to show the effect of the GMP concentration on the total thermal conductivity, the absolute change in thermal conductivity coefficient (comparing the EPS C and B products), was defined as difference:

$$\Delta\lambda = \lambda_C - \lambda_B \tag{3}$$

It is shown as a vertical down arrow in Figure 4b. Also, the relative change in thermal conductivity coefficient was defined as:

$$\frac{\Delta\lambda}{\lambda_B} = \frac{\lambda_C - \lambda_B}{\lambda_B} \tag{4}$$

Both $\Delta\lambda = -6.5 \times 10^{-3}$ W·m^{-1}·K^{-1} and $\Delta\lambda/\lambda_B = -0.172$ have constant negative values within the whole range of panel thickness. Thus, the total thermal conductivity of the "grey" EPS C was about 17.2% smaller than that of the "dotted" EPS B. This result was very close to the literature data, comparing "grey" and "pure" EPS [37,43] of bulk density ca. 14–17 kg·m^{-3}. Cautiously comparing the "grey" EPS C and "pure" EPS from [37] of comparable bulk density ca. 14 kg·m^{-3} results in −25% change. Thus, the total thermal conductivity of the "dotted" EPS B was about 7.8% smaller than of the "pure" EPS from [37].

The results are discussed further in Section 4.1.

3.4. The Thermal Resistance Results

The declared thermal resistance, R_D, as given for each tested EPS product in the TDS, was compared with the corrected resistance $R(d)$. The requirements of Standards [31,32] were satisfied.

The $R(d)$ values for the EPS A and B panels were calculated by converging the corrected $\lambda(d)$ data (see Supplementary Material part 1). Yet, for the EPS C panels, the $R(d)$ were directly assigned to the $R'(d)$, as measured on the HFM Fox 600 (see Section 2.4.3).

In Figure 5a, only the EPS B and C products were compared, as the EPS A and B panels differed too much in average bulk density (Table 2), which made their comparison not precise. In EPS C, resistance increased faster with thickness. Thus, considerable improvement in thermal insulation performance appears evident, when comparing the R-values of "grey" EPS C to "dotted" EPS B.

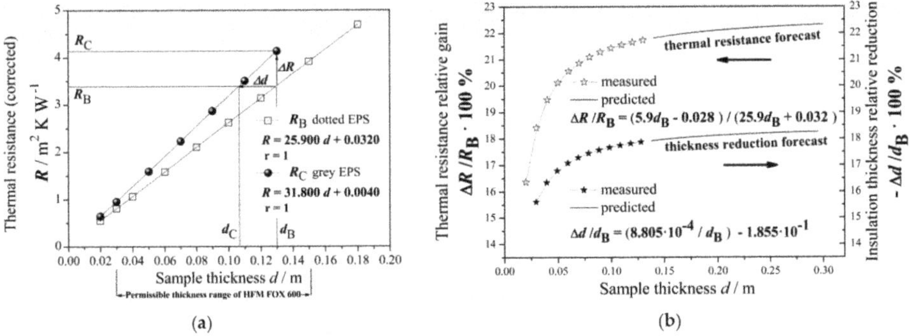

Figure 5. (**a**) The thermal resistance, R, at Tm = 10 °C versus the panel thickness d for the EPS B and C products. The equations resulting from the linear fit (1.9) from Supplementary Material part 1 are shown together with the linear correlation coefficient r (solid lines). ΔR shows the difference between B and C panels of the same thickness and Δd shows the difference between the B and C panels of the same R-value. (**b**) The thermal resistance relative gain $\Delta R/RB$ (left axis) and the insulation thickness relative reduction $-\Delta d/dB$ (right axis) for EPS C with respect to EPS B, versus the panel thickness d. The graph is reflected horizontally, since Equation (8) gives negative values. The experimental points are extrapolated (solid lines) based on the functions shown, corresponding to Equations (9) and (11).

To account for the data, the linear model (1.9), from Supplementary Material part 1, was applied to the $R(d)$ points. After the linear fit, in order to examine the effect of GMP, ΔR was defined as the difference between the EPS B and C panel's R-values at a given thickness d_B:

$$\Delta R = R_C - R_B \tag{5}$$

This is shown as a vertical arrow, ΔR, in Figure 5a. Hence, the relative gain $\Delta R/R_B$ in thermal resistance of the EPS C, with respect to EPS B of the same panel thickness d, could be defined as:

$$\frac{\Delta R}{R_B} = \frac{R_C - R_B}{R_B} \tag{6}$$

which is plotted in Figure 5b as a function of d.

Maintaining a constant given thermal resistance level, the corresponding change in the insulating layer thickness could be defined, for the EPS B and C panels, by:

$$\Delta d = d_C - d_B \tag{7}$$

which is shown as a horizontal arrow, Δd, in Figure 5a. As can be seen, greater the thermal resistance $R(d)$, the greater is the difference in thickness Δd. Hence, the relative change $\Delta d/d_B$ in thickness of the EPS C, with respect to EPS B of the same R-value, can be defined as:

$$\frac{\Delta d}{d_B} = \frac{d_C - d_B}{d_B} \tag{8}$$

which is also plotted in Figure 5b as a function of d.

From Equation (6) and the system of equations (Figure 5a), one may derive the analytical expression for the percentage increase (gain) in thermal resistance:

$$\frac{\Delta R}{R_B} \cdot 100\% = \frac{5.9d - 0.028}{25.9d + 0.032} \cdot 100\% \tag{9}$$

where $d = d_B = d_C$ may range from $4.7 \cdot 10^{-3}$ to 0.30 m. From Equation (9), one may calculate the maximum asymptotic value of about 22.8% (as $d_B \to \infty$). In practice, it is possible to achieve the maximum value of 22.3% only, for panels of the highest available thickness (0.30 m).

By using the linear fit equations reported in Figure 5a, one may find the practical equation:

$$d_C = 0.814d_B + 0.0008805 \tag{10}$$

where d_B is the thickness (in m) of EPS B and d_C is the thickness of the EPS C panel of the same R-value ($R_B = R_C$).

From Equations (7) and (10), one may derive the analytical expression for the percentage change in thickness:

$$\frac{\Delta d}{d_B} \cdot 100\% = \left(\frac{0.0008805}{d_B} - 0.1855 \right) \cdot 100\% \tag{11}$$

which is valid for d_B ranging from 4.7×10^{-3} to 0.30 m. Expression (11) allows negative values; thus, the magnitude of thickness reduction increases with the insulation thickness, d_B. The thicker the insulating layer required, the greater is the benefit in terms of material and cost savings when replacing EPS B with EPS C. As calculated from Equation (11), the outermost theoretical value of the percentage change could be achieved as -18.55% (as $d \to \infty$). In practice, it is possible to get only -18.26% for panels of the highest available thickness (0.30 m).

Finally, based on Equations (9) and (11), the additional analytical expression relating the thickness reduction and the resistance gain can be rewritten as:

$$-\frac{\Delta d}{d_B} = \frac{0.001006}{d_B} \frac{\Delta R}{R_B} + 0.8145 \tag{12}$$

It is worthy to highlight that $|-\Delta d/d_B| \neq |\Delta R/R_B|$, so the relative resistance gain does not directly determine the relative thickness reduction, especially for thicker EPS panels. If $d = 0.30$ m, then $-\Delta d/d_B \approx 0.82\ \Delta R/R_B$. In other words, when considering EPS B and C, the relative thickness reduction $|-\Delta d/d_B|$ may reach only 82% of the relative resistance gain, $\Delta R/R_B$, when increasing thickness up to 0.300 m. When decreasing the EPS B panel thickness, this relation achieves equivalence, such that $-\Delta d/d_B \approx 1.00\ \Delta R/R_B$ at the theoretical value $d \approx 0.005$ m.

3.5. Micro-Raman Spectra of Tested Products

The Raman spectra were separately collected for the "white" part selected only from the "dotted" EPS B and for the "grey" EPS C products. The "white" part (no black dotted isles) cut from the EPS B, one may consider as the equivalent of "white" EPS material (pure). Raman spectra were registered for two excitation wavelengths to distinguish phonon and luminescent peaks.

According to the above indicated, for the "white" EPS material (red lines in Figure 6a,b), without GMP, in both Raman spectra, the bands characteristic for the pure polystyrene matrix were observed. There are intensive phonon modes related to phenyl ring: ~650 cm^{-1}, ~1100 cm^{-1}, and ~1600 cm^{-1} and hydrocarbon chain modes in the range of 2900–3000 cm^{-1} [56,57]. In low frequencies, below 200 cm^{-1}, an increasing band corresponding to the Boson peak was detected [58]. The bosonic peak is derived from acoustic phonons and appears in glasses and amorphous materials, where the selection rules for Raman scattering have been broken. It is observed as low-shaped, often asymmetrical, broad peak occurring in the low frequency region of the Raman spectrum (below 200 cm^{-1}).

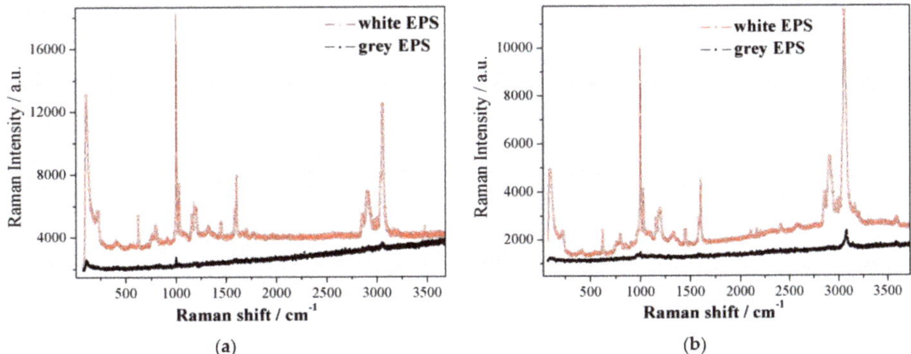

Figure 6. (a) Raman spectra for selected "white" part of the EPS B (red line) and for the "grey" EPS C (black line) products. The 633 nm He–Ne-ion laser line was used as the excitation wavelength. (b) Raman spectra for selected "white" part of the EPS B (red line) and for the "grey" EPS C (black line) products. The 514 nm Ar-ion laser line was used as the excitation wavelength.

For the "grey" EPS C (black lines in Figure 6a,b), in both Raman spectra, a significant decrease in Raman spectral intensity was observed. Such suppression of phonon spectra is usually associated with structural deformation of the molecules or molecule chains, which results in transitions to more disordered structural forms of polymer matrix [59]. The suppression effect was observed for both optical and acoustic phonons for the polystyrene matrix with GMP. The attenuation of acoustic phonons for the polystyrene matrix with GMP is indirectly visible due to the absence of boson peak in the Raman spectrum. In polymeric insulation where there are no free to move carriers, the polymer matrix thermal conduction is determined by phonons, especially by acoustic phonons. The fact that acoustic and optical phonons are suppressed in the matrix with GMP results in lowering the matrix thermal conduction and, thus, the total thermal conductivity of the EPS C insulation.

Moreover, the Raman spectra of the "grey" EPS C show a decrease in the background signal, resulting from luminescence. Such a process may also indicate an increased electromagnetic radiation absorption coefficient for graphite-containing EPS insulation.

It should be emphasized that Raman studies for the "grey" EPS C do not show graphite-specific peaks. This is different than in the case of carbon nanotubes-containing polystyrene samples [60]. Therefore, the addition of GMP to the polystyrene matrix results in modified/disturbed matrix and the EPS C cannot be treated as a simple mixture of polystyrene and graphite.

3.6. Thermal Analysis of Tested Products

The TGA and DSC thermograms were separately collected for the "white" part selected only from the "dotted" EPS B and for the "grey" EPS C products. One may consider the "white" part (no black dotted isles) as the equivalent of "white" EPS material (pure).

According to the above indicated data, the TGA thermograms indicate an improvement on the thermal stability of the "grey" EPS C compared with the "white" EPS material, as shown in Figure 7a,b. The initial mass losses of 3% occurred as follows: 311 °C for the "white" EPS, 337 °C for the EPS C, in the insulation samples with mass ranging between 1.994 and 1.968 mg, respectively. One may attribute the considerable increase in thermal stability of the EPS C to homogenous GMP dispersion in polystyrene matrix. Additionally, the presence of GMP impedes the burning process by reducing the oxygen diffusion towards bulk. The resulting maximum degradation rates (calculated from percentage mass change derivative) were as follows: 418.5 °C for the "white" EPS and 418.8 °C for the EPS C samples. The mass loss of the insulation samples occurred at only one stage and finally reached at 550 °C as follows: 0.8% and 5.1%, respectively. Thus, the graphite content calculated as the difference between the EPS C and "white" residue masses, 0.100 mg and 0.016 mg respectively, was up to 4.3% of the total mass of the EPS C sample.

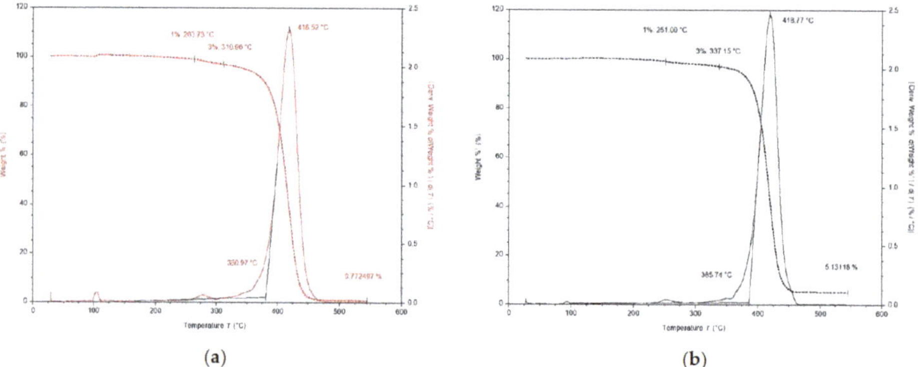

Figure 7. TGA thermograms for: (**a**) selected "white" part of the EPS B (red lines) and (**b**) the "grey" EPS C (black lines) products.

During the first conventional DSC heating scan, the resulting midpoint of T_g was as follows: 107.8 °C for the "white" EPS material and 109.7 °C for the "grey" EPS C, in the insulation samples with mass ranging between 2.274 and 2.248 mg, respectively. The melting temperature (T_{cm}), typical for the crystalline phase, was not observed in the EPS C, even up to the degradation temperature occurring at 350 °C (see degradation onsets at 381 °C and 386 °C in Figure 7). In the case of the "white" EPS, one may detect a small exothermic peak due to the cold crystallization at 194.5 °C and the melting peak at $T_{cm} = 282.9$ °C; yet, the calculated crystallinity $X_C = 0.7\%$, was extremely low. The increase in T_g of the EPS C can be explained, as the effect of intermolecular interactions between GMP and the closest polystyrene matrix chains, thereby reducing the mobility of the polymer chains and thus increasing the T_g value.

4. Discussion

4.1. Thermal Conductivity Analysis

4.1.1. Relationship between Thickness Effect and Density

For industrially produced "pure" EPS panels, the literature has reported that bulk density is the dominant controlling variable determining the mechanical and thermal properties [54]. The $\lambda(\rho)$

function has been observed in the range of 10–45 kg·m^{-3}, where total thermal conductivity decreased with increasing bulk density [6,27,37,41,43,54]. In particular, the coefficient λ can slightly decrease from 0.041 to 0.038 W·m^{-1}·K^{-1}, whereas bulk density increases from 16 to 18 kg·m^{-3}. The average value $<\lambda_A> \approx 0.039$ W·m^{-1}·K^{-1} at $<\rho_A> \approx 17$ kg·m^{-3} found in this study was in agreement with the literature [37,43]. Furthermore, the average values for EPS B and C agreed with the literature as well [42].

Comparing Figures 2 and 3, one may observe the impact of varying bulk density gradients on thermal conductivity changing along thickness. The density function $\rho_A(d)$, visibly "waving" around its average level, $<\rho_A>$, seems to be synchronized with conductivity function, $\lambda_A(d)$, which simultaneously "waves" around its average level, $<\lambda_A>$, yet, opposite in phase (Figure 4b). In the case of EPS B and C, $\lambda_B(d)$ and $\lambda_C(d)$ did not display such unsteadiness; instead, these materials revealed more uniform packing of the EPS beads or better structural homogeneity (in terms of cell morphology in the beads). These effects are understandable as, in practice, thermal conductivity (either corrected or not) is the composite function λ of ρ versus d, such that:

$$\lambda(\rho,d) = (\lambda \circ \rho)(d) = \lambda(\rho(d)) \tag{13}$$

As reported in [37,43], the thickness effect is much more visible for very low bulk density "pure" EPS than in greater density panels. The more significant the effect, the longer the curvature and the further the thickness limit position d_L, the value of which decreases as the bulk density increases. The results for the EPS A and B panels were in good agreement with the literature, as $d_{LA} < d_{LB}$ while $\rho_A > \rho_B$ (Tables 2 and 3). Yet, unlike the results explained in [37], the thickness effect may not have originated from the experimental setup. According to [37], radiation can be blocked by reflection from the colder black plate and might not be absorbed by a thinner "pure" EPS panel before returning to the "hot" black plate; this was expected to reduce the electric power required by the heating system in HFM and, hence, lower measured conductivity of the thermal insulation. Alternatively, in the light of the relation (13), the thickness effect could be explained as a simple consequence of the structural differences between the rough sample surface region (EPS panel interfacing with the "hot" or "cold" HFM black plate) and the deeper bulk core region (of slightly lower bulk density). Thus, this effect, which is common in lighter EPS products, may result from the density gradient (normal to the panel surface). Moreover, it is due to conduction (of the matrix component), rather than radiation, which is shown in Section 4.1.3.

4.1.2. GMP Effect

Another outcome from the literature is that, when comparing EPS foams with and without GMP, the nearly constant levels reached by $\lambda'(d)$ (as in Figure 4a) differ; furthermore, this difference is greater with lower EPS density [37,43]. Dependence on GMP is also evident from Figure 4b, where the $<\lambda>$ level drops gradually with increasing GMP content. To understand the effect of GMP on the total thermal conductivity levels, one may look at the impact of GMP on the individual components of total thermal conductivity, which can be assumed to be additive. As air cannot flow through the EPS closed-cell structure, the convection component can be neglected in this particular case [1,27]. Therefore, total thermal conductivity of the EPS should be resolved into its three main components:

- radiation (through both solid matrix and air),
- solid matrix conduction, and
- gas conduction (air thermal conductivity without radiation).

To this end, an interesting analysis was done by compilation and comparison of the experimental results and data reported in the literature [42] (indicated in Table 4), by combining the HFM Fox 600 and Fox 314 measurements (with and without the two parallel 10 μm Al-foil layers at the bottom and on the top of the sample). All EPS products were of comparable, very low bulk density, from 14.0 to

15.0 kg·m^{-3}. Figure 8 presents the combined data, together with an additionally simulated Al-foil effect on EPS B and C.

Table 4. A brief comparison of EPS B and C (tested) with the corresponding EPS ("dotted" and "grey") from [42] (marked as literature data) and the list of HFM instruments with test setup and output data.

EPS Materials	Low Industrial Concentration of GMP in the "dotted" EPS Materials			
	"dotted" EPS (Adapted)—Literature Data [42]		"dotted" EPS B (Tested)—Measured and Simulated Data	
	HFM FOX 314		HFM FOX 600	
HFM instrument and test setup	Without Al-foil	With Al-foil	Without Al-foil	With Al-foil
HFM output as the apparent thermal conductivity coefficient	Measured λ'_{dotted} (d)	Measured $\lambda''_{\text{dotted}}$ (d)	Measured λ'_{B} (d)	Simulated λ''_{B} (d)
EPS Materials	High Industrial Concentration of GMP in the "grey" EPS Materials			
	"grey" EPS (Adapted)—Literature Data [42]		"grey" EPS C (Tested)—Measured and Simulated Data	
	HFM FOX 314		HFM FOX 600	
HFM instrument and test setup	Without Al-foil	With Al-foil	Without Al-foil	With Al-foil
HFM output as the apparent thermal conductivity coefficient	Measured λ'_{grey} (d)	Measured λ''_{grey} (d)	Measured λ'_{C} (d)	Simulated λ''_{C} (d)

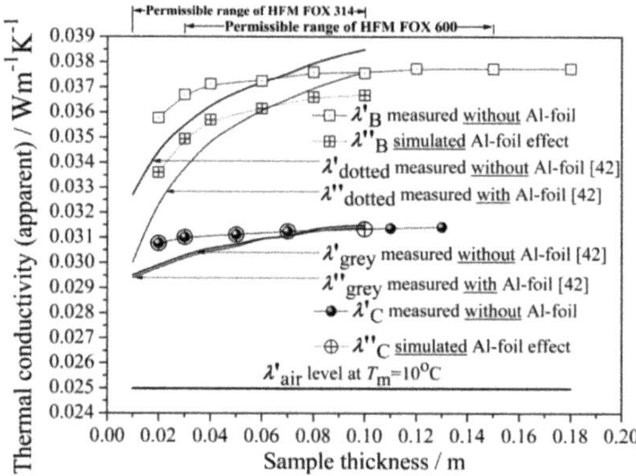

Figure 8. The effects of Al-foil and GMP on the thermal conductivity of EPS. The dotted squares and black spheres show the apparent thermal conductivity $\lambda'(d)$ at T_{m} = 10 °C for EPS B and C, respectively. The crossed squares and circles show the simulated data of $\lambda''(d)$ for EPS B and C with Al-foil. The solid lines are plotted based on data from [42]. The thin and thick lines indicate tests with and without Al-foil, respectively.

Based on this data set (Table 4) and the collected curves (Figure 8), one could carry out the quantitative estimation of each thermal component contribution by applying the procedure described in Supplementary Material part 3. The calculated results, in terms of percentage contributions, are listed in Table 5 (for the first time) as well as visualized in Figure 9 for the tested EPS B and C, in terms of the resolved components of total thermal conductivity.

Table 5. Percentage contributions of the total thermal conductivity components (radiation, air conduction, and polymer matrix conduction) presented for the apparent $\lambda'(d)$ and corrected $\lambda(d)$ (in brackets). The calculated results compare EPS B and C (tested) with the corresponding EPS ("dotted" and "grey") from [42].

Thickness (m)	Radiation	Air Conduction	Polymer Matrix Conduction	Radiation	Air Conduction	Polymer Matrix Conduction
	Contribution (%)			Contribution (%)		
	Low Industrial Concentration of GMP in the "dotted" EPS Materials					
	"dotted" EPS (adapted)—Based on Literature Data [42]			"dotted" EPS B (tested)—Based on Measured and Simulated Data (in Brackets Corrected below $d_{LB}=0.060$ m)		
0.01	8.3	76.4	15.3	-	-	-
0.02	6.1	72.2	21.7	6.1 (10.8)	69.8 (66.3)	24.1 (22.9)
0.03	4.8	70.4	24.8	4.8 (7.4)	68.1 (66.2)	27.1 (26.4)
0.04	3.9	68.8	27.3	3.9 (5.4)	67.3 (66.2)	28.8 (28.4)
0.05	3.5	67.9	28.6	-	-	-
0.06	3.0	67.1	29.9	3.0 (3.9)	67.1 (66.4)	29.9 (29.7)
0.07	2.7	66.6	30.7	-	-	-
0.08	2.6	65.7	31.7	2.6 (2.6)	66.4 (66.4)	31.0 (31.0)
0.09	-	-	-	-	-	-
0.10	2.3	64.9	32.8	2.3 (2.3)	66.5 (66.5)	31.2 (31.2)
0.11	-	-	-	-	-	-
Thickness (m)	High Industrial Concentration of GMP in the "grey" EPS Materials					
	"grey" EPS (Adapted)—Calculated Based on Literature Data [42]			"grey" EPS C (Tested)—Calculated Based on Measured and Simulated Data		
0.01	0	84.7	15.3	0	–	–
0.02	0	83.5	16.5	0	81.2	18.8
0.03	0	82.4	17.6	0	80.6	19.4
0.04	0	81.6	18.4	0	–	–
0.05	0	81.4	18.6	0	80.3	19.7
0.06	0	80.6	19.4	0	–	–
0.07	0	80.6	19.4	0	79.9	20.1
0.08	0	79.6	20.4	–	–	–
0.09	–	–	–	–	–	–
0.10	0	79.3	20.7	0	≈ 79.7	≈ 20.3
0.11	–	–	–	0	79.6	20.4

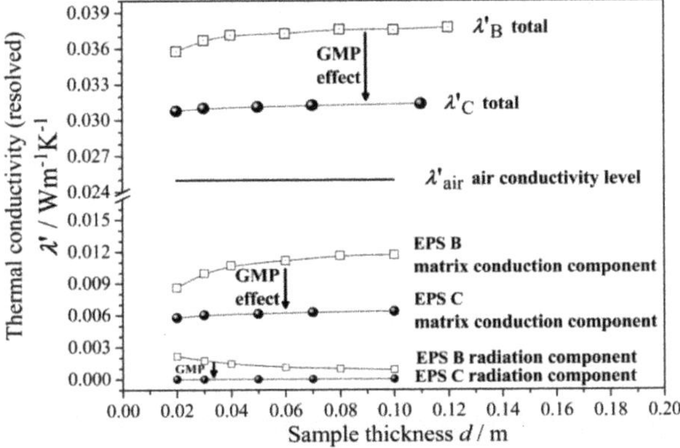

Figure 9. The GMP effect on the total thermal conductivity, resolved into its components. Each component value is calculated by multiplying the contribution fraction (Table 5) by the total thermal conductivity. Notice the impact of GMP on the thickness effect.

As can be seen from Table 5, the numerical change of each component (radiation, air conduction, and matrix conduction) could be best observed for thicknesses from 0.01 to 0.10 m, simultaneously for the two GMP industrial concentrations (EPS B and C tested or the corresponding EPS "dotted" and "grey" from [42]). For thickness up to 0.10 m, the percentage contributions are shown, which describe the evolution of all components of total thermal conductivity (before and after correction) versus thickness and GMP content.

As one may notice from Table 5, the radiation contribution to the total thermal conductivity of the "dotted" EPS from [42] decreased from 8.3% to 2.3%, whereas the matrix conduction contribution increased from 15.3% to 32.8%, with thickness increasing from 0.01 to 0.10 m. Comparing the obtained percentage contribution at each thickness, it appears that the matrix conduction may play a greater role than the radiation in the overall heat transport through EPS. For the "dotted" EPS panels of the lowest thickness, the radiation contribution appeared to be no more than 8.3%, whereas the matrix contribution result was 15.3%—nearly twice as big. On the other hand, in the thicker 0.10 m EPS panels, the radiation contribution dropped down to 2.3% and the matrix contribution reached 32.8%, such that the thicker sample had a much greater matrix contribution. For the EPS B tested panels, similar trends can be observed.

For the "grey" EPS from [42], the radiation contribution was zero and the matrix contribution revealed an increasing trend, from 15.3% to 20.7% (at the expense of the air contribution), with thickness increasing from 0.01 to 0.10 m. For EPS C, the radiation contribution was also zero and the matrix contribution revealed a similar increasing trend—from 18.8% to 20.5%—with thickness increasing from 0.02 to 0.11 m. For the EPS C tested panels, similar trends can be observed.

Thus, comparing from Table 5 (as mentioned above), the calculated results for the corresponding EPS "dotted" and "grey" from [42] with EPS B and C, one may notice that there was a good agreement, in terms of the observed trends and the calculated values.

Interestingly, it may be noticed from Figure 8 that, after applying Al-foil, the λ''_{dotted} (or λ''_B) did not drop down to the λ''_{grey} (or λ''_C) level for any d. In order to explain the apparent gap between the "grey" and "dotted" EPS material's, i.e., the total thermal conductivity levels (both with Al-foil and thus, without radiation), one must take into account the polymer matrix conduction component, which, besides radiation, can also be reduced by the GMP. The latter fact seems to have been neglected in the literature [42–50]. One may try to explain that the λ''_{dotted} (or λ''_B) may not have dropped more, due to insufficient blocking (cutting off) of radiation by the Al-foil or by some "secondary" radiation generated internally (some flux of the phonons could be converted into photons). Yet, one must notice that the Al-foil emissivity value is no more than 0.04 and that above 96% of the photon flux was reflected back by the upper Al-foil, either "primary" or "secondary". Hence, the concept of such "secondary" radiation cannot explain such an evident gap. Facing the facts that radiation can be efficiently blocked by reflection from the Al-foil and that the conductivity gap appears, the effect of GMP on the polymer matrix conduction becomes evident; that is, after addition of the GMP, conduction of the polymer matrix is dramatically reduced. The latter effect is also visualized in Figure 9. In particular, the absolute negative change in the matrix conductive component (e.g., from about 0.012 down to 0.006 W·m^{-1}·K^{-1} at 0.10 m) can be one order of magnitude greater than the radiative one (from about 0.001 down to 0.000 W·m^{-1}·K^{-1} at 0.10 m).

As observed in Figure 9, the total thermal conductivity of the "grey" EPS C, relative to the "dotted" EPS B, was reduced, which might be due to both the thermal radiation (through the whole foam) drop and conduction (only through the polymer matrix with graphite, not air) drop. This might be caused by several physical phenomena. In particular, the interfacial effects that are induced by incorporation of the GMP at the highest concentration could be responsible [61–64]. From a microscopic view, GMP is a specific opacifier of proper size (to prevent agglomeration), evenly distributed in the polymer matrix and located between the solid (matrix) and gas (air) phase, hence forming additional interfaces (GMP/air and GMP/matrix).

A question which arose was: why does GMP cause such a dramatic decrease in the thermal conduction component of the polymer matrix, the lower-than-expected, when comparing the conduction to radiation drop, observed in Figure 9, for both EPS-GMP industrial systems? First, phonons are hindered at the GMP/air and matrix/GMP interfaces as well as at the GMP exterior and interior regions. At the interfaces, they are either (strongly) scattered or (much less probable) blocked by reflection on GMP. The presence of various scattering processes for phonons leads to a reduction in their lifetime and thus, also to slow down the heat transport process taking place with their participation. At the GMP exterior region, phonons can also be temporary blocked or delayed, due to the intermolecular interactions between GMP and the closest polystyrene matrix chains causing modified/disturbed matrix, as it results from the increase of T_g value (Section 3.6) and supressed Raman spectra (Figure 6). At the

GMP interior region of high thermal capacity, further (strong) delay is caused by absorption–emission in random direction (by the delocalized nonbonded electrons of the sp^2 carbon atoms in GMP), simultaneously to (rare) refraction of the phonons. The above may produce a local thermally isolative barrier, increasing resistance. As a result, the presence of GMP significantly improves the insulating qualities of EPS materials.

4.1.3. Further Explanation of the Thickness Effect

As mentioned above, the thickness effect can be related to bulk density [37,43], however, it is also related to the GMP concentration changes. As shown in Figure 4a, the results for the EPS B and C were in good agreement with the literature, as $d_{LC} < d_{LB}$, while the GMP concentration was greater in EPS C than in EPS B at comparable bulk densities. The maximum of $\lambda'_C(d)$ differed from the minimum value by only 2%, which indicates a dramatic reduction in the observable thickness effect due to the addition of GMP. Other researchers have also found a negligible thickness effect in such "grey" EPS C-like products [37,42,43].

On one hand, the thermal conductivity of "grey" EPS, λ'_{grey} (or λ'_C), almost does not depend on thickness and, thus, practically does not reveal the thickness effect, whereas the conductivity of "dotted" EPS, λ''_{dotted} (or λ''_B), even measured with Al-foil, still does cutting off radiation by Al-foil did not remove the conductivity curvature (compare the curves for "dotted" EPS in Figure 8). Thus, distributing GMP with a higher industrial concentration in the polymer matrix may significantly reduce the thickness effect, as compared to the "pure" or "dotted" EPS with or without Al-foil.

On the other hand, applying Al-foil to the "grey" EPS practically did not cause any further decrease in the measured thermal conductivity λ'_{grey} (Figure 8). Therefore, GMP and Al-foil seem to have a similar effect in terms of blocking the thermal radiation flux.

Besides the relation between density and the thickness effect, the above observations from the experiment with Al-foil (which effectively blocks radiation though permitting conduction) suggest that the thickness effect is caused by matrix conduction (increasing with depth when crossing the EPS surface region of higher density) rather than radiation.

Thus, both observed phenomena—reduction of the thickness effect and the significant drop of total thermal conductivity after addition of GMP—might be caused by a stronger decrease in polymer matrix conduction (e.g., from about 0.012 to 0.006 $W \cdot m^{-1} \cdot K^{-1}$ at 0.10 m) than a decrease in the thermal radiation component (0.001 to 0.000 $W \cdot m^{-1} \cdot K^{-1}$ at 0.10 m, based on Table 5 and Figure 9). Note that radiation reveals a very poor contribution at the applied temperature difference.

Comparing the curves in Figure 9, the matrix conduction clearly reveals responsibility for the observed thickness effect on the EPS thermal conductivity (i.e., the λ'_B total and the EPS B matrix conduction are convex, while the EPS B radiation component is concave). As also seen in Figure 9, the EPS C matrix conduction component does not reveal the thickness effect at all. By this feature, one may discover that GMP manifests a strong effect on the polymer matrix. This must be stronger than the effect of the density gradient (normal to surface) as well as stronger than interfacial effect at the EPS panel rough surface (matrix/air).

5. Conclusions and Evaluation

This study analyses heat transfer in practical closed-cell EPSs insulation. Some conclusions and evaluation derived from the experimental and simulated findings are summarized below.

1. Initial testing of EPS product quality should be a homogeneity assessment, which can be based either on bulk density or thermal conductivity measurements versus thickness. This is possible due to the observed $\rho(d)$ and corrected $\lambda(d)$ synchronizing and the experimental relationship of $\lambda(\rho)$ between the density and corrected conductivity functions. Absence of data scattering and constant level indicates good quality. The worst homogeneity was found for the "white" EPS A product of poor quality, possibly due to recycling process used during production. The "dotted"

EPS B and "grey" C products revealed good homogeneity. As the poor homogeneity may have a great impact on all thermal measurements and material characteristics, the EPS A product had to be excluded from further analysis.

2. The analysis and evolution of the total thermal conductivity components versus the EPS panel thickness in the range of 0.02–0.1 m, for two different GMP concentrations (which are applied industrially): low ("dotted") and high ("grey") were described. The EPS materials from which the panels were made had comparable and very low bulk densities, from 14 to 15 kg·m^{-3}. The simulated data for the "dotted" EPS B and "grey" EPS C products are presented in Table 5 and plotted in Figure 8. The analysis was carried out by combining experimental measurements (HFM Fox 600) and literature data (HFM Fox 314) [42]. Simulation of the thermal radiation component was carried out through the above data processing, which was used to separate all thermal conductivity components (radiation, air conduction, and matrix conduction), as plotted in Figure 9. A lack of convection was assumed, due to the EPS closed-cell structure. The percentage contributions of all thermal components were then calculated.

3. In EPS materials that differ in GMP concentration ("dotted" and "grey"), the percentage contribution of the polymer matrix thermal conduction component and the thermal radiation component in the total thermal conductivity vary with the thickness of the thermal insulating layer in both product types. In detail, we noticed the following main points (Table 5):

   a. In the "dotted" EPS B at the smallest panel thickness (up to the thickness limit value), the thermal radiation component reached its highest percentage contribution in the total heat transport. At the smallest thickness (of 0.02 m), the thermal radiation contribution was 10.8% (corrected data in brackets). The thermal radiation contribution decreased with an increase of the panel thickness. Above the thickness limit, the contribution of the thermal radiation component was negligible and, at the highest available thickness of 0.1 m, it was only 2.3% (corrected data in brackets). On the contrary, the contribution of the solid matrix thermal conduction component increased with an increase of panel thickness. The contributions of the polymer matrix thermal conduction component were 22.9% and 31.2% (corrected data in brackets) for the 0.02 and 0.1 m panels, respectively; and

   b. In the "grey" EPS C, regardless of the panel thickness, the thermal radiation component was negligible. The percentage contribution of the polymer matrix thermal conduction component was 18.8% in the 0.02 m panels. The contribution increased with thickness and reached 20.4% at the highest panel thickness (0.1 m).

4. As resulted, adding GMP in high industrial concentrations as in "grey" EPS material may force a change in the radiative–conductive heat transfer mechanism; yet, it does not cause a perceivable decrease of the air conduction contribution. Based on the analysis results presented in Table 5, unfortunately, the percentage contributions in both the "dotted" EPS B and "grey" C products at the smallest panel thickness (0.02 m) can reach up to 70% and 81% and at the highest panel thickness (0.1 m), to about 66% and 80%, respectively. In order to reduce air conduction contribution, one may apply volume compression during foam manufacturing, as in the case of XPS foam production [65]. Such volume compression may be realized in combination with cell morphology regulation by altering the cell orientation (in one preferred spatial direction) and cell anisotropy (of 3D form), as compared with substantially round celled materials [66]. Additionally, one may reduce the cell size to obtain nanocellular PS foams [67,68].

5. The comparison of EPS materials ("dotted" and "grey"), regarding their distributions of percentage contributions of thermal conductivity components (Table 5, e.g., the "dotted" EPS B and "grey" C products) at the highest panel thickness (0.10 m), showed a dramatic effect of change in thermal radiation, by nearly −100% (i.e., (0–0.023)/0.023 × 100%). Furthermore, the polymer matrix thermal conduction was reduced strongly, by c.a. 35% (i.e., (0.203–0.312)/0.312 × 100%). One may conclude that the incorporation of GMP implicates elimination of the thermal radiation. It also

considerably weakens the polymer matrix thermal conduction, especially for large thickness panels, as the contribution of the matrix conduction becomes substantial for panels above the thickness limit. In general, the results indicate that the higher the thickness, the greater the reduction effect of matrix thermal conduction.

6. The apparent evolution of all thermal conductivity components was found in the analysis, based on measured and simulated data for EPS materials of two different GMP industrial concentrations ("dotted" and "grey"). In order to confirm the observed effects, verification may be required in terms of additional measurements. Yet, the trends revealed in this experiment are not expected to radically change.

7. As shown in Figure 6, the GMP addition to the polystyrene matrix (as in "grey" EPS C) leads to polymer matrix structural modification processes, resulting in significant attenuation of phonon spectra characteristic of pure matrix (as in "white" EPS material). This directly supports the observed drop in matrix thermal conduction component (Figure 9) and thereby explains the decrease in total thermal conductivity of EPS insulation (Figure 8). It is well known that the graphite's thermal conductivity is very high. However, based on Raman spectra, we can conclude that the addition of GMP does not lead to a simple mixture of graphite and polystyrene. In the Raman spectrum of the matrix of the EPS C, there are no modes characteristic for graphite. It should be assumed that we are dealing with particular intermolecular interactions between graphite particles and polystyrene, leading to a structurally modified/disturbed polymer matrix.

8. The thermal isolation of required resistance can be designed, regarding EPS "grey", rather than EPS "dotted" or EPS "white" panels, of reduced thickness (0.18–0.30 m) and at comparable density to EPS materials. In building practice, this means that the highest achievable reduction of at least 18.3% in the EPS insulating layer thickness is possible, referring to the thickest 0.30 m "dotted" EPS B or even "white" EPS panels.

Supplementary Materials: The following are available online at http://www.mdpi.com/1996-1944/13/11/2626/s1.

Author Contributions: Conceptualization, A.B. and M.W.; data curation, A.B.; methodology, A.B. and M.W.; validation, A.B. and C.J.; formal analysis, A.B.; investigation, A.B. and C.J.; resources, A.B.; data curation, A.B.; writing—original draft preparation, A.B.; writing—review and editing, A.B. and C.J.; visualization, A.B. and C.J.; project administration, A.B.; and funding acquisition, A.B. All authors have read and agreed to the published version of the manuscript.

Funding: This research received no external funding.

Acknowledgments: The research was supported by Laboratory Water Center, SGGW (Poland). The authors are grateful to TA Instruments (subsidiary in Poland), for help in thermal data analyses during the preparation of this work. They also need to express their gratitude to J.A. Glaeser, U.S. Forest Service, Madison, WI USA, for valuable remarks to the edition of the manuscript.

Conflicts of Interest: The authors declare no conflicts of interest.

The List of Symbols and Abbreviations

Symbols and Abbreviations	Meanings	Units
< … >	– distinguishing symbol for a physical quantity averaged over thickness range	
A	– "white" (pure) EPS product, tested herein or elsewhere in literature	
B	– "dotted" EPS product, tested herein	
dotted	– "dotted" EPS product, tested elsewhere [42]	
C	– "grey" EPS product, tested herein	
grey	– "grey" EPS product, tested elsewhere [42]	
C	– correction parameter (Table 1)	$(W{\cdot}m^{-1}{\cdot}K^{-1})$
CEN	– European Committee for Standardization	
DSC	– differential scanning calorimetry	
TGA	– thermogravimetric analysis	
d	– thickness (height) of the panel or sample	(m)
d_A, d_B, d_C	– thickness of the panel or sample referring to product A, B, or C, respectively	
d_{max}	– maximum permissible thickness of the sample assigned to the chamber of specified dimensions of the plate instrument	(m)
d_{min}	– minimum permissible thickness of the sample assigned to the chamber of specified dimensions of the plate instrument	(m)
d_L	– the panel thickness limit of the R' and λ' (non-)linearity	(m)
d_{LA}, d_{LB}, d_{LC}	– the thickness limit referring to product A, B, or C, respectively	
EPS/XPS	– expanded/extruded polystyrene	

Symbol	Description	Units
GFMs	– gas-filled materials	
GHP	– guarded hot plate instrument	
GMP	– graphite microparticles	
HFM	– heat flow meter instrument	
k	– coverage factor used in statistical analysis	
L	– thickness effect parameter	
μ-RS	– micro-Raman spectroscopy	
m	– measured mass	(kg)
NIMs	– nanoinsulation materials	
PCMs	– phase change materials	
PhF	– phenolic foam	
PIR	– polyisocyanurate	
PS	– polystyrene	
PUR	– polyurethane	
q	– density of heat flux through an insulation panel	$(\mathrm{W \cdot m^{-2}})$
R_0	– extrapolated thermal resistance corresponding to panel thickness $d = 0$,	
R_{0A}, R_{0B}, R_{0C}	in particular referring to product A, B, or C, respectively (after correction)	$(\mathrm{m^2 \cdot K \cdot W^{-1}})$
R'_0	– extrapolated thermal resistance corresponding to panel thickness $d = 0$, in particular referring to product A, B, or C,	
$R'_{0A}, R'_{0B}, R'_{0C}$	respectively (before correction)	$(\mathrm{m^2 \cdot K \cdot W^{-1}})$
R	– corrected thermal resistance (obtained by converging the corrected λ for a given panel thickness d)	
R_A, R_B, R_C	– corrected thermal resistances referring to product A, B, or C, respectively	$(\mathrm{m^2 \cdot K \cdot W^{-1}})$
R'	– apparent thermal resistance (as measured for a given panel thickness d)	
R'_A, R'_B, R'_C	– apparent thermal resistance referring to product A, B, or C, respectively	$(\mathrm{m^2 \cdot K \cdot W^{-1}})$
R_D	– declared thermal resistance at $T_m = 10\,°C$ (assigned by a manufacturer to each panel of thickness d)	
R_{DA}, R_{DB}, R_{DC}	– declared thermal resistance referring to product A, B, or C, respectively	$(\mathrm{m^2 \cdot K \cdot W^{-1}})$
SIMs	– super insulating materials	
SRM	– standard reference material	
TDS	– technical data sheet provided by the manufacturer	
T_{cm}	– crystalline melting temperature	$(°C)$
T_g	– glass transition temperature	$(°C)$
T_m	– average measurement (test) temperature	$(°C)$
U-value	– thermal transmittance (overall heat transfer coefficient)	$(\mathrm{W \cdot m^{-2} \cdot K^{-1}})$
VIMs	– vacuum insulation materials	
x, y	– panel dimensions: length, width	(m)
ΔT	– temperature difference between the "hot" and "cold" plates	$(°C)$
λ	– corrected thermal conductivity coefficient (resulted after correction of λ' for a given panel of thickness d)	
$\lambda_A, \lambda_B, \lambda_C$	– corrected thermal conductivity coefficient referring to product A, B, or C, respectively	$(\mathrm{W \cdot m^{-1} \cdot K^{-1}})$
λ'	– apparent thermal conductivity coefficient (as measured without Al-foil for a given panel thickness d)	
$\lambda'_A, \lambda'_B, \lambda'_C, \lambda'_{dotted}, \lambda'_{grey}$	– apparent thermal conductivity coefficient for a given product (as measured without Al-foil)	$(\mathrm{W \cdot m^{-1} \cdot K^{-1}})$
λ''	– apparent thermal conductivity coefficient (as simulated or measured with Al-foil for a given panel thickness d)	
$\lambda''_A, \lambda''_B, \lambda''_C, \lambda''_{dotted}, \lambda''_{grey}$	– apparent thermal conductivity coefficient for a given product (as simulated or measured with Al-foil)	
λ_D	– thermal conductivity coefficient at $T_m = 10\,°C$, as declared by a manufacturer (independent on thickness d)	
$\lambda_{DA}, \lambda_{DB}, \lambda_{DC}$	– declared thermal conductivity coefficient referring to product A, B, or C, respectively	$(\mathrm{W \cdot m^{-1} \cdot K^{-1}})$
λ^C_{SRM}	– SRM thermal conductivity coefficient (given in the Certificate for $T_m = 10\,°C$)	$(\mathrm{W \cdot m^{-1} \cdot K^{-1}})$
λ^M_{SRM}	– SRM thermal conductivity coefficient (measured with HFM FOX 600 at $T_m = 10\,°C$)	$(\mathrm{W \cdot m^{-1} \cdot K^{-1}})$
λ_t	– thermal transmissivity, horizontal asymptote of $\lambda(d)$, reciprocal gradient of the $R(d)$ linear fit (after correction)	
$\lambda_{tA}, \lambda_{tB}, \lambda_{tC}$	– thermal transmissivity, referring to product A, B, or C, respectively	$(\mathrm{W \cdot m^{-1} \cdot K^{-1}})$
λ'_t	– thermal transmissivity, horizontal asymptote of $\lambda'(d)$, reciprocal gradient of the $R'(d)$ oblique asymptote (before correction)	
$\lambda'_{tA}, \lambda'_{tB}, \lambda'_{tC}$	– thermal transmissivity referring to product A, B, or C, respectively	$(\mathrm{W \cdot m^{-1} \cdot K^{-1}})$
ρ	– bulk density of polymeric insulation material (very low or low)	
ρ_A, ρ_B, ρ_C	– bulk density value of product A, B, or C, respectively	$(\mathrm{kg \cdot m^{-3}})$
$\mathfrak{I}$	– heat transfer factor, the effective conductivity coefficient (Supplementary Material part 1)	
$\mathfrak{I}_A, \mathfrak{I}_B, \mathfrak{I}_C$	– heat transfer factor referring to product A, B, or C, respectively	$(\mathrm{W \cdot m^{-1} \cdot K^{-1}})$
Uncertainty		
Δ	– absolute error, accuracy or absolute change of a quantity	
$U(C)$	– expanded uncertainty of the correction parameter in the calculation of C	$(\mathrm{W \cdot m^{-1} \cdot K^{-1}})$
$U(d)$	– expanded uncertainty of the sample thickness measurement in HFM FOX 600	(m)
$U(L)$	– expanded uncertainty of the thickness effect parameter calculation	
$U(m)$	– expanded uncertainty of the mass measurement	(kg)
$U(q)$	– expanded uncertainty of the heat flux density measurement	$(\mathrm{W \cdot m^{-2}})$
$U(x)$ or $U(y)$	– expanded uncertainty of the x, y dimensions measurement	(m)
$U(\Delta T)$	– expanded uncertainty of the temperature difference measurement	$(°C)$
$U(\lambda)$	– expanded uncertainty in determination of the thermal conductivity coefficient λ	$(\mathrm{W \cdot m^{-1} \cdot K^{-1}})$
$U(\lambda')$	– expanded uncertainty in measurement of the thermal conductivity coefficient λ'	$(\mathrm{W \cdot m^{-1} \cdot K^{-1}})$
$U(\lambda)/\lambda$	– relative expanded uncertainty in determination of the thermal conductivity coefficient λ	(%)
$U(\lambda')/\lambda'$	– relative expanded uncertainty in measurement of the thermal conductivity coefficient λ'	(%)
$U(\lambda^C_{SRM})$	– expanded uncertainty in determination of the SRM thermal conductivity coefficient λ^C_{SRM} (given in the Certificate for $T_m = 10\,°C$)	$(\mathrm{W \cdot m^{-1} \cdot K^{-1}})$
$U(\lambda^M_{SRM})$	– expanded uncertainty in measurement of the SRM thermal conductivity coefficient λ^M_{SRM} (measured with HFM FOX 600 at $T_m = 10\,°C$)	$(\mathrm{W \cdot m^{-1} \cdot K^{-1}})$
$U(\rho)$	– expanded uncertainty of the bulk density measurement	$(\mathrm{kg \cdot m^{-3}})$
$U(\rho)/\rho$	– relative expanded uncertainty of the bulk density measurement	(%)
σ	– standard deviation	

References

1. Suh, K.W. Polystyrene and Structural Foam. In *Handbook of Polymeric Foams and Foam Technology: Polystyrene and Structural Foam*, 2nd ed.; Klempner, D., Sendijarevic, V., Eds.; Hanser: Munich, Germany, 2004; pp. 189–225.
2. Pfundstein, M.; Gellert, R.; Spitzner, M.; Rudolphi, A. *Detail Practice: Insulating Materials: Principles, Materials, Applications*, 1st ed.; Birkhäuser: Basel, Switzerland, 2008.
3. Gray, J.E. *Polystyrene: Properties, Performance and Applications*, 1st ed.; Nova Science Publishers Inc.: New York, NY, USA, 2011.
4. Polystyrene. Available online: https://www.sciencedirect.com/topics/materials-science/polystyrene (accessed on 1 September 2019).
5. Gibson, L.J.; Ashby, M.F. *Cellular Solids: Structure and Properties*, 2nd ed.; Gibson, L.J., Ashby, M.F., Eds.; Cambridge University Press: Cambridge, UK, 1997.
6. Simmler, H.; Brunner, S.; Heinemann, U.; Schwab, H.; Kumaran, K.; Mukhopadhyaya, P.; Quénard, D.; Sallée, H.; Noller, K.; Kücükpinar–Niarchos, E.; et al. Vacuum Insulation Panels. Study on VIP—Components and Panels for Service Life Prediction of VIP in Building Applications (Subtask A), IEA/ECBCS Annex 39 HiPTI—High Performance Thermal Insulation (2005). Available online: https://www.osti.gov/etdeweb/biblio/21131463 (accessed on 1 September 2019).
7. Jelle, B.P.; Gustavsen, A.; Baetens, R. The path to the high performance thermal building insulation materials and solutions of tomorrow. *J. Build. Phys.* **2010**, *34*, 99–123. [CrossRef]
8. Fantucci, S.; Lorenzati, A.; Kazas, G.; Levchenko, D.; Serale, G. Thermal Energy Storage with Super Insulating Materials: A Parametrical Analysis. *Energy Procedia* **2015**, *78*, 441–446. [CrossRef]
9. Al–Homoud, M.S. Performance characteristics and practical application of common building thermal insulation materials. *Build. Environ.* **2005**, *40*, 353–366. [CrossRef]
10. Papadopoulos, M.A. State of the art in thermal insulation materials and aims for future developments. *Energy Build.* **2005**, *37*, 77–86. [CrossRef]
11. Domínguez–Muñoz, F.; Anderson, B.; Cejudo–López, J.; Carrillo–Andrés, A. Uncertainty in the thermal conductivity of insulation materials. *Energy Build.* **2010**, *42*, 2159–2168. [CrossRef]
12. Jelle, P.B. Traditional, state–of–the–art and future thermal building insulation materials and solutions—Properties, requirements and possibilities. *Energy Build.* **2011**, *43*, 2549–2563. [CrossRef]
13. Directive 2010/31/EU of the European Parliament and of the Council of 19 May 2010 on the Energy Performance of Buildings. Available online: https://eur-lex.europa.eu/legal-content/en/TXT/?uri=celex%3A32010L0031 (accessed on 1 September 2019).
14. Chung, D.D.L. Review Graphite. *J. Mater. Sci.* **2002**, *37*, 1475–1489. [CrossRef]
15. Muzyka, R.; Drewniak, S.; Pustelny, T.; Chrubasik, M.; Gryglewicz, G. Characterization of Graphite Oxide and Reduced Graphene Oxide Obtained from Different Graphite Precursors and Oxidized by Different Methods Using Raman Spectroscopy. *Materials* **2018**, *11*, 1050. [CrossRef] [PubMed]
16. Jara, A.D.; Betemariam, A.; Woldetinsae, G.; Kim, J.Y. Purification, application and current market trend of natural graphite: A review. *Int. J. Min. Sci. Technol.* **2019**, *29*, 671–689. [CrossRef]
17. Causin, V.; Marega, C.; Marigo, A.; Ferrara, G.; Ferraro, A. Morphological and structural characterization of polypropylene/conductive graphite nanocomposites. *Eur. Polym. J.* **2006**, *42*, 3153–3161. [CrossRef]
18. Tu, H.; Ye, L. Thermal conductive PS/graphite composites. *Polym. Adv. Technol.* **2009**, *20*, 21–27. [CrossRef]
19. Glueck, G.; Batscheider, K.-H.; Riethues, M.; Schmitt, H. Polystyrene Foam Beads, Useful for the Production of Drainage Sheets for Cellar and Structural Walls. DE 19828250A1. 30 December 1999. Available online: https://worldwide.espacenet.com/patent/search/family/007871933/publication/DE19828250A1?q=DE%2019828250A1 (accessed on 1 September 2019).
20. Glueck, G.; Hahn, K.; Batscheider, K.-H.; Naegele, D.; Kaempfer, K.; Husemann, W.; Hohwiller, F. Method for Producing Expandable Styrene Polymers Containing Graphite Particles. US6130265A, 10 October 2000.
21. Glück, G.; Hahn, K.; Kaempfer, K.; Naegele, D.; Braun, F. Expandable Styrene Polymers Containing Graphite Particles. US6340713B1, 22 January 2002.
22. Gluck, G. Expandable Styrene Polymers Containing Carbon Particles. US20040039073A1, 26 February 2004.
23. Datko, A.; Hahn, K.; Allmendinger, M. Expanded Styrene Polymers Having A Reduced Thermal Conductivity. US8173714B2, 8 May 2012.

24. Zhang, C.; Li, X.; Chen, S.; Yang, R. Effects of polymerization conditions on particle size distribution in styrene-graphite suspension polymerization process. *J. Appl. Polym. Sci.* **2016**, *133*, 44270. [CrossRef]
25. Kondratowicz, F.L.; Rojek, P.; Mikoszek-operchalska, M.; Utrata, K. Combination of Silica and Graphite and Its Use for Decreasing the Thermal Conductivity of Vinyl Aromatic Polymer foam. US20180030231A1, 1 February 2018.
26. Hohwiller, F. Neopor ®, A New EPS-Generation. In Proceedings of the Conference Particle Foam 2000, Wiesloch, Germany, 10–11 May 2000; p. 169. Available online: https://www.tib.eu/de/suchen/id/TIBKAT% 3A312103158/Particle-foam-2000/ (accessed on 1 September 2019).
27. Schellenberg, J.; Wallis, M. Dependence of Thermal Properties of Expandable Polystyrene Particle Foam on Cell Size and Density. *J. Cell. Plast.* **2010**, *46*, 209–222. [CrossRef]
28. ISO:EN 8301:1991 Thermal insulation–Determination of Steady–State Thermal Resistance and Related Properties–Heat Flow Meter Apparatus. Available online: https://www.iso.org/standard/15421.html (accessed on 1 September 2019).
29. ISO:EN 8301:2010 Amendment 1 Thermal insulation–Determination of Steady—State Thermal Resistance and Related Properties—Heat Flow Meter Apparatus. Available online: https://www.iso.org/standard/15421.html (accessed on 1 September 2019).
30. EN 1946-3:1999 Thermal Performance of Building Products and Components—Specific Criteria for the Assessment of Laboratories Measuring Heat Transfer Properties—Part 3: Measurement by Heat Flow Meter Method. Available online: https://shop.bsigroup.com/ProductDetail?pid=000000000019973656 (accessed on 1 September 2019).
31. CEN:EN 13172:2012 Thermal insulation Products–Evaluation of Conformity. Available online: https://standards.cen.eu/dyn/www/f?p=204:110:0::::FSP_PROJECT,FSP_ORG_ID:34678,6071&cs= 1AE67E7D596D1D5F55D18830615383105 (accessed on 1 September 2019).
32. CEN:EN 13163:2012 and CEN: EN 13163:2008 Thermal Insulation Products for Buildings—Factory Made Expanded Polystyrene (EPS) Products–Specification. Available online: https://standards.cen.eu/dyn/www/f? p=204:110:0::::FSP_PROJECT,FSP_ORG_ID:62941,6071&cs=13DF8DF21FF2A47C751908A6DC15AFAA4 (accessed on 1 September 2019).
33. ISO:EN 9288:1996 Thermal Insulation—Heat Transfer by Radiation—Physical Quantities and Definitions (ISO 9288:1989). Available online: https://www.iso.org/standard/16943.html (accessed on 1 September 2019).
34. Tleoubaev, A. Conductive and Radiative Heat Transfer in Insulators, TA Instruments Portal. Available online: http://www.tainstruments.com/pdf/literature/Conductive%20and%20Radiative%20Heat% 20Transfer%20in%20Insulators.pdf (accessed on 1 September 2019).
35. CEN:EN 12939:2000 Thermal Performance of Building Materials and Products—Determination of Thermal Resistance by Means of Guarded Hot Plate and Heat Flow Meter Methods—Thick Products of High and Medium Thermal Resistance. Available online: https://standards.cen.eu/dyn/www/f?p=204:110:0::::FSP_ PROJECT,FSP_ORG_ID:2296,6072&cs=1745C3AD11A87A7F7BC47527DA6D98D88 (accessed on 1 September 2019).
36. CEN:EN 12667:2001 Thermal Performance of Building Materials and Products—Determination of Thermal Resistance by Means of Guarded Hot Plate and Heat Flow Meter Methods—Products of High and Medium Thermal Resistance. Available online: https://standards.cen.eu/dyn/www/f?p=204:110:0::::FSP_PROJECT, FSP_ORG_ID:2314,6072&cs=1D562318E1F7D2D1B6702D7070657863D (accessed on 1 September 2019).
37. Firkowicz–Pogorzelska, K. Transport ciepła przez promieniowanie w lekkich materiałach termoizolacyjnych/in Polish/, Materiały konferencyjne. In Proceedings of the VI Ogólnopolska Konferencja Naukowo–Techniczna Energodom 2002. Problemy projektowania, realizacji i eksploatacji budynków o niskim zapotrzebowaniu na energię, Kraków–Zakopane, Poland, 13–16 October 2002; pp. 83–90.
38. Coquard, R.; Baillis, D. Modeling of Heat Transfer in Low-Density EPS Foams. *J. Heat Transf.* **2006**, *128*, 538–549. [CrossRef]
39. Baillis, D.; Coquard, R. Radiative and Conductive Thermal Properties of Foams. In *Cellular and Porous Materials: Thermal Properties Simulation and Prediction*, 1st ed.; Öchsner, A., Murch Graeme, E., de Lemos, M.J.S., Eds.; Wiley-VCH Verlag GmbH & Co. KGaA: Weinheim, Germany, 2008; pp. 343–384. [CrossRef]
40. Coquard, R.; Baillis, D.; Quenard, D. Radiative Properties of Expanded Polystyrene Foams. *J. Heat Transf.* **2009**, *131*, 012702. [CrossRef]

41. Jeong, Y.-S.; Choi, H.-J.; Kim, K.-W.; Choi, G.-S.; Kang, J.-S.; Yang, K.-S. A study on the thermal conductivity of resilient materials. *Thermochim. Acta* **2009**, *490*, 47–50. [CrossRef]

42. Kondrot–Buchta, A.; Koniorczyk, P.; Zmywarczyk, J. Badania eksperymentalne efektu redukcji przewodności cieplnej w piance polistyrenowej (styropianie)/in Polish/—Experimental Investigations of Thickness effect curve in Polystyrene foam (Styrofoam). *Ciepłownictwo Ogrzew. Went.* **2013**, *44*, 509–512. Available online: http://www.sigma-not.pl/publikacja-81057-badania-eksperymentalne-efektu-redukcji-przewodno%C5%9Bci-cieplnej-w-piance-polistyrenowej-(styropianie)-cieplownictwo-ogrzewnictwo-wentylacja-2013-12.html (accessed on 1 September 2019).

43. Lakatos, Á.; Kalmár, F. Investigation of thickness and density dependence of thermal conductivity of expanded polystyrene insulation materials. *Mater. Struct.* **2013**, *46*, 1101–1105. [CrossRef]

44. Baillis, D.; Coquard, R.; Randrianalisoa, J.H.; Dombrovsky, L.A.; Viskanta, R. Thermal radiation properties of highly porous cellular foams. *Spec. Topopics Rev. Porous Media* **2013**, *4*, 111–136. [CrossRef]

45. Akolkar, A.; Rahmatian, N.; Unterberger, S.; Petrasch, J. Modeling the Effect of Infrared Opacifiers on Coupled Conduction-Radiation Heat Transfer in Expanded Polystyrene. *J. Heat Transf.* **2018**, *140*, 112005. [CrossRef]

46. Coquard, R.; Quenard, D.; Baillis, D. Numerical and experimental study of the IR opacification of Polystyrene Foams for Thermal Insulation enhancement. *Energy Build.* **2018**, *183*, 54–63. [CrossRef]

47. Hasanzadeh, R.; Azdast, T.; Doniavi, A.; Eungkee Lee, R. Multi-objective optimization of heat transfer mechanisms of microcellular polymeric foams from thermal-insulation point of view. *Therm. Sci. Eng. Prog.* **2019**, *9*, 21–29. [CrossRef]

48. Hasanzadeh, R.; Azdast, T.; Doniavi, A.; Eungkee Lee, R. Thermal Conductivity of Low-Density Polyethylene Foams Part I: Comprehensive Study of Theoretical Models. *J. Therm. Sci.* **2019**, *28*, 745–754. [CrossRef]

49. Hasanzadeh, R.; Azdast, T.; Doniavi, A. Thermal Conductivity of Low-Density Polyethylene Foams Part II: Deep Investigation using Response Surface Methodology. *J. Therm. Sci.* **2020**, *29*, 159–168. [CrossRef]

50. Solórzano, E.; Rodriguez-Perez, M.A.; Lázaro, J.; de Saja, J.A. Influence of solid phase conductivity and cellular structure on the heat transfer mechanisms of cellular materials: Diverse case studies. *Adv. Eng. Mater.* **2009**, *11*, 818–824. [CrossRef]

51. Reglero Ruiz, J.A.; Saiz-Arroyo, C.; Dumon, M.; Rodríguez-Perez, M.A.; Gonzalez, L. Production, cellular structure and thermal conductivity of microcellular (methyl methacrylate)—(butyl acrylate)—(methyl methacrylate) triblock copolymers. *Polym. Int.* **2011**, *60*, 146–152. [CrossRef]

52. Zhang, H.; Fang, W.Z.; Li, Y.M.; Tao, W.Q. Experimental study of the thermal conductivity of polyurethane foams. *Appl. Therm. Eng.* **2017**, *115*, 528–538. [CrossRef]

53. Aram, E.; Mehdipour-Ataei, S. A review on the micro- and nanoporous polymeric foams: Preparation and properties. *Int. J. Polym. Mater. Polym. Biomater.* **2016**, *65*, 358–375. [CrossRef]

54. Mihlayanlar, E.; Dilmac, S.; Güner, A. Analysis of the effect of production process parameters and density of expanded polystyrene insulation boards on mechanical properties and thermal conductivity. *Mater. Des.* **2008**, *29*. [CrossRef]

55. Schellenberg, J.; Wallis, M. Dependence of properties of expandable polystyrene particle foam on degree of fusion. *J. Appl. Polym. Sci.* **2010**, *115*, 2986–2990. [CrossRef]

56. Palm, A. Raman Spectrum of Polystyrene. *J. Phys. Chem.* **1951**, *55*, 1320–1324. [CrossRef]

57. Sears, W.M.; Hunt, J.L.; Stevens, J.R. Raman scattering from polymerizing styrene. I. Vibrational mode analysis. *J. Chem. Phys.* **1981**, *75*, 1589–1598. [CrossRef]

58. Kirillov, S.A.; Perova, T.S.; Faurskov Nielsen, O.; Praestgaard, E.; Rasmussen, U.; Kolomiyets, T.M.; Voyiatzis, G.A.; Anastasiadis, S.H. Fitting the low-frequency Raman spectra to boson peak models: Glycerol, triacetin and polystyrene. *J. Mol. Struct.* **1999**, *479*, 271–277. [CrossRef]

59. Menezes, D.B.; Reyer, A.; Marletta, A.; Musso, M. Glass transition of polystyrene (PS) studied by Raman spectroscopic investigation of its phenyl functional groups. *Mater. Res. Express* **2017**, *4*, 015303. [CrossRef]

60. Chipara, D.M.; Macossay, J.; Ybarra, A.V.R.; Chipara, A.C.; Eubanks, T.M.; Chipara, M. Raman spectroscopy of polystyrene nanofibers—Multiwalled carbon nanotubes composites. *Appl. Surf. Sci.* **2013**, *275*, 23–27. [CrossRef]

61. Bartz, A.M.; Hitchcock, M.K. Methods of Insulating with Plastic Structures Containing Thermal Grade Carbon Black. US 5373026A, 13 December 1994.

62. Cahill, D.G.; Braun, P.V.; Chen, G.; Clarke, D.R.; Fan, S.; Goodson, K.E.; Keblinski, P.; King, W.P.; Mahan, G.D.; Majumdar, A.; et al. Nanoscale thermal transport. II. 2003–2012. *Appl. Phys. Rev.* **2014**, *1*, 011305. [CrossRef]

63. Vo, T.Q.; Kim, B.H. Transport Phenomena of Water in Molecular Fluidic Channels. *Sci. Rep.* **2016**, *1*, 1–8. [CrossRef]

64. bin Saleman, A.R.; Chilukoti, H.K.; Kikugawa, G.; Shibahara, M.; Ohara, T. A molecular dynamics study on the thermal energy transfer and momentum transfer at the solid-liquid interfaces between gold and sheared liquid alkanes. *Int. J. Therm. Sci.* **2017**, *120*, 273–288. [CrossRef]

65. Vo, C.V.; Bunge, F.; Duffy, J.; Hood, L. Advances in Thermal Insulation of Extruded Polystyrene Foams. *Cell. Polym.* **2011**, *30*, 137–155. [CrossRef]

66. Miller, L.; Breindel, R.; Weekley, M.; Cisar, T. To Enhance the Thermal Insulation of Polymeric Foam by Reducing Cell Anisotropic Ratio and the Method for Production Thereof. US20050192368A1, 15 December 2013.

67. Liu, S.; Duvigneau, J.; Vancso, G.J. Nanocellular polymer foams as promising high performance thermal insulation materials. *Eur. Polym. J.* **2015**, *65*, 33–45. [CrossRef]

68. Forest, C.; Chaumont, P.; Cassagnau, P.; Swoboda, B.; Sonntag, P. Polymer nano-foams for insulating applications prepared from CO_2 foaming. *Prog. Polym. Sci.* **2015**, *41*, 122–145. [CrossRef]

 *materials*

Article

A Design of Experiment Approach for Surface Roughness Comparisons of Foam Injection-Moulding Methods

Gethin Llewelyn [1,*] , **Andrew Rees** [1] , **Christian Griffiths** [1] **and Martin Jacobi** [2]

1   College of Engineering, Swansea University, Swansea, Wales SA1 8EN, UK;
    Andrew.Rees@Swansea.ac.uk (A.R.); c.a.griffiths@swansea.ac.uk (C.G.)
2   Trexel GmbH, Ahlefelderstr. 64, D-51645 Gummersbach, Germany; m.jacobi@trexel.com
*   Correspondence: gethin.llewelyn@gmail.com

Received: 28 April 2020; Accepted: 19 May 2020; Published: 20 May 2020

Abstract: The pursuit of polymer parts produced through foam injection moulding (FIM) that have a comparable surface roughness to conventionally processed components are of major relevance to expand the application of FIM. Within this study, 22% talc-filled copolymer polypropylene (PP) parts were produced through FIM using both a physical and chemical blowing agent. A design of experiments (DoE) was performed whereby the processing parameters of mould temperatures, injection speeds, back-pressure, melt temperature and holding time were varied to determine their effect on surface roughness, Young's modulus and tensile strength. The results showed that mechanical performance can be improved when processing with higher mould temperatures and longer holding times. Also, it was observed that when utilising chemical foaming agents (CBA) at low-pressure, surface roughness comparable to that obtained from conventionally processed components can be achieved. This research demonstrates the potential of FIM to expand to applications whereby weight saving can be achieved without introducing surface defects, which has previously been witnessed within FIM.

Keywords: polypropylene; talc; TecoCell®; MuCell®; foam injection moulding

---

1. Introduction

Recent demands from lowering polymer consumption and making lightweight parts has seen the rise of foam injection moulding (FIM) through different foaming techniques [1–3]. FIM can be performed through either physical blowing agents (PBA) or chemical blowing agents (CBA) [4]. PBA is used by injecting a super critical gas through the moulding barrel while the polymer is being metered, in order to form a single-phase solution [5]; whilst CBA are added to the parent material in small amounts prior to processing [6]. The introduction of FIM does result in component weight saving, however it also lowers the mechanical properties and introduces the surface defect of swirl marks [7]. The swirl marks can be attributed to the mould being filled by the polymer/gas solution. In particular, cell nucleation has been initiated at this stage due to the rapid pressure drop at the injection point. Following this, the fountain flow affect freezes and stretches the cells at the mould/polymer interface resulting in swirl marks [8].

Traditionally, FIM removes the packing phase which is witnessed in conventional injection moulding (IM). This is removed as essentially the packing phase is completed by the foaming of the polymer. This method is referred to as low-pressure FIM and it was the original technique used when the technology was first developed [9]. The technique has introduced weight savings of up to 15% within thermoplastics such as polypropylene (PP) [10]. At present, this technique is the most utilised in an industrial capacity, however recent developments in alternative techniques have seen the position challenged.

Recent developments in FIM have seen the introduction of high-pressure processing. During high-pressure processing, the packing phase is not removed from the conventional IM process, but instead reduced or kept constant. Like low-pressure FIM, the single-phase solution begins to nucleate during the filling stage of the injection cycle through pressure changes. However, the introduction of the packing stage causes the pre-nucleated cells to re-dissolve back into the melt. This occurs if the packing pressure is kept above the solubility pressure of the PBA, the pre-nucleated cells can be re-dissolved into the polymer and nucleated in situ within the mould [11]. Finally, the holding pressure is stopped, and the cells nucleate through either thermal shrinkage through normal cooling or by a secondary pressure drop; usually by mould opening [12]. Although the resulting part weight reductions cannot meet the level of low-pressure FIM, the technique has been shown to improve the cellular properties [11,12].

The use of semi-crystalline polymers, such as PP, are extensively used within the plastics industry due to their high thermal stability, moisture resistance, excellent chemical and corrosion resistance, ease of processability and low density [13]. However, with the additional functionality requirements in recent years, it is becoming more difficult to use PP in its neat state. Instead, fillers are added to increase structural properties [14]. Moreover, unfilled PP experiences poor foaming behaviour due to it being a linear hydro-carbon polymer which has poor melt strength and low viscoelastic properties, leading to cell coalescence and resulting in poor cellular structures [15–17]. Therefore, fillers such as silica, carbon black, calcium carbonate and talc have all been included in the FIM of polymers with poor foaming properties to improve the isothermal crystallisation process which improves the nucleation and resulting cellular structures [18–20].

Within the current knowledge base, limited work has been performed on the effect of high-pressure FIM on the resulting mechanical properties and the surface roughness of the final part. Shaayegan et al. investigated high-pressure foaming, whereby cellular properties were dramatically increased if the mould opening is utilised to initiate a secondary pressure drop [21]. Further research has demonstrated that with an increase in packing time from 4 s to 8 s in FIM, polystyrene (PS) with CO_2 as the blowing agent caused the average cell size to drop and cell density to increase [22]. In addition, it has been proven that through high pressure FIM, nano cellular structures are obtainable [23–25]. Costeux has highlighted the levels at which nanofoams have been produced in recent years within the foam industry [26]. In a further study, Ameli et al. achieved a minimum cell size of 70 nm and maximum cell density of 2×10^{14} with a PP/Multi-Walled Carbon Nanotubes mixture [27]. However, most of these studies are through laboratory-based procedures, such as batch foaming, and have yet to be achieved through industrial foaming routes, like injection moulding.

Obtaining FIM parts that have comparable surface roughness as their conventionally processed solid counterpart has been investigated. In particular, various methods to improve these surface finish defects have been attempted through co-injection moulding [28], gas counter pressure [29] and vario-thermal moulding [30,31]. All process variants have shown to improve the FIM parts surface appearance. However, the improvement in surface finish has introduced detrimental effects with regards to cycle time and environmental aspects. Therefore, these technologies have seen limited impact within an industrial context and have led researchers to continue the pursuit of improvements in surface finish when processing through FIM.

Lee et al. have achieved near perfect surface finish polyethylene (PE) parts with FIM. This was achieved by optimising processing whereby 0.173 wt.% of supercritical N_2 is applied [32]. More recently, Guo et al., proposed that through in-mould decoration, foamed PP could be produced that had the same surface appearance as solid parts [33]. In another study, Wang et al. used a multi stage process to create a defect-free surface with a nanocellular microstructure through incorporating polytetrafluoroethylene (PTFE) into the PP matrix through in situ nanofibrillation [23]. Although research in FIM has advanced greatly in recent years, there is yet to be a study that has demonstrated that FIM can achieve surface roughness comparable to conventional IM when moulding components in PP with high weight savings.

In this research, a 22% talc-filled PP was processed initially through conventional injection moulding (IM) then produced using variations of FIM process. For the PBA, super critical N_2 was used (through the MuCell® system) and the CBA used was TecoCell® H1. In addition, for the FIM processing a design of experiment (DoE) varying the 5 main processing parameters of mould temperature, injection speed, back-pressure, melt temperature and holding time were performed to investigate the effect of varying processing parameters on the resulting tensile strength and surface roughness.

2. Materials and Methods

2.1. Materials

This research has used a commercial grade copolymer polypropylene (PP), synthesised with 22% talc and black masterbatch as the base material. It has a melt flow rate (MFR) of 35 g/10 min and a density (ρ) of 1.05 g/cm^3. The pressure-volume-temperature (PvT) data of this polymer can be seen in Figure 1.

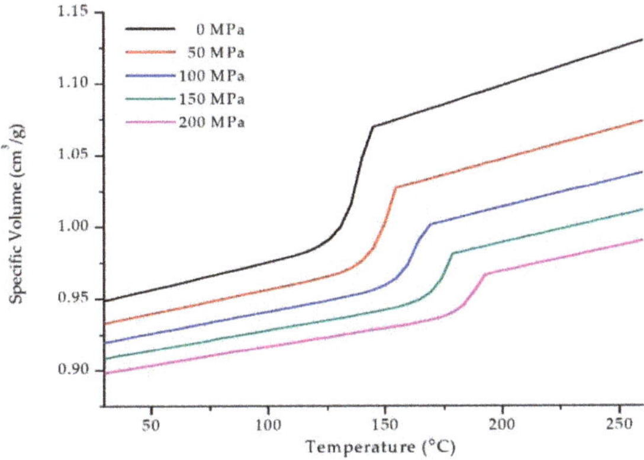

Figure 1. Pressure-Volume-Temperature data of the copolymer polypropylene used in this research.

For the CBA experiments, an endothermic CBA was used: TecoCell® H1 (Trexel GmbH, Gummersbach, Germany), which was added to the base material. This CBA creates an endothermic reaction when heated within the injection moulding barrel above 200 °C; between monosodium citrate ($C_6H_7NaO_7$) and calcium carbonate ($CaCO_3$) [6]; releasing carbon dioxide (CO_2) into the IM barrel.

For the PBA experiments, a gas dosing unit (T100, Trexel GmbH, Gummersbach, Germany) was used with nitrogen (N_2) of 99.998% purity, to produce MuCell® injection moulded parts. For all of the experiments performed in this research, the polymer was dried for 4 h at 80 °C prior to processing. In each moulding cycle, 2 tensile bars complying to type A1 within BS EN ISO 2073:2014 were produced (Figure 2) and mould layout can be seen in the authors' previous publication [19]. The Injection Moulding machine used for this research was a 40 mm screw diameter IM machine (e-Victory 120, ENGEL, Warwick, UK); with a maximum clamping force of 1200 kN.

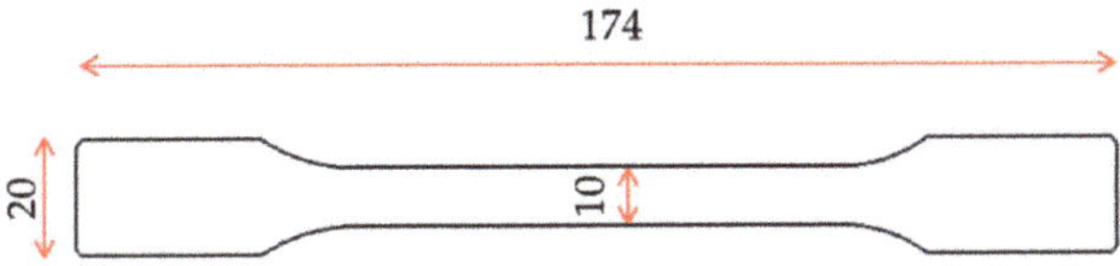

Figure 2. Moulded part geometry, thickness is 4 (dimensions in millimetres).

2.2. Experimental Procedure

To evaluate the resulting tensile strength and surface roughness when using a PBA, a full 2^5 factorial DoE was used by varying the mould temperatures, injection speeds, barrel back-pressure, melt temperature and injection packing time (factorial design Table A1, maximum and minimum values displayed in Table 1).

Table 1. 2^5 Full factorial design of experiment (DoE).

Input Variable	−1	0	1
Mould Temperature (°C)	25	57.5	90
Injection Speed (mm/s)	100	183.9	267.8
Back-Pressure (MPa)	12	14	16
Average Melt Temperature (°C)	155	207.5	260
Holding Time (s)	0	3.5	7

Similar to other research, preliminary experiments were performed to obtain the maximum and minimum input values when utilising a PBA [34]. The '1' values shown Table 1 for mould temperature and injection speeds are limited to the equipment available, while the other maximum values are limited by the polymer. The '−1' values for all the parameters were limited by the polymer; recommended by the material supplier. Also, conventional injection moulded parts were moulded as an experimental reference in terms of the weight savings of the foamed parts. The target weight savings of the final parts in relation to the conventional IM, were 12.6% and 8.8% for the 0 s and 7 s holding time, respectively. For all experiments the clamping force was kept consistent at 1000 kN. In addition, for the PBA experiments 40 mg of supercritical N_2 was used. All of the experiment settings were repeated 25 times to ensure the process was repeatable; of which 5 of these samples were tested at random all testing procedures used in this research,

Following the characterisation of the optimum processing condition for minimum achievable surface roughness when utilising a PBA, the same parameters were used for the CBA experiments.

2.3. Characterisation Methods

2.3.1. Tensile Properties

Tensile tests were performed on a mechanical testing unit (H25 KS, Hounsfield, Surrey, UK) with BS EN ISO 527–1:2012 compliance, to obtain the maximum stress and subsequently to calculate the Young's modulus (E). The ultimate tensile stress (S_u) was obtained by dividing the maximum force value by cross-sectional area. While the Young's modulus (E) was obtained by taking the strain values between 0.0005 and 0.0025, collected using an axial extensometer (3542, Epsilon, WY, USA), and using the chord slope method of the stress obtained from the force values within this range. This was run using 1 mm/min until a strain of 0.0025 had been met, then increased to 10 mm/min until fracture.

2.3.2. Part Surface Roughness

The resulting surface characteristics were quantified through surface roughness measurements. This was completed using a surface profilometer (Dektak 150, Veeco UK, St. Ives, UK) using 4 mg of force along 10 mm (L) with the resulting height being y. The calculated surface roughnesses were the arithmetical average deviation from the mean line (R_a), and the root mean squared value of the roughness (R_q); seen below in Equations (1) and (2), respectively.

$$R_a = \frac{1}{L} \int_{x=0}^{x=L} |y| \, dx \tag{1}$$

$$R_q = \sqrt{\int_{x=0}^{x=L} y^2(x)\, dx} \tag{2}$$

3. Results and Discussion

Initially within this section the main findings from the DoE activities for the processing with a PBA are presented. Then, the results from the two CBAs are compared to the best surface finish appearance PBA parts and the conventionally moulded parts. The DoE data is presented as simplified regression models using the analysis of variance (ANOVA) model. This is a method to identify which of the processing parameters have the highest statistical influence on the process. The values are generated using the experimental values in a sum of squares and divided by the numbers of degrees of freedom of each error: the variance for each parameter is then compared with the variance of the error [35]. This DoE data is also shown in interaction plots; whereby the input processing parameters are compared against each other to determine whether there is an interaction between processing parameters which alter the final results [36]. Finally, the regression models derived from the DoE data can be seen in Table A2 in Appendix A.

3.1. Tensile Strength

3.1.1. Modulus of Elasticity (E)

Figure 3 displays the resulting mean Young's modulus (E) against the five variable processing parameters from the DoE: mould temperature, injection speeds, back-pressure, melt temperature and holding time.

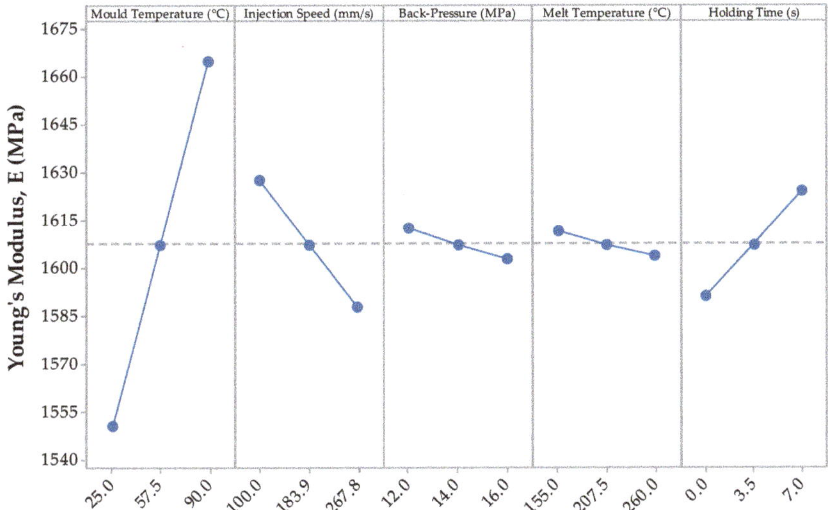

Figure 3. Main effects plot for the mean Young's modulus (E) results from the DoE.

The DoE results for the Young's modulus clearly identify the effect of each of the five input processing parameters. In particular, with increasing mould temperature and holding time, the Young's modulus of the part increases. However, with an increase in the injection speed, back-pressure and melt temperature, the Young's modulus reduces. The most significant input processing parameter on the resulting Young's modulus is that of mould temperature, a range of 116.7 MPa while the least significant input processing parameter range is 7.6 MPa from the melt temperature. The information within Table 2 confirms the viability of this data with a P-value below 0.05; showing that the confidence level of this parameter having an effect on the outcome being highly significant [37].

Table 2. Analysis of variance model for Young's modulus against processing parameters.

Source	*df*	Sum of Squares	Mean Squares	F-Value	*p*-Value
Model	9	187,000	20,778	14.18	0.000
Linear	5	138,934	27,787	18.97	0.000
Mould Temperature	1	114,005	114,005	77.83	0.000
Injection Speed	1	11,190	11,190	7.64	0.011
Back-Pressure	1	27	27	0.02	0.893
Melt Temperature	1	1444	1444	0.99	0.331
Holding Time	1	12,268	12,268	8.37	0.008
2-Way Interactions	4	48,066	12,016	8.2	0.000
Injection Speed × Melt Temperature	1	12,591	12,591	8.6	0.007
Injection Speed × Holding Time	1	11,461	11,461	7.82	0.010
Back-Pressure × Melt Temperature	1	7558	7558	5.16	0.033
Melt Temperature × Holding Time	1	16,456	16,456	11.23	0.003
Residual	23	33,692	1465		
Curvature	1	0	0	0.00	0.994
Lack-of-Fit	22	33,692	1531		
Total	32	220,692			

Within this research, all tensile strength factors and interactions with a P-value lower than that of the confidence level ($\alpha = 0.05$) are deemed significant. Therefore, they are seen to have a great effect on the response when the testing level is moved from low to high or vice versa [38].

The data in Table 2 display the significant data from the analysis of variance model which shows all of the input processing parameters, along with any other interaction between these parameters which show a P-value lower than 0.05. The data validates the analysis from Figure 1 that the mould temperature is the most significant variable as it has a very low P-value of less than 0.001. The only significant factors were that of 2-way interactions; with 3- to 5-way interactions having very high P values and hence, not being of any significance. Holding time and injection speed were the other 2 variables that showed significant effects with p-values of 0.003 and 0.011, respectively. Finally, as shown in both Table 2 and Figure 3, the least significant variables were those of back-pressure and melt temperature with p-values of 0.893 and 0.331, respectively. As for the 2-way interaction plots, the most significant was that of melt temperature with holding time, followed closely by that of injection speed with melt temperature. Injection speed combined with holding time also showed significance effects, along with back-pressure combined with melt temperature.

It is plausible that mould temperature having such a large effect on the Young's modulus can be attributed to the presence of shish-kebab microstructures typically witnessed at higher cooling rate in the sample interior as it inhibits the relaxation of the crystalline structure prior to crystallisation [39].

3.1.2. Ultimate Tensile Strength (S_u)

Figure 4 shows the results of the ultimate tensile strength (S_u) from the DoE of the PBA.

Unlike Figure 3, the mean S_u data shows a weaker trend, compared to the Young's Modulus data, due to the input processing parameters due to the midpoint DoE processing setting being 15.64 MPa; a higher mean value than all the processing parameters. However, the increase and decrease trends are in line with Young's modulus data. In particular, mould temperature and holding time increase the mean S_u whilst injection speed, back-pressure and melt temperature reduce S_u. However, the input processing parameter setting with the greatest effect on the S_u, are that of the melt temperature with 0.85 MPa. Table 3 confirms the observation as the P-value of this processing parameter is close to 0: showing a great significance on the S_u.

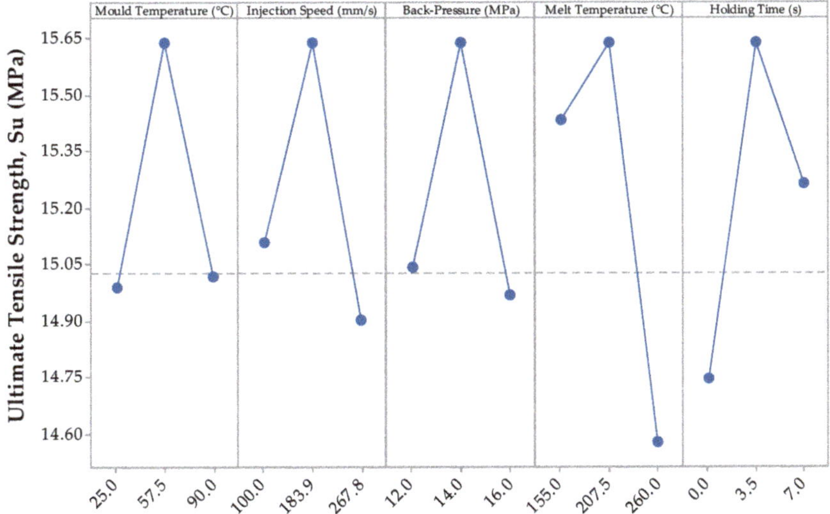

Figure 4. Main effects plot for the mean maximum tensile strength (S_u) results from the DoE.

Table 3. Analysis of variance linear model for ultimate tensile strength against processing parameters.

Source	df	Sum of Squares	Mean Squares	F-Value	*p*-Value
Model	7	9.5998	1.3714	31.07	0.000
Linear	5	8.4735	1.69469	38.39	0.000
Mould Temperature	1	0.0107	0.01068	0.24	0.627
Injection Speed	1	0.3843	0.38433	8.71	0.007
Back-Pressure	1	0.0369	0.03691	0.84	0.369
Melt Temperature	1	5.9321	5.9321	134.38	0.000
Holding Time	1	2.1094	2.10944	47.78	0.000
2-way interactions	1	0.7347	0.73466	16.64	0.000
Melt Temperature × Holding Time	1	0.7347	0.73466	16.64	0.000
Curvature	1	0.3917	0.39165	8.87	0.006
Residual	25	1.1036	0.04415		
Total	32	10.7034			

It has been previously reported that excessive melt temperatures of the single-phase solution causes excessive growth of cells, leading to poor cell microstructure and hence, a lower mechanical strength [40]. This is evident with this research and highlights the importance of cooling rates with FIM for tensile strength.

3.2. Part Surface Roughness

The main effects plot for the mean R_a values from the DoE can be seen in Figure 5.

As the mould temperature increases from 25 °C to 90 °C, the mean R_a also increases from 0.819 μm to 1.550 μm. This result can be explained whereby the higher mould temperatures delay the formation of the frozen front; hence the polymer/gas solution continues to nucleate towards the moulding surface causing a pitted surface which causes an increase in surface roughness. A further explanation is that this is post-blow: whereby the gas diffuses out of the part after the moulding process as the polymer continues to balance its thermodynamic boundaries [41,42]. Furthermore, when injection speed is increased from 100 mm/s to 267.8 mm/s, the mean R_a decreased from 1.297 μm to 1.072 μm. This reduction in surface roughness with an increase in injection speed can be attributed to the cavity being filled earlier hence the frozen layer may have started to freeze thus mitigating the polymer/gas solution from reaching the moulding surface [43,44]. When processing with a higher melt temperature, the R_a decreased from 1.324 µm to 1.045 µm. The processing parameters with the least effect on the mean R_a are the back-pressure and the holding time, whereby a contribution of 0.035 µm and 0.032 µm respectively were witnessed on the resulting moulding components. Tables 4 and 5 show the analysis of variance model results for the R_a and R_q respectively.

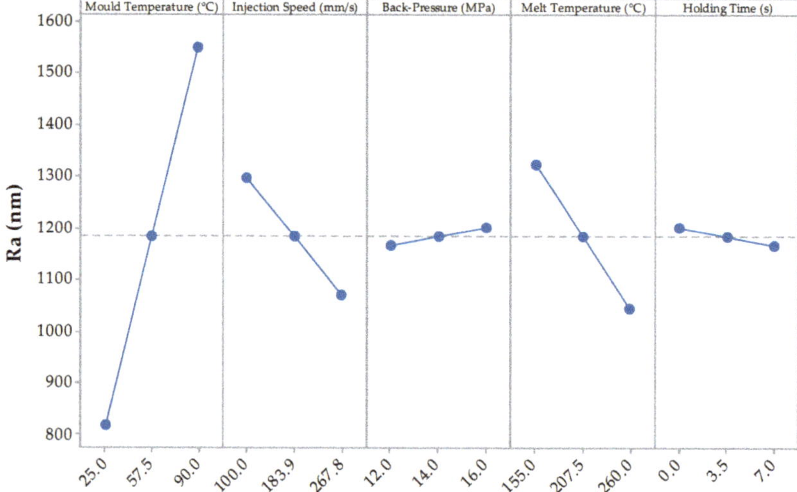

Figure 5. Main effects plot for the R_a results from the DoE.

Table 4. Analysis of variance linear model for R_a against processing parameters.

Source	df	Sum of Squares	Mean Squares	F-Value	*p*-Value
Model	4	7,334,374	1,833,593	2.01	0.12
Linear	3	4,691,865	1,563,955	1.72	0.186
Mould Temperature	1	4,277,081	4,277,081	4.7	0.039
Injection Speed	1	404,925	404,925	0.44	0.51
Back-pressure	1	9858	9858	0.01	0.918
2-Way Interactions	1	2,642,509	2,642,509	2.9	0.1
Injection Speed × Back pressure	1	2,642,509	2,642,509	2.9	0.1
Residual	28	25,507,360	910,977		
Curvature	1	0	0	0	1
Lack-of-Fit	27	25,507,359	944,717		
Total	32	32,841,733			

Table 5. Analysis of variance linear model for R_q against processing parameters.

Source	df	Sum of Squares	Mean Squares	F-Value	*p*-Value
Model	4	12,792,135	3,198,034	2.31	0.082
Linear	3	8,473,030	2,824,343	2.04	0.13
Mould Temperature	1	7,381,443	7,381,443	5.34	0.028
Injection Speed	1	1,046,061	1,046,061	0.76	0.392
Back Pressure	1	45,527	45,527	0.03	0.857
2-Way Interactions	1	4,319,105	4,319,105	3.13	0.088
Injection Speed × Back Pressure	1	4,319,105	4,319,105	3.13	0.088
Residual	28	38,689,194	1,381,757		
Curvature	1	1523	1523	0	0.974
Lack-of-Fit	27	38,687,671	1,432,877		
Total	32	51,481,329			

Again, the *p*-values for each source are similar on both tables; with the mould temperature having the most significance with 0.039 and 0.028, respectively; the only processing parameters that have a major significance (<0.05) on the part surface. The melt temperature and holding time *p*-values demonstrate that they have a negligible effect on the resulting surface roughness. Furthermore, from the results obtained from the linear model, none of the 2- to 5-way interactions have any significance on the R_a and R_q values. However, when considering the significance of a 2-way interaction, injection speed and back-pressure had *p*-values of 0.1 and 0.088 respectively.

The root mean squared height deviation (R_q) of the DoE for PBA is presented in Figure 6, whereby the data corresponds to the results presented in Figure 5 except for the mid-point DoE setting not lying on the trend line. However, all the processing parameters have the same effect, with the mould temperature having the greatest effect on the R_q and the back-pressure and holding time having negligible effect.

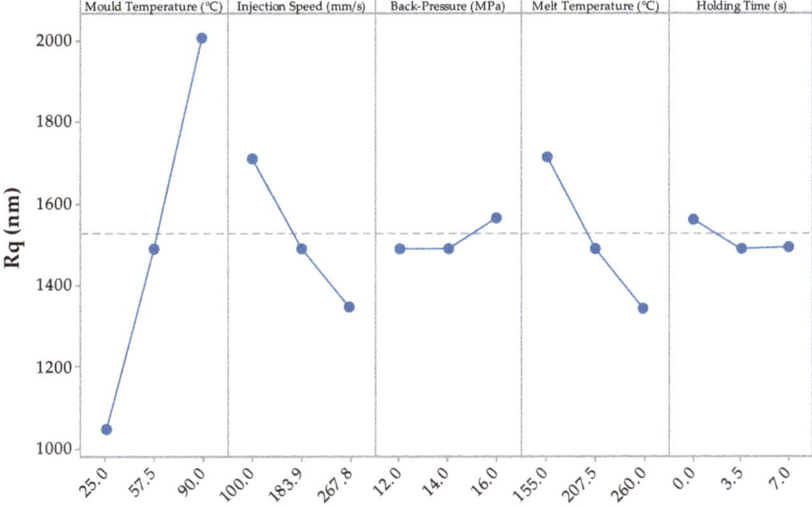

Figure 6. Main effects plot for the R_q results from the DoE.

The R_a and R_q values of the conventionally moulded part were 0.367 µm and 0.439 µm, respectively. As seen in Figures 5 and 6, many of the mean results from the DoE when processing with FIM yield higher surface roughness profiles. The fast injection speeds typically applied in FIM aided in reducing surface roughness as the resulting filling time is reduced and, therefore, allows the cells to nucleate earlier during the filling stage [10]. The mould temperature had a large effect on the surface profiles as this alters the crystallisation of the polymer/gas solution. With a higher mould temperature, this increases the cooling rate (due to further time required in order to meet the crystallisation temperature), and hence cause different cellular microstructures [32].

3.3. Comparison of Physical Blowing Agents to Chemical Blowing Agents

Previous work performed by Llewelyn et al. showed that with unfilled and talc-filled PP, superior mechanical properties were witnessed in parts processed with CBA [19]. This trend continues, as shown Figure 7.

For the CBAs, the processing parameters that resulted in the lowest R_a and R_q were then used to create FIM parts through a CBA using no holding pressure and holding pressure with 1 wt.%. The main reason why the chemically foamed parts result in higher E and S_u values, are due to the thicker skin wall and can be correlated to the modelling theory derived by Xu and Kishbaugh where a thicker skin results in a stronger tensile and flexural part [45]. The physically foamed parts without/with holding pressure in Figure 7 have a skin thickness (T_w) of 549 µm and 394 µm, respectively. The thinnest T_w exhibited in the chemically foamed parts is 775 µm while the thickest is 1060 µm.

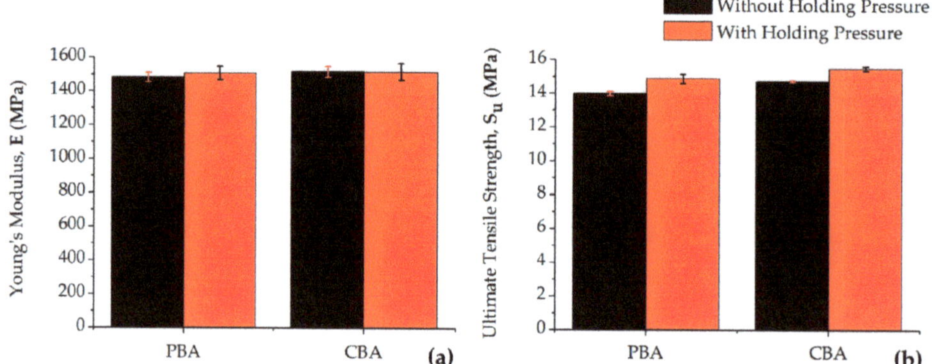

Figure 7. (**a**) Young's modulus (E) and (**b**) ultimate tensile strength (S$_u$) data of a physical blowing agent (PBA) compared to a similarly processed chemical blowing agent (CBA).

The mean R$_a$ and R$_q$ values for the chemically foamed parts are shown in Figure 8. From the results it can be concluded that when processing with a CBA the addition of holding pressure does not lead to any improvement in surface roughness. Also, the results demonstrate that with the addition of CBA when processed with low pressure, surface roughness can be achieved that ae directly comparable to conventionally moulded parts.

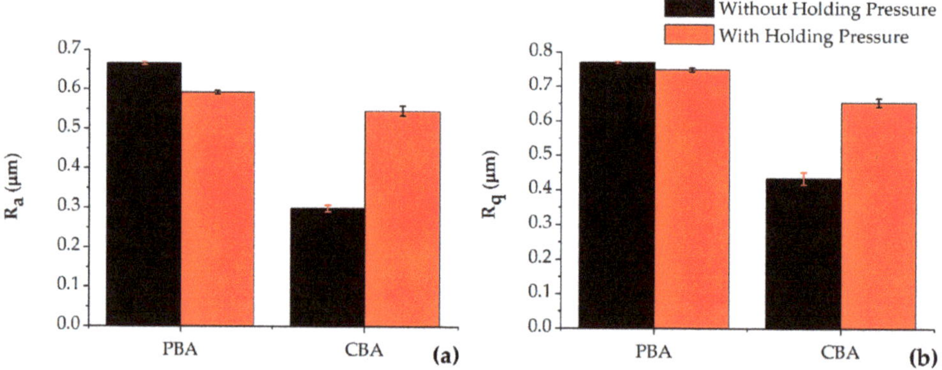

Figure 8. (**a**) R$_a$ and (**b**) R$_q$ data of a physical blowing agent compared to a similarly processed chemical blowing agent.

Figure 9 shows a visual representation of the parts compared in this section of the research. The parts foamed through the PBA are distinct as they have a much greyer appearance than the other parts; a regular problem exhibited through the PBA. Finally, the parts foamed using the CBA can be seen to have a similar appearance to conventionally moulded part. This is confirmed by the surface roughness measurements as the conventionally moulded parts have an average R$_a$ and R$_q$ of 0.367 µm and 0.439 µm, while the CBA parts (low-pressure foaming), have 0.299 µm and 0.434 µm, respectively.

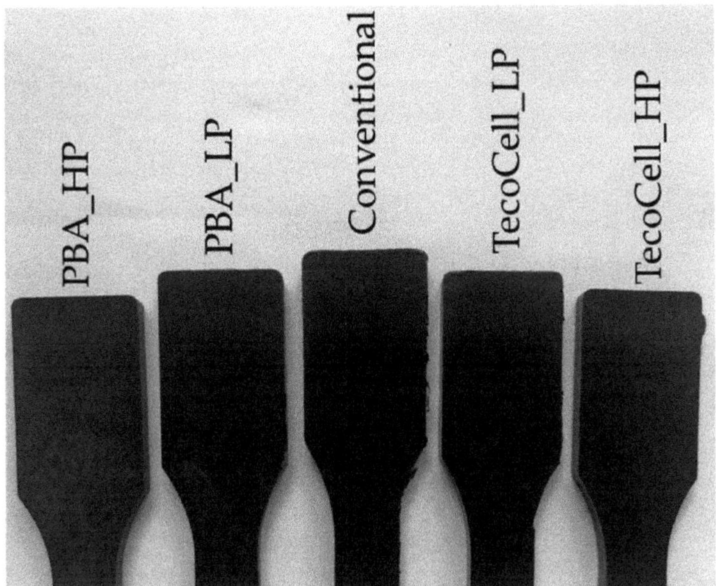

Figure 9. Visual comparison between the conventional, PBA and CBA parts.

4. Conclusions

This work took a talc-filled PP and used low gas dosing in a microcellular injection moulding setup. The initial research performed was a design of experiments with 5 varying input processing parameters: mould temperature, injection speed, back-pressure, melt temperature and holding time. The physically foamed parts with the best surface finishes from this research were compared to similarly manufactured parts through a chemical foaming agent. The major findings from this research are:

- The Young's modulus is heavily affected by mould temperature, whilst melt temperature has the lowest statistical effect on the process.
- The resulting surface roughness is affected by all 5 of the input processing parameters. The mould temperature was found to have the greatest influence. This can be attributed to the high mould temperatures causing post-blow (gas diffusing out of the part after moulding) [41].
- The parts produced through chemical blowing agents were superior to the physically foamed parts with regards to both mechanical properties and the surface roughness. However, using holding pressure with the chemical foaming resulted in poorer mechanical and surface properties.
- Finally, the parts produced through the chemical foaming agent had resulting surface roughnesses that are comparable to conventionally processed components. This demonstrates the great potential for this technology to be applied in engineering applications where lightweight components are required, as surface roughness will not be compromised.

Author Contributions: Conceptualisation, G.L., M.J. and C.G.; Formal analysis, G.L.; Investigation, G.L.; Supervision, A.R.; Writing—original draft, G.L.; Writing—review and editing, A.R., C.G. and M.J. All authors have read and agreed to the published version of the manuscript.

Funding: This research was part funded by the EU's European Funding Office and the Engineering and Physical Sciences Research Council (EPRSC) and the Welsh European Funding Office (WEFO).

Acknowledgments: The authors would like to acknowledge the financial support of the Materials and Manufacturing Academy (M2A) and the access of AIM facilities at the College of Engineering.

Conflicts of Interest: The authors declare no conflict of interest.

Appendix A

Table A1 is the DoE used in this research. It has 5 input processing parameters; with a low and high settings for each along with randomised run order to ensure repeatability (refer to Table 1). Table A2 shows the regression models for each of the 4 output parameters, using the 5 main input processing parameters; derived from the DoE data used in this research.

Table A1. 2^5 DoE for the polymer used in this research.

Experimental Number	Run Order	Mould Temperatures (°C)	Injection Speeds (mm/s)	Back-Pressure (MPa)	Melt Temperature (°C)	Holding Time (s)
30	1	90.0	100.0	16.0	260.0	7.0
32	2	90.0	267.8	16.0	260.0	7.0
18	3	90.0	100.0	12.0	155.0	7.0
10	4	90.0	100.0	12.0	260.0	0.0
22	5	90.0	100.0	16.0	155.0	7.0
2	6	90.0	100.0	12.0	155.0	0.0
13	7	25.0	100.0	16.0	260.0	0.0
14	8	90.0	100.0	16.0	260.0	0.0
24	9	90.0	267.8	16.0	155.0	7.0
28	10	90.0	267.8	12.0	260.0	7.0
11	11	25.0	267.8	12.0	260.0	0.0
4	12	90.0	267.8	12.0	155.0	0.0
12	13	90.0	267.8	12.0	260.0	0.0
8	14	90.0	267.8	16.0	155.0	0.0
27	15	25.0	267.8	12.0	260.0	7.0
15	16	25.0	267.8	16.0	260.0	0.0
20	17	90.0	267.8	12.0	155.0	7.0
29	18	25.0	100.0	16.0	260.0	7.0
26	19	90.0	100.0	12.0	260.0	7.0
33	20	57.5	183.9	14.0	207.5	3.5
31	21	25.0	267.8	16.0	260.0	7.0
17	22	25.0	100.0	12.0	155.0	7.0
1	23	25.0	100.0	12.0	155.0	0.0
25	24	25.0	100.0	12.0	260.0	7.0
23	25	25.0	267.8	16.0	155.0	7.0
3	26	25.0	267.8	12.0	155.0	0.0
6	27	90.0	100.0	16.0	155.0	0.0
16	28	90.0	267.8	16.0	260.0	0.0
19	29	25.0	267.8	12.0	155.0	7.0
9	30	25.0	100.0	12.0	260.0	0.0
5	31	25.0	100.0	16.0	155.0	0.0
21	32	25.0	100.0	16.0	155.0	7.0
7	33	25.0	267.8	16.0	155.0	0.0

Table A2. Regression models derived from the DoE factorial design analysis.

Response	Regression model (A: Mould Temperature, B: Injection Speed, C: Back-Pressure, D: Melt Temperature and E: Holding Time
Young's Modulus (E)	$E = 1434 + 1.837A - 1.383B + 29.9C + 0.661D - 31.87E + 0.00450B \times D + 0.0644B \times E - 0.1464C \times D + 0.1234D \times E$
Ultimate Tensile Strength (S_u)	$S_u = 17.495 + 0.00056A - 0.001306B - 0.0170C - 0.01109D - 0.0977E + 0.000825D \times E$
Arithmetic Mean Deviation (R_a)	$R_a = -3747 + 11.25A + 22.6B + 324C - 1.71B \times C$
Root Mean Deviation (R_q)	$R_q = -4826 + 14.78A + 28.5B + 421C - 2.19B \times C$

References

1. Yusa, A.; Yamamoto, S.; Goto, H.; Uezono, H.; Asaoka, F.; Wang, L.; Ando, M.; Ishihara, S.; Ohshima, M. A new microcellular foam injection-molding technology using non-supercritical fluid physical blowing agents. *Polym. Eng. Sci.* **2016**, *57*, 105–113. [CrossRef]

2. Gómez-Monterde, J.; Hain, J.; Sánchez-Soto, M.; Maspoch, M.L. Microcellular injection moulding: A comparison between MuCell process and the novel micro-foaming technology IQ Foam. *J. Mater. Process. Technol.* **2019**, *268*, 162–170. [CrossRef]

3. Standau, T.; Zhao, C.; Castellón, S.M.; Bonten, C.; Altstädt, V. Chemical Modification and Foam Processing of Polylactide (PLA). *Polymers* **2019**, *11*, 306. [CrossRef]

4. Wu, H.; Zhao, G.; Wang, G.; Zhang, W.; Li, Y. A new core-back foam injection molding method with chemical blowing agents. *Mater. Des.* **2018**, *144*, 331–342. [CrossRef]

5. Realinho, V.; Arencón, D.; Antunes, M.; Velasco, J.I. Effects of a Phosphorus Flame Retardant System on the Mechanical and Fire Behavior of Microcellular ABS. *Polymers* **2018**, *11*, 30. [CrossRef] [PubMed]

6. Trexel. TecoCell®—CFA. Available online: http://www.trexel.com/en/tecocell-cfa (accessed on 9 June 2019).

7. Zhao, P.; Zhao, Y.; Kharbas, H.; Zhang, J.; Wu, T.; Yang, W.; Fu, J.; Turng, L.-S. In-situ ultrasonic characterization of microcellular injection molding. *J. Mater. Process. Technol.* **2019**, *270*, 254–264. [CrossRef]

8. Wang, G.; Zhao, G.-Q.; Wang, J.-C.; Zhang, L. Research on formation mechanisms and control of external and inner bubble morphology in microcellular injection molding. *Polym. Eng. Sci.* **2014**, *55*, 807–835. [CrossRef]

9. Llewelyn, G.; Rees, A.; Griffiths, C.A.; Scholz, S.G. Advances in microcellular injection moulding. *J. Cell. Plast.* **2020**, *56*. [CrossRef]

10. Xu, J. *Microcellular Injection Molding*, 1st ed.; Xu, J., Ed.; John Wiley & Sons, Inc.: Hoboken, NJ, USA, 2010; Volume 1.

11. Shaayegan, V.; Wang, G.; Park, C.B. Study of the bubble nucleation and growth mechanisms in high-pressure foam injection molding through in-situ visualization. *Eur. Polym. J.* **2016**, *76*, 2–13. [CrossRef]

12. Wang, C.; Shaayegan, V.; Ataei, M.; Costa, F.; Han, S.; Bussmann, M.; Park, C.B. Accurate theoretical modeling of cell growth by comparing with visualized data in high-pressure foam injection molding. *Eur. Polym. J.* **2019**, *119*, 189–199. [CrossRef]

13. Muñoz-Pascual, S.; Lopez-Gonzalez, E.; Saiz-Arroyo, C.; Rodriguez-Perez, M.A. Effect of Mold Temperature on the Impact Behavior and Morphology of Injection Molded Foams Based on Polypropylene Polyethylene-Octene Copolymer Blends. *Polymers* **2019**, *11*, 894. [CrossRef] [PubMed]

14. Świetlicki, M.; Chocyk, D.; Klepka, T.; Prószyński, A.; Kwaśniewska, A.; Borc, J.; Gładyszewski, G. The Structure and Mechanical Properties of the Surface Layer of Polypropylene Polymers with Talc Additions. *Materials* **2020**, *13*, 698. [CrossRef] [PubMed]

15. Wang, G.; Zhao, G.; Dong, G.; Mu, Y.; Park, C.B. Lightweight and strong microcellular injection molded PP/talc nanocomposite. *Compos. Sci. Technol.* **2018**, *168*, 38–46. [CrossRef]

16. Yan, K.; Guo, W.; Mao, H.; Yang, Q.; Meng, A.Z. Investigation on Foamed PP/Nano-CaCO$_3$ Composites in a Combined in-Mold Decoration and Microcellular Injection Molding Process. *Polymers* **2020**, *12*, 363. [CrossRef]

17. Fu, L.; Shi, Q.; Ji, Y.; Wang, G.; Zhang, X.; Chen, J.; Shen, C.; Park, C.B. Improved cell nucleating effect of partially melted crystal structure to enhance the microcellular foaming and impact properties of isotactic polypropylene. *J. Supercrit. Fluids* **2020**, *160*, 104794. [CrossRef]

18. Lee, S.H.; Zhang, Y.; Kontopoulou, M.; Park, C.B.; Wong, A.; Zhai, W. Optimization of Dispersion of Nanosilica Particles in a PP Matrix and Their Effect on Foaming. *Int. Polym. Process.* **2011**, *26*, 388–398. [CrossRef]

19. Llewelyn, G.; Rees, A.; Griffiths, C.A.; Jacobi, M. A Novel Hybrid Foaming Method for Low-Pressure Microcellular Foam Production of Unfilled and Talc-Filled Copolymer Polypropylenes. *Polymers* **2019**, *11*, 1896. [CrossRef]

20. Kaltenegger-Uray, A.; Rieß, G.; Lucyshyn, T.; Holzer, C.; Kern, W. Physical Foaming and Crosslinking of Polyethylene with Modified Talcum. *Polymers* **2019**, *11*, 1472. [CrossRef]

21. Shaayegan, V.; Wang, C.; Park, C.B.; Costa, F.; Han, S. Mechanism of Cell Nucleation in High-Pressure Foam Injection Molding Followed by Precise Mold-Opening. In Proceedings of the SPE-ANTEC, Indianapolis, IN, USA, 23–25 May 2016; pp. 1151–1155.

22. Tromm, M.; Shaayegan, V.; Wang, C.; Heim, H.-P.; Park, C.B. Investigation of the mold-filling phenomenon in high-pressure foam injection molding and its effects on the cellular structure in expanded foams. *Polymers* **2019**, *160*, 43–52. [CrossRef]

23. Wang, G.; Zhao, G.; Zhang, L.; Mu, Y.; Park, C.B. Lightweight and tough nanocellular PP/PTFE nanocomposite foams with defect-free surfaces obtained using in situ nanofibrillation and nanocellular injection molding. *Chem. Eng. J.* **2018**, *350*, 1–11. [CrossRef]

24. Bernardo, V.; León, J.M.-D.; Pinto, J.; Verdejo, R.; Rodriguez-Perez, M.A. Modeling the heat transfer by conduction of nanocellular polymers with bimodal cellular structures. *Polymers* **2019**, *160*, 126–137. [CrossRef]

25. Tiwary, P.; Park, C.B.; Kontopoulou, M. Transition from microcellular to nanocellular PLA foams by controlling viscosity, branching and crystallization. *Eur. Polym. J.* **2017**, *91*, 283–296. [CrossRef]

26. Costeux, S. CO_2-blown nanocellular foams. *J. Appl. Polym. Sci.* **2014**, *131*, 131. [CrossRef]

27. Ameli, A.; Nofar, M.; Park, C.; Pötschke, P.; Rizvi, G. Polypropylene/carbon nanotube nano/microcellular structures with high dielectric permittivity, low dielectric loss, and low percolation threshold. *Carbon* **2014**, *71*, 206–217. [CrossRef]

28. Suhartono, E.; Chen, S.-C.; Lee, K.-H.; Wang, K.-J. Improvements on the tensile properties of microcellular injection molded parts using microcellular co-injection molding with the material combinations of PP and PP-GF. *Int. J. Plast. Technol.* **2017**, *21*, 351–369. [CrossRef]

29. Li, S.; Zhao, G.; Wang, G.; Guan, Y.; Wang, X. Influence of relative low gas counter pressure on melt foaming behavior and surface quality of molded parts in microcellular injection molding process. *J. Cell. Plast.* **2014**, *50*, 415–435. [CrossRef]

30. Llewelyn, G.; Rees, A.; Griffiths, C.A.; Scholz, S.G. Advances in Near Net Shape Polymer Manufacturing Through Microcellular Injection Moulding. In *Materials Forming, Machining and Tribology*; Springer Science and Business Media LLC: Cham, Switzerland, 2019; pp. 177–189.

31. Hopmann, C.; Lammert, N.; Zhang, Y. Improvement of foamed part surface quality with variothermal temperature control and analysis of the mechanical properties. *J. Cell. Plast.* **2019**, *55*, 507–522. [CrossRef]

32. Lee, J.; Turng, L.-S.; Dougherty, E.; Gorton, P. A novel method for improving the surface quality of microcellular injection molded parts. *Polymers* **2011**, *52*, 1436–1446. [CrossRef]

33. Guo, W.; Yang, Q.; Mao, H.; Meng, A.Z.; Hua, L.; He, B. A Combined In-Mold Decoration and Microcellular Injection Molding Method for Preparing Foamed Products with Improved Surface Appearance. *Polymers* **2019**, *11*, 778. [CrossRef]

34. Bellantone, V.; Surace, R.; Trotta, G.; Fassi, I. Replication capability of micro injection moulding process for polymeric parts manufacturing. *Int. J. Adv. Manuf. Technol.* **2012**, *67*, 1407–1421. [CrossRef]

35. Ryu, Y.; Sohn, J.S.; Kweon, B.C.; Cha, S.W. Shrinkage Optimization in Talc- and Glass-Fiber-Reinforced Polypropylene Composites. *Materials* **2019**, *12*, 764. [CrossRef] [PubMed]

36. Gómez-Monterde, J.; Sánchez-Soto, M.; Maspoch, M.L. Influence of injection molding parameters on the morphology, mechanical and surface properties of ABS foams. *Adv. Polym. Technol.* **2018**, *37*, 2707–2720. [CrossRef]

37. Montgomery, D.C. *Design and Analysis of Experiments*, 8th ed.; John Wiley & Sons, Inc.: New Jersey, NJ, USA, 2017; p. 724.

38. Orooji, Y.; Ghasali, E.; Emami, N.; Noorisafa, F.; Razmjou, A. ANOVA Design for the Optimization of TiO_2 Coating on Polyether Sulfone Membranes. *Molecules* **2019**, *24*, 2924. [CrossRef]

39. Mi, D.; Xia, C.; Jin, M.; Wang, F.; Shen, K.; Zhang, J. Quantification of the Effect of Shish-Kebab Structure on the Mechanical Properties of Polypropylene Samples by Controlling Shear Layer Thickness. *Macromolecules* **2016**, *49*, 4571–4578. [CrossRef]

40. Xi, Z.; Sha, X.; Liu, T.; Zhao, L. Microcellular injection molding of polypropylene and glass fiber composites with supercritical nitrogen. *J. Cell. Plast.* **2014**, *50*, 489–505. [CrossRef]

41. Turng, L.-S.; Kharbas, H. Development of a Hybrid Solid-Microcellular Co-injection Molding Process. *Int. Polym. Process.* **2004**, *19*, 77–86. [CrossRef]

42. Pomerleau, J.; Sanschagrin, B. Injection molding shrinkage of PP: Experimental progress. *Polym. Eng. Sci.* **2006**, *46*, 1275–1283. [CrossRef]

43. Dong, G.; Zhao, G.; Guan, Y.; Li, S.; Wang, X. Formation mechanism and structural characteristics of unfoamed skin layer in microcellular injection-molded parts. *J. Cell. Plast.* **2015**, *52*, 419–439. [CrossRef]

44. Lohr, C.; Beck, B.; Henning, F.; Weidenmann, K.A.; Elsner, P. Process comparison on the microstructure and mechanical properties of fiber-reinforced polyphenylene sulfide using MuCell technology. *J. Reinf. Plast. Compos.* **2018**, *37*, 1020–1034. [CrossRef]

45. Xu, J.; Kishbaugh, L. Simple Modeling of the Mechanical Properties with Part Weight Reduction for Microcellular Foam Plastic. *J. Cell. Plast.* **2003**, *39*, 29–47. [CrossRef]

Article

Static Mechanical Properties of Expanded Polypropylene Crushable Foam

Przemysław Rumianek [1,*], Tomasz Dobosz [2], Radosław Nowak [1], Piotr Dziewit [3] and Andrzej Aromiński [1]

1   Faculty of Automotive and Construction Machinery Engineering, Warsaw University of Technology, 02-524 Warsaw, Poland; radoslaw.nowak@pw.edu.pl (R.N.); andrzej.arominski@pw.edu.pl (A.A.)
2   Faculty of Mechanical Engineering, Department of Machine and Vehicle Design and Research, Wroclaw University of Science and Technology, 50-370 Wrocław, Poland; tomasz.dobosz@pwr.edu.pl
3   Faculty of Mechatronics and Aerospace, Military University of Technology, 00-908 Warsaw, Poland; piotr.dziewit@wat.edu.pl
*   Correspondence: przemyslaw.rumianek@pw.edu.pl; Tel.: +48-22-234-87-72

Abstract: Closed-cell expanded polypropylene (EPP) foam is commonly used in car bumpers for the purpose of absorbing energy impacts. Characterization of the foam's mechanical properties at varying strain rates is essential for selecting the proper material used as a protective structure in dynamic loading application. The aim of the study was to investigate the influence of loading strain rate, material density, and microstructure on compressive strength and energy absorption capacity for closed-cell polymeric foams. We performed quasi-static compressive strength tests with strain rates in the range of 0.2 to 25 mm/s, using a hydraulically controlled material testing system (MTS) for different foam densities in the range 20 g/dm^3 to 220 g/dm^3. The above tests were carried out as numerical simulation using ABAQUS software. The verification of the properties was carried out on the basis of experimental tests and simulations performed using the finite element method. The method of modelling the structure of the tested sample has an impact on the stress values. Experimental tests were performed for various loads and at various initial temperatures of the tested sample. We found that increasing both the strain rate of loading and foam density raised the compressive strength and energy absorption capacity. Increasing the ambient and tested sample temperature caused a decrease in compressive strength and energy absorption capacity. For the same foam density, differences in foam microstructures were causing differences in strength and energy absorption capacity when testing at the same loading strain rate. To sum up, tuning the microstructure of foams could be used to acquire desired global materials properties. Precise material description extends the possibility of using EPP foams in various applications.

Keywords: compressive deformation; EPP foam; foam; microstructure; strain rate

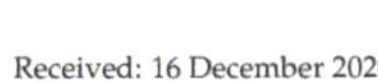

Citation: Rumianek, P.; Dobosz, T.; Nowak, R.; Dziewit, P.; Aromiński, A. Static Mechanical Properties of Expanded Polypropylene Crushable Foam. *Materials* **2021**, *14*, 249. https://doi.org/10.3390/ma14020249

Received: 16 December 2020
Accepted: 2 January 2021
Published: 6 January 2021

Publisher's Note: MDPI stays neutral with regard to jurisdictional claims in published maps and institutional affiliations.

1. Introduction

For designing protective structures or energy absorbing elements, the closed-cell material, such as expanded polypropylene (EPP), is the most reasonable choice. Energy-absorbing structures have the ability to take over the kinetic energy of impacts. This energy is equivalent to the work of force destroying the material (crushing, breaking). Closed-cell EPP foam is commonly used in automobile bumpers for absorbing impact energy. In this application, foam energy absorbing characteristics are related to the loading received by the bumper reinforcement beam and body frame. Varying foam density and thickness changes the energy-absorbing capabilities, enabling further optimization of bumper systems. Low production costs, low weight and high energy-absorbing capabilities enable foam sandwich structures to be used in aerospace and automotive industries. Foam materials are often manufactured as steam chest moulding, using a sintering-like process (heat and pressure). The final performance of manufactured parts may be affected by many variables such as: manufacturing method, gas used for closed cell foams, cell geometry, constituent material.

Kinetic energy-absorbing capability on the right level is very important, because values of the peak force and acceleration over the threshold could cause injury or damage. Energy absorbed during single or multiple impacts can be described in several ways [1–3]. For example, as a relation between peak impact deceleration of the real foam and "ideal" foam, and the ability to completely absorb impact energy. A lot of empirical data is needed in order to assess this factor, as it is dependent for example on material thickness and impact energy. Alternatively, this factor can be evaluated by obtaining curves representing the absorbed energy as a function of peak stress. Such curves can be derived from stress–strain dependence. Research and analyses conducted by Rusch [4] lead to the construction of peak stress vs. specific energy absorption curves. This method still relies on empirical evaluation of foam characteristics at different levels of impact energy, but it provides better generality than the J-factor. Burgess et al. [5] and Castiglioni et al. [6] attempted to model foam cushioning behaviour. The so-called Maiti diagrams [7] are a more refined approach. In those diagrams, both normalized to foam modulus, energy per volume unit is plotted against stress level for different foam densities. Selecting proper foam material and density for particular application can be done by finding a set of points corresponding to densification onset for given foam density. An extensive data set acquired at different strain rates is needed to cover application requirements for particular material use. The material structure and the strain rate of polymer foams was considered by Luca Andena et al. They have used Nagy's phenomenological model and determined the material stress–strain behaviour at a reference strain rate [3]. To predict foam behaviour under specific conditions and optimizing design of energy absorption devices, various models were developed, trying to capture actual material mechanics. The Gibson and Ashby [8] model is commonly used to gain basic understanding of foam behaviour. The Gibson and Ashby model assumes regular cell structure. Foams with irregular structures can be described by this model, but with limited predicting capabilities. Different models, such as Kelvin [9], Shulmeister [10] or Roberts-Garboczi [11,12] try to overcome limitations of the Gibson and Ashby model. All of the above models are unable to realistically represent real foam structure and all of them have limited accuracy. Mechanical properties and energy absorption capabilities of the foams change with temperature levels. The influence of temperature on foam characteristics was investigated in several studies [13–15]. According to Zhang et al. [16] EPP foam energy absorption capability of the component was found to be highly dependent on temperature [17].

The aim of this study was to investigate the impact of the structure of the material on energy dissipation in various conditions. The verification of the material properties was carried out on the basis of the results of experimental tests and simulations performed using the finite element method (FEM). The method of describing foam properties is used for analysis of protective elements in vehicles. Analyses take into account the foam materials in a manufacturing process, where we define: basic material and technologies (gas pressure used during foaming). Experimental tests were performed for various loads and at various initial temperatures of the tested sample. The presented method of material analysis allows for the selection of an appropriate hyper-elastic model and its modification, resulting in even better determination of material properties. During experimental tests, compressive stress test was performed with 10%, 60% and 80% relative deformation. Deformation values ware based on the methods specified in the standards [18,19].

Experimental, analytical and FEM simulation results were compared and presented in a graphical manner. FEM material model analysis takes into account the experimental results and modifications of the existing material models.

2. Materials and Methods

2.1. Experimental Methods

2.1.1. Materials and Specimens

The material under investigation is a polypropylene polymer foam. EPP foam components are made from small beads, moulded into shape. At first polypropylene beads are expanded, forming larger, foamed beads with a diameter of 0.25–5 mm. The next stage is further bead expansion in a mould. At this stage the temperature is higher than in the first stage, causing melting EPP and sintering of the foamed beads into the final component. Process parameters of polypropylene expansion determine the degree of expansion, foam density and cellular structure. The complex cellular structure of each EPP bead and its interaction with adjoining beads allows any energy exerted in the executed part to be managed [17]. A typical cellular structure of an individual EPP foam bead is shown in Figure 1a–d. The research structure was determined using Phenom G2 scanning electron microscope (SEM) (Phenom-World BV, Eindhoven, The Netherlands).

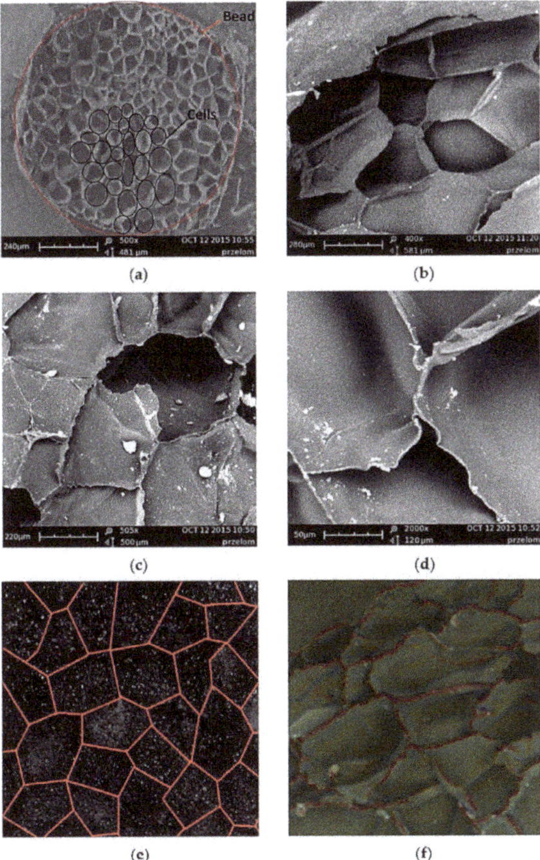

Figure 1. Scanning electron microscope (SEM) microphotographs cellular structure of expanded polypropylene (EPP) of density $\rho = 20\,\mathrm{g/dm^3}$ (**a–d**) microphotograph structure bead (**e**) diagram and photograph structure bead (**f**) morphology structure closed cell.

The cross section images were taken by notching one side of a rectangular sample, cooling sample in liquid nitrogen and then breaking by bending while frozen. Breaking frozen samples was done to reduce potential plastic deformation in the breaking region of the sample. Then the samples were coated with gold using a sputter deposition method and imaged by a SEM device. Cellular structure can be seen in the inside of the single EPP bead. After extracting samples from the component, foam density was estimated by weighing specimens and measuring specimen dimensions using a calliper. The average density of samples taken from a component was 20 g/dm³ with a standard deviation of 0.5 g/dm³.

Detailed structure tests were performed on rigid polypropylene foam of different densities 20 g/dm^3, 30 g/dm^3, 80 g/dm^3, 120 g/dm^3 and 200 g/dm^3. 3D reconstruction software (Phenom ProSuite v2.3, Phenom-World BV, Eindhoven, The Netherlands) was used to measure structure of the selected areas. The software is based on the ProSuite system.

Due to the wide structural diversity of material (Figure 1e,f), which affect the mechanical properties, studies were carried out on averaged sizes of the closed cell of the materials, which is described by Equation (1). The mechanical tests for randomly selected cells show that for different foam densities, we obtain different values of stresses and strains, which can be seen in quasi-static compressive strength experiment tests.

The closed cell size test method is already used for the analysis of metallic foams [20–23]. The individual closed cell size is presented as D_m [20,24]. More than 100 cells from each sample were selected for size calculation. The mean cell size D_m is determined by the following equation D_m,

$$D_m = \frac{S_i}{\sum_{i=1}^{n} S_i} \times \sqrt{\frac{S_i}{\pi}} \tag{1}$$

where n is the total cell number, S_i is the area of the i th pore, which was captured by image processing software (Phenom ProSuite v2.3, Phenom-World BV, Eindhoven, The Netherlands) [22]. Figure 2 shows a representative measurement of average cell dimension.

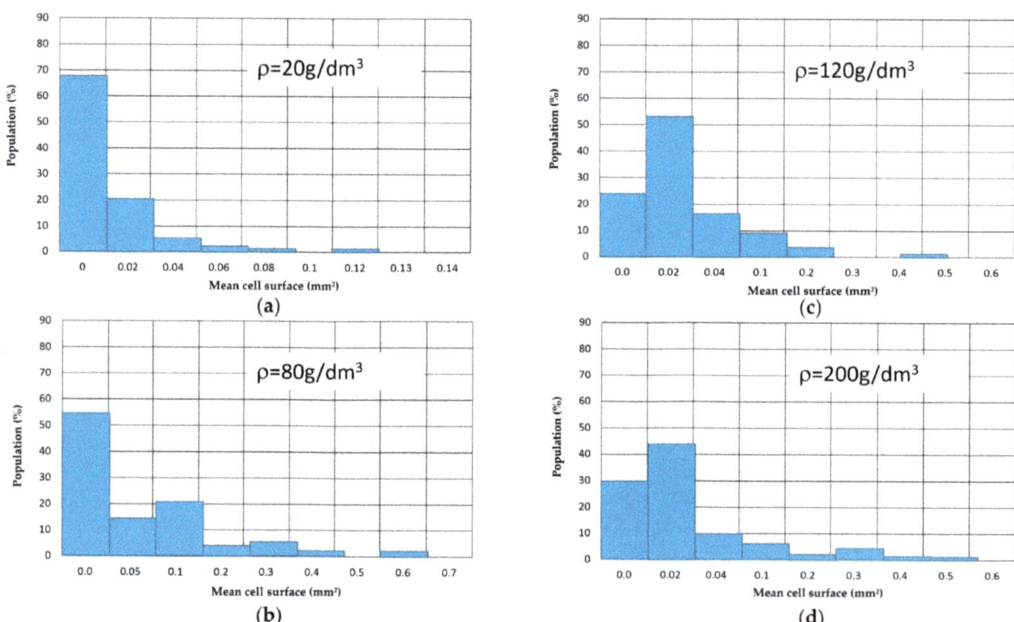

Figure 2. Representative measurement of average cell dimension (**a**) cell surface ρ = 20 g/dm^3 (**b**) ρ = 80 g/dm^3 (**c**) ρ = 120 g/dm^3 (**d**) ρ = 2000 g/dm^3.

Due to the requirements of the test machines used for research, samples were prepared in two sizes: 80 mm × 80 mm, 40 mm height and 20 mm × 20 mm, 30 mm height. Samples were prepared from foams with a density range from 20 g/dm^3 to 220 g/dm^3 and initial dimensions of moulded rectangular blocks (500 mm × 800 mm × 80 mm), which was made by automotive EPP manufacture (Izoblok, Chorzów, Poland). The specimens used for compression tests are shown in Figure 3.

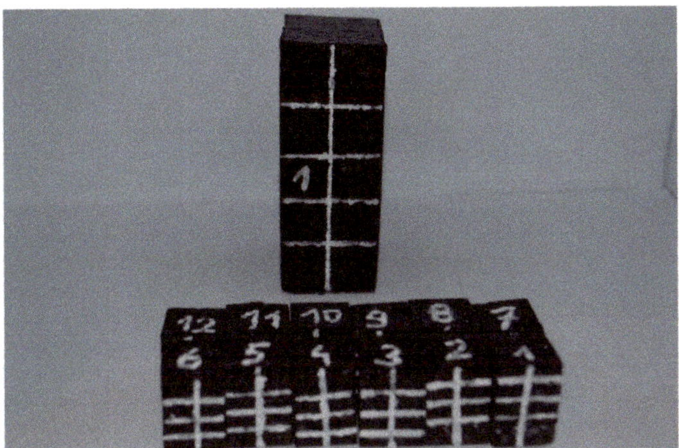

Figure 3. Tested specimens EPP foam.

Tests were performed on samples of different sizes, which allowed the estimate to influence the stress–strain of a tested material [25,26]. Tests using big and small samples were performed because of the needs of the automotive industry, where the energy absorption elements with cuboidal shape have a wide range of cross-section areas. These samples were made in various sizes in order to understand the geometric effect [27]. Experimental compression tests were performed at different strain rates on identical samples to allow direct comparison of quasi-static responses [28]. Exact dimensions were measured using a calliper prior to each test. A typical static stress–strain curve, with each region identified, for EPP foam is presented in Figure 4.

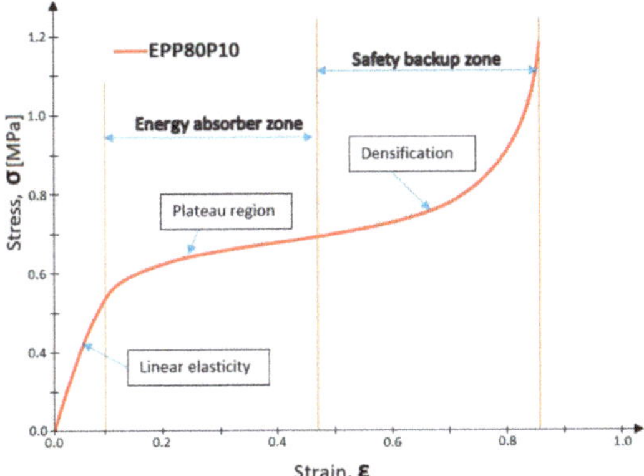

Figure 4. Regions of EPP static stress-strain curve.

A typical stress–strain curve contains linear elasticity and densification regions. The area between those regions is characterized by a slowly rising plateau. From 0% to 5–10% strain the material is in the linear elasticity region, which defines foam Young's modulus at specific density. The plateau region refers to the material's elastoplastic behavior. In this region, cellular structures dissipate the applied load by transferring energy through the cellular structures of both individual beads and between individual beads. Moulded EPP has an anisotropic nature, this enables it to transfer energy in the most efficient manner giving EPP foam superior resilience. Due to this, the stress–strain curve can be maintained after repeated impacts and at quasi-static test speeds (<25 mm/s). When all cells have collapsed, the densification area is reached.

2.1.2. Quasi-Static Test

Compressive strength tests of foam samples were conducted using a testing machine located in the laboratory of the Institute of Machine Design Fundamentals at the Faculty of Automotive and Construction Machinery Engineering at Warsaw University of Technology and Department of Materials Science and Engineering at Warsaw University of Technology. Compressive strength was determined using Q-test 10 material testing system (MTS Systems GmbH, Berlin, Germany) and Zwick/Roell Z005 testing machines (Zwick-Roell GmbH & Co., Ulm, Germany) (Figure 5). Compression force between static and moving pistons was recorded during the tests. This was the first part of the experimental testing of samples. Graphs showing dependence between stress and strain were prepared based on acquired data. Samples were prepared in the form of cuboids using foams of different density, varying from 20 g/dm^3 to 220 g/dm^3. Table 1 shows the static strength of used materials delivered by the material manufacturer.

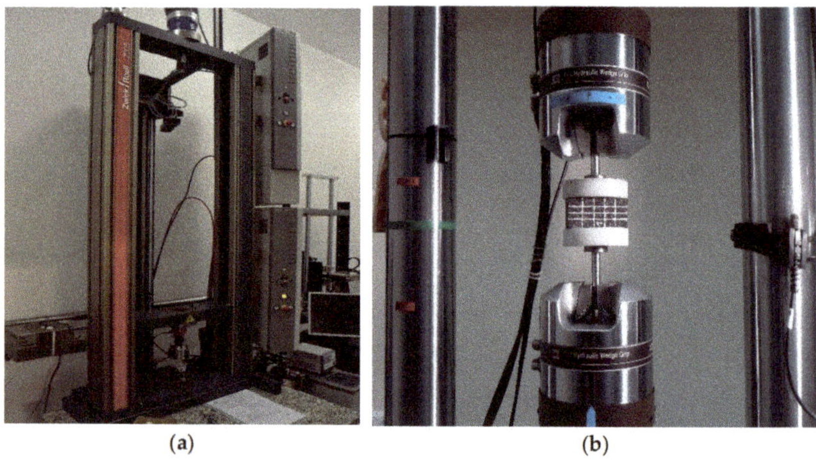

(a)   (b)

Figure 5. Testing machines (a) Zwick/Roell Z005, (b) Q-test 10 material testing system (MTS).

Table 1. Material properties of the specimens.

Specimen	ρ_1	ρ_2	ρ_3	ρ_4	ρ_5
Density (g/dm^3)	20	30	80	120	200
Strength (10^{-2} MPa)	23.52	42.14	95.4	98	110
Failure strain (%)	12	17	16	30	35
Compressive plastic strain (%)	8.5	8.6	9.2	11	16

Research was conducted on samples with temperatures of $-30\ °C$, $23\ °C$, $80\ °C$. The environmental chamber was used to keep the appropriate and constant temperature level in a single test and, of course, this facilitated an easy change to the temperature value, thus it was possible to test the influence of the aforementioned parameters. A pyrometer was used for control of temperature measurement during the tests. The strain rate compression test includes uniaxial compression of tested samples at variable loading rates on cube specimens. Actuator position vs. time for different sample densities is shown on Figure 6. Velocity decreases with time, which corresponds to changes in strain rate from 0.2 to 25 mm/s throughout the test.

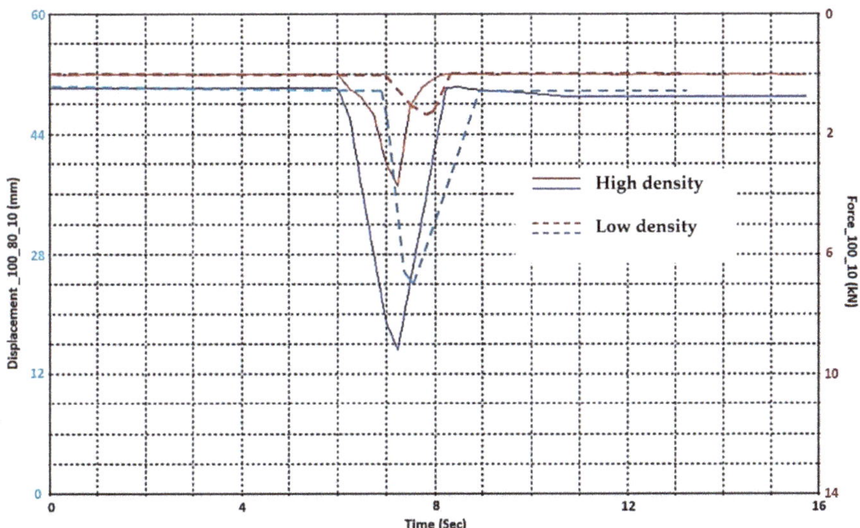

Figure 6. Actuator position vs. time for different sample densities, low (20 g/dm^3) and high (200 g/dm^3) density EPP foams.

The stress–strain curve is perceived to be dependent on foam density. Tests were repeated three times, observed spread in obtained test curves was minimal for all tested foam densities. Figure 7 shows three samples of different densities low-, medium- and high-comparison strain of representative stress at strain rate 25 mm/s.

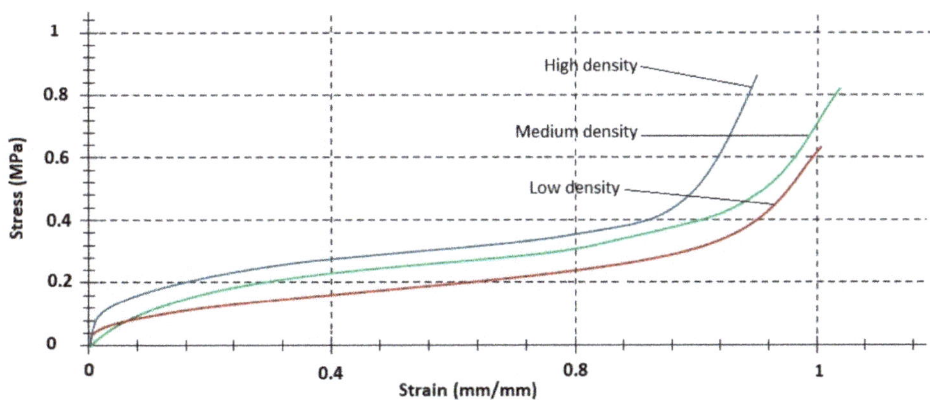

Figure 7. Comparison strain of representative stress–strain curves at strain rate 25 mm/s for low (20 g/dm^3), medium (80 g/dm^3) and high (200 g/dm^3) density EPP foams.

Figure 8 shows three repeated varying strain rate compression tests for low-density EPP foam.

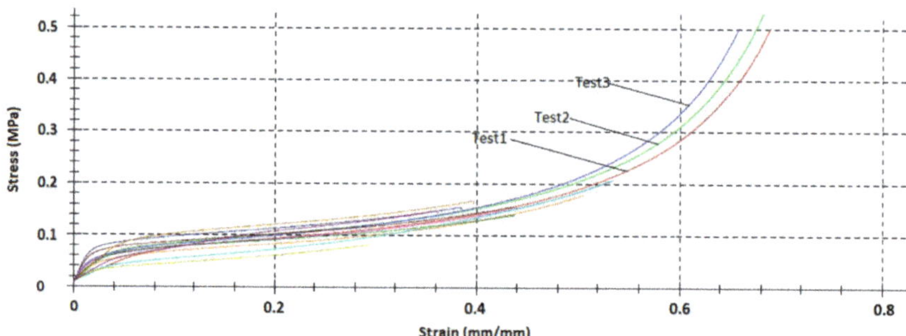

Figure 8. Varying strain rate compression tests three repeats (Tests 1–3) for low-density (20 g/dm^3) EPP foams, sample size 20 mm × 20 mm, 30 mm height.

2.2. Numerical Implementation

The finite element method (FEM) was used for model calculations. Numerical simulations of the EPP foam were performed in Abaqus software version 6.12 (Dassault Systemes, Vélizy-Villacoublay, France). Foam was appropriately modelled for numerical analysis. Simulations were performed using Abaqus/Standard and Abaqus/Explicit module as the incorporated thermo-mechanical task (due to temperature changes and associated changes in stress).

Samples of the same dimensions as in the experimental tests were used for simulation of the compression tests. Stress–strain curves were used to estimate levels of correlation. In the simulations, FEM foam specimens were compressed by modeling a rigid compression steel plate. Plate velocity during the test was specified as the same as the experiment by using the velocity vs. time curve and it is dependent on a required strain rate to be simulated. Contact between the compression plate and the tested specimen was defined as automatic, with general constrain formulation [29]. As shown on Figure 9a, one of the nodes of the specimen was fully constrained to simulate rigid substrate.

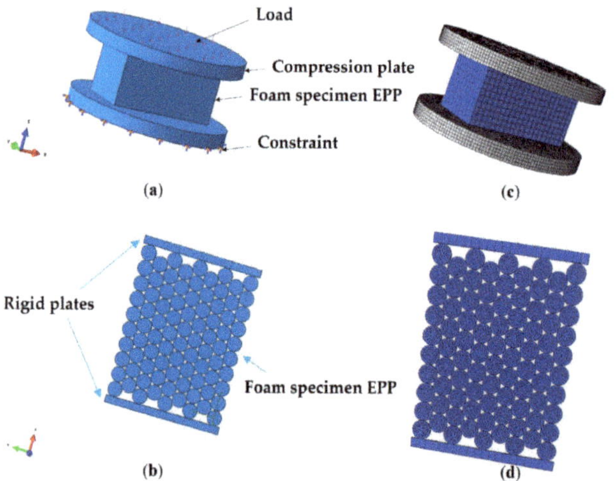

Figure 9. Uniaxial compression EPP specimen (**a**) CAD model (**b**) structure small beads: section view of CAD model; (**c**) finite element model of specimen; (**d**) structure small beads: section view of Finite element model.

Simulation analysis allowed investigation of the phenomena occurring during deformation of the foam structure–tightening pores (separately for open and closed pores) in hyperelastic materials. Additionally, a specific issue that was considered was friction in the fixed joints, traditionally named construction friction, widely regarded as the contact task, taking into account making connection and workloads. The finite element method was used for validation and prediction of the developed material models capabilities during complex load cases.

While describing a material's properties, all of the relevant assumptions for this kind of material was used: theory of hyper-elastic materials [30], model of elastic foam, elastic-plastic and plastic foam and other causes of energy dissipation–internal dampening and material friction. Simulations were carried out using models for crushable foam. The volumetric hardening model assumes that the hydrostatic compression strength evolves as a result of compaction or dilation [29]. Simulations were performed in a simplified system (modeled the plate without fixing bars). Table 2 shows material and element properties of the numerical simulations of the small sample.

Table 2. Material and element properties for the model used in the numerical simulations.

Density of Foam EPP (g/dm^3)	Elastic Modules E (MPa)	Poisson Ratio υ	Tested Elements	Mesh Element Size (m)	Mesh Refinement	Coefficient of Static Friction	Coefficient of Kinetic Friction	Coefficient of Decay
20	3.5	0	C3D8R, CPS4R	0.002	2000	0.6	0.5	0.1
30	4.3	0.2	C3D8R, CPS4R	0.002	2000	0.6	0.5	0.1
80	8.2	0.3	C3D8R, CPS4R	0.002	2000	0.6	0.5	0.1
120	16.0	0.3	C3D8R, CPS4R	0.001	16,000	0.6	0.5	0.1
200	92.1	0.3	C3D8R, CPS4R	0.001	16,000	0.6	0.5	0.1

The structures were also modelled as small beads Figure 9b making it possible to introduce various degrees of structural irregularity.

The mesh was created at the part level, since all parts used in this investigation are dependent. Figure 9c shows a finite element model of specimen and Figure 9d shows a section view of finite element model of the structure of small beads.

In the numerical simulation the temperatures of the analyzed samples were defined [29]. Temperatures were adopted for the tested samples $-30\,°C$, $23\,°C$ and $80\,°C$. Numerical analysis was carried out in the field of static calculations in strain rate from 0.2 to 25 mm/s throughout the test.

3. Results

3.1. Experiment

In Figure 10 the yield curve for quasistatic compression for three types of samples that have different densities but the same dimensions can be seen. During the tests, it was observed that the stress value becomes higher with the increase of density.

It can be observed that a higher unit weight of the granules and a higher packing index in a high-density sample resulted in a high modulus of elasticity. Due to the lower strain rate, the material strengthening phenomenon resulting from gas compression in cavities is negligible. Figure 11 shows the yield curves for quasistatic compression tests of samples of various shapes.

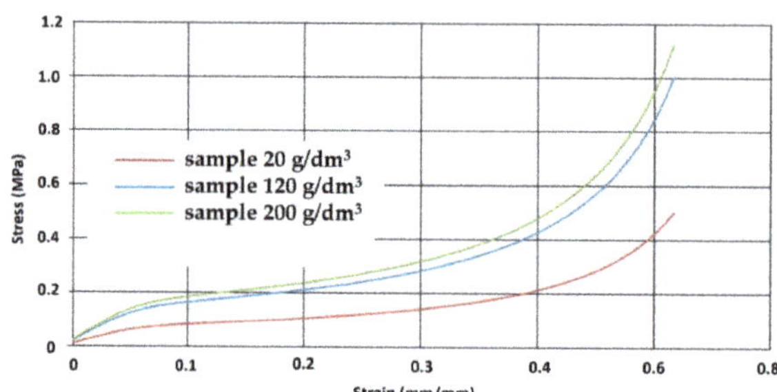

Figure 10. The stress–strain curves of three kinds of EPP foam at a quasi-static loading rate of 2 mm/s.

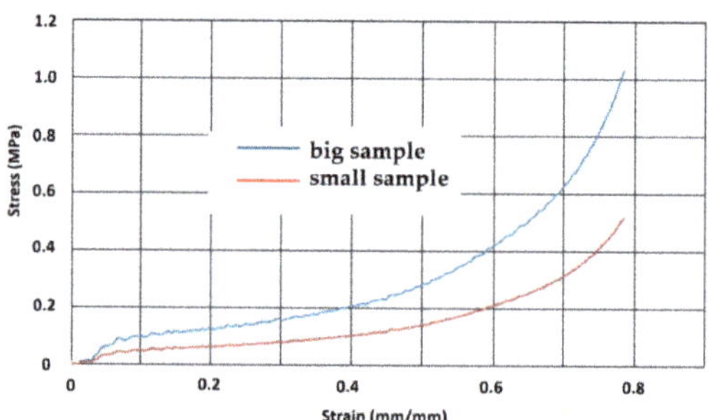

Figure 11. The stress–strain curves of big (80 mm × 80 mm, 40 mm height) and small (20 mm × 20 mm, 30 mm height) sample of EPP foam at a quasistatic loading strain rate of 2 mm/s.

The research showed a significant influence of temperature on the samples tested. This has been confirmed by determining the Young's modulus in EPP materials. With increasing temperature, stresses significantly decrease, the material loses its elastic properties. Figure 12 shows the differences in stresses and strains depending on the temperature of a sample of the same density. The sample at 80 degrees was damaged with a strain value of 0.46.

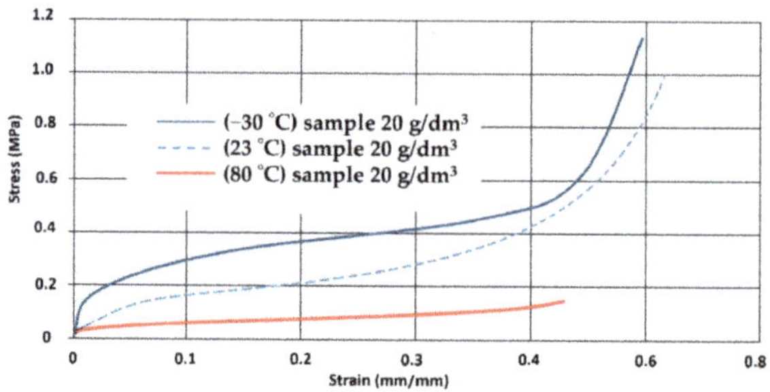

Figure 12. The stress–strain curves of EPP foam (density 20 g/dm^3) at a quasistatic loading, strain rate of 2 mm/s and different temperature, sample size 20 mm × 20 mm, 30 mm height.

Figure 13 shows the differences in stress and strain depending on the quasistatic loading of the different strain rate in the sample of 20 g/dm^3.

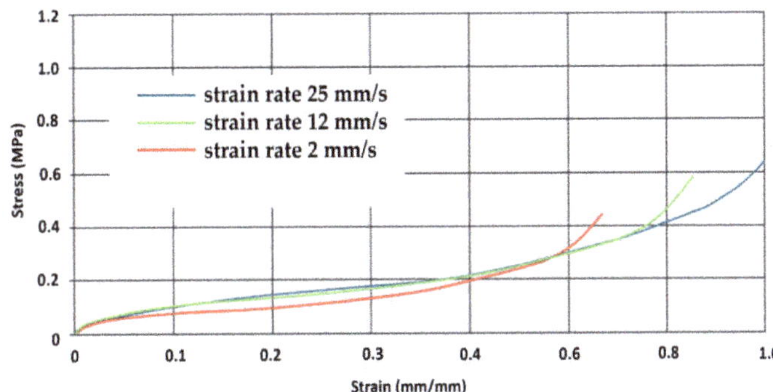

Figure 13. The stress–strain curves of EPP foam (density 20 g/dm^3) at a quasistatic loading, different strain rate.

3.2. Numerical Simulations

Due to the presence of large volumetric deformations considered in this work, there is the need for description of the strength other than the most commonly used Huber–Mises–Hencky model of plasticity condition. The model used is called "crumbling foam" (crushable foam). Conducted research and analysis have shown that elastic deformation occurred up to 60% load, over 60% there were elasto-plastic deformations. The examples of reduced stress results for one of the simulations in the Abaqus program depending on the compression are shown in Figure 14a–c.

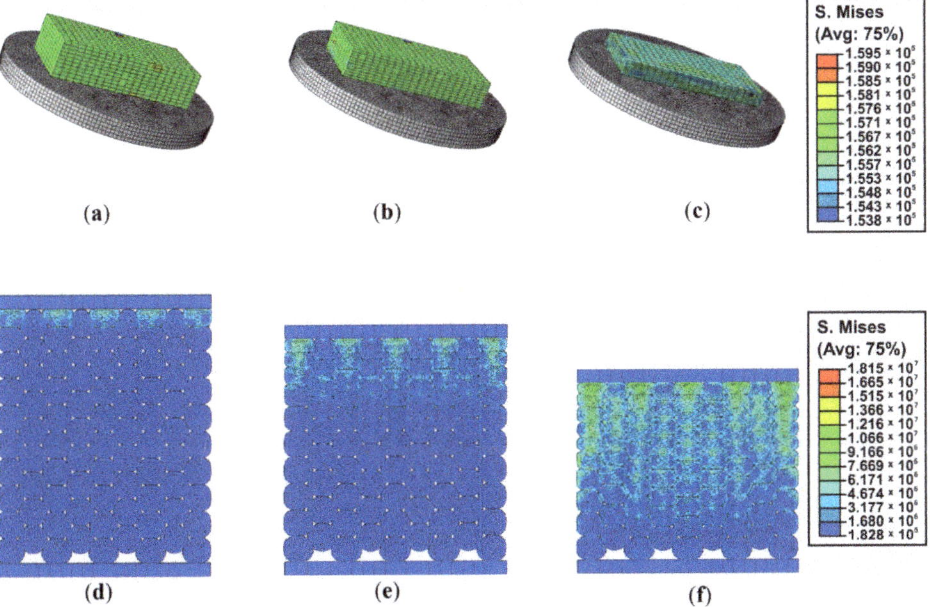

Figure 14. Reduced stresses of relative deformation at (**a**) 10% (**b**) 60% (**c**) 80% (**d**) 10% (**e**) 30% (**f**) 60%.

To investigate a structure of the material simulations of interconnected EPP granules were carried the detailed influence of the cellular material structure. Figure 14d–f shows examples of reduced stress of the sample.

Simulation analysis of different material models was conducted. By comparing them to the experimental studies, we have been able to make a series of comparisons for different types of models. Except for the values of the coefficient α_i (α_i represents the shape of the yield ellipse in the stress plane and can be calculated using the initial yield stress in uniaxial compression [29]) others have led to a non-linear model, which allows for the description of materials and compressibility. The values of the coefficients were determined by approximation based on experimentally defined stress–strain. When using the FEM application, hyperelastic models can be chosen, for which the set of properties have been described. The correct choice of the applied material model should be completed with the comparison of the results of the FE model and experiment. In the case of acceptance of the description of the coefficients $\alpha_i = 2, 4, 6,...$, then we have a polynomial model, including various special cases: the models of Mooney–Rivlin [31,32] and Neo-Hookean [33]. Figure 15 shows a series of compression curves set experimentally by Ogden models for foams with a density of 20 g/dm^3.

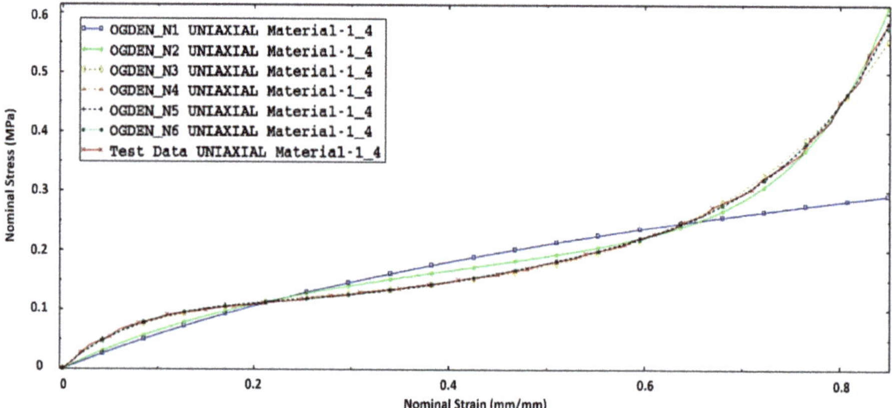

Figure 15. The simulations nominal stress-nominal strain curves of the Ogden model and experimental test data at a quasistatic loading rate of 2 mm/s.

By introducing the equation of α_i coefficients with fractional values, non-linear models could be obtained already in the first approximation. Final results of the process revealed that the consistent stresses waveforms in the real and numerical studies have occurred for the Ogden model. The material has been described by the third row of Ogden's model [29,34]. According to simulation studies, Ogden's model claims that resilience potential can be described as:

$$U = \sum_{i=1}^{N} \frac{2\mu_i}{\alpha_i^2} \left(\bar{\lambda}_1^{\alpha_i} + \bar{\lambda}_2^{\alpha_i} + \bar{\lambda}_1^{\alpha_i} - 3 \right) + \sum_{i=1}^{N} \frac{1}{D_i} (J_{el} - 1)^{2i} \tag{2}$$

where:

$$\bar{\lambda}_i = J^{-\frac{1}{3}} \lambda_i \rightarrow \bar{\lambda}_1 \bar{\lambda}_2 \bar{\lambda}_3 = 1 \tag{3}$$

$\lambda_1^{\alpha_i}$ —invariant of deformation state; J_{el}—sample volume change dependent parameter, μ_i, α_i, D_i—experimentally determined coefficients.

Hence, the first part of Ogden's strain energy function depends only on $\bar{I}_1$ and $\bar{I}_2$. Ogden's energy function cannot be written explicitly in terms of $\bar{I}_1$ and $\bar{I}_2$. It is, however, possible to obtain closed-form expressions for the derivatives of U with respect to $\bar{I}_1$ and $\bar{I}_2$.

The value of N and tables giving the values as functions of temperature are specified by the user.

In the Ogden form the initial shear modulus, μ_o, depends on all coefficients:

$$\mu_o = \sum_{i=1}^{N} \mu_i \tag{4}$$

$$k_o = \sum_{i=1}^{N} 2(\frac{1}{3} + D_i)\mu_i, \tag{5}$$

and the initial bulk modulus, k_o, depends on D_i as before. The user can request that Abaqus calculate the μ_i and α_i values from measurements of nominal stress and strain.

$$\sigma_i = 2(\lambda_i)^{-1} \sum_{i=1}^{N} \frac{\mu_i}{\alpha_i} \left(\lambda_i{}^{\alpha_i} - J^{-\alpha_i D_i} \right), \tag{6}$$

where the material parameters μ_i, α_i, and D_i ($i = 1, 2, 3$) can be determined by fitting the experimental nominal stress–strain curve.

The coefficients of Ogden's material model, used in the numeric analysis FEM, are presented in Table 3.

Table 3. Material properties of the specimens. Coefficients of Ogden's model.

i	μ_i	α_i	D_i
1	$-1{,}062{,}310.500$	7.7010	8.6448
2	781,324.564	8.0848	0
3	776,736.530	-11.4395	0

For modeling foam, the Ogden model was modified with the introduction of the real exponent in the second part of the equation that describes the volume deformations (in this case, also non-linear dependencies), Table 4.

Table 4. Material properties of the specimens modify the coefficients of Ogden's model.

i	μ_i	α_i	D_i
1	$-832{,}131.490$	16.209480	8.935334
2	831,230.727	16.211660	0
3	$-2{,}528{,}968.430$	-1.172925	0

4. Conclusions

Tests were conducted to characterize the material structure and the mechanical response of EPP foam. During the research on the structure, SEM photographs showed that depending on the density of the foam there are finer cell structures with smaller cells. Measurements of compression and responses from the foam were presented at different speeds of deformation. Samples were tested in compression tests with strain rates in the range of 0.2 to 25 mm/s. Convergence of results for different strain rates conditions implies no gas strengthening. The research for samples having different temperatures show large effects of high temperatures on the test material. The material reach the yield point and the sample was damaged. A huge influence was observed on the shape due to stresses and strains. There were noticeable differences in the mechanical responses between the foams of similar density as confirmed by the lack of reproducibility of the structure. The main idea of this work was to identify and propose types of hyperelastic models for specific groups of material, which are used for increasing safety levels in the automotive industry. The analysis performed allowed us to extend the possibilities of using modern construction materials, plastics and composites. It is impossible to perform such analysis without precise material description. Usage of known hyperelastic material models–polynomial, Ogden's normal and reduced non-linear allowed us to describe the properties of the tested foam material (EPP) correctly. The use of the modified Ogden's model makes it possible to accurately determine the description of the material, which makes it possible to increase the accuracy and effectiveness of the simulation. The research carried out allows an even better selection of the material and its properties.

Author Contributions: Conceptualization, P.R., R.N. and T.D.; methodology, P.R., R.N., P.D.; software, P.R. and T.D.; validation, P.R., A.A., R.N. and P.D.; formal analysis, P.R., T.D., P.D.; investigation, R.N., P.R.; resources, P.R., T.D.; data curation, P.R. and P.D.; writing—original draft preparation, P.R., A.A.; writing—review and editing, P.R., T.D.; visualization, A.A., P.R.; supervision, P.R., R.N.; project administration, P.R.; funding acquisition, P.R., T.D. All authors have read and agreed to the published version of the manuscript.

Funding: This research received no external funding.

Data Availability Statement: Data sharing not applicable. No new data were created or analyzed in this study. Data sharing is not applicable to this article.

Conflicts of Interest: The authors declare no conflict of interest.

References

1. Zhai, W.; Kim, Y.-W.; Park, C.B. Steam-Chest Molding of Expanded Polypropylene Foams. 1. DSC Simulation of Bead Foam Processing. *Ind. Eng. Chem. Res.* **2010**, *49*, 9822–9829. [CrossRef]
2. Zhai, W.; Kim, Y.-W.; Jung, D.W.; Park, C.B. Steam-Chest Molding of Expanded Polypropylene Foams. 2. Mechanism of Interbead Bonding. *Ind. Eng. Chem. Res.* **2011**, *50*, 5523–5531. [CrossRef]
3. Andena, L.; Caimmi, F.; Leonardi, L.; Nacucchi, M.; De Pascalis, F. Compression of polystyrene and polypropylene foams for energy absorption applications: A combined mechanical and microstructural study. *J. Cell. Plast.* **2019**, *55*, 49–72. [CrossRef]
4. Rusch, K.C. Load–compression behavior of flexible foams. *J. Appl. Polym. Sci.* **1969**, *13*, 2297–2311. [CrossRef]
5. Burgess, G. Consolidation of cushion curves. *Packag. Technol. Sci.* **1990**, *3*, 189–194. [CrossRef]
6. Castiglioni, A.; Castellani, L.; Cuder, G.; Comba, S. Relevant materials parameters in cushioning for EPS foams. *Colloids Surf. A Physicochem. Eng. Asp.* **2017**, *534*, 71–77. [CrossRef]
7. Maiti, S.K.; Gibson, L.J.; Ashby, M.F. Deformation and energy absorption diagrams for cellular solids. *Acta Metall.* **1984**, *32*, 1963–1975. [CrossRef]
8. Gibson, L.J.; Ashby, M.F. *Cellular Solids*; Cambridge University Press: Cambridge, UK, 1997; ISBN 9780521499118.
9. Thomson, W. On the division of space with minimum partitional area. *Acta Math.* **1887**, *11*, 121–134. [CrossRef]
10. Shulmeister, V. *Modelling of the Mechanical Properties of Low-Density Foams*; Shaker Publishing B.V.: Maastricht, The Netherlands, 1998; ISBN 9042300256.
11. Roberts, A.P.; Garboczi, E.J. Elastic moduli of model random three-dimensional closed-cell cellular solids. *Acta Mater.* **2001**, *49*, 189–197. [CrossRef]
12. Lee, Y.S.; Park, N.H.; Yoon, H.S. Dynamic Mechanical Characteristics of Expanded Polypropylene Foams. *J. Cell. Plast.* **2010**, *46*, 43–55. [CrossRef]
13. Mills, N. *Polymer Foams Handbook*; Elsevier: Amsterdam, The Netherlands, 2007; ISBN 9780750680691.
14. Krundaeva, A.; De Bruyne, G.; Gagliardi, F.; Van Paepegem, W. Dynamic compressive strength and crushing properties of expanded polystyrene foam for different strain rates and different temperatures. *Polym. Test.* **2016**, *55*, 61–68. [CrossRef]
15. Arezoo, S.; Tagarielli, V.L.; Siviour, C.R.; Petrinic, N. Compressive deformation of Rohacell foams: Effects of strain rate and temperature. *Int. J. Impact Eng.* **2013**, *51*, 50–57. [CrossRef]
16. Zhang, J.; Kikuchi, N.; Li, V.; Yee, A.; Nusholtz, G. Constitutive modeling of polymeric foam material subjected to dynamic crash loading. *Int. J. Impact Eng.* **1998**, *21*, 369–386. [CrossRef]
17. Morton, D.T.; Reyes, A.; Clausen, A.H.; Hopperstad, O.S. Mechanical response of low density expanded polypropylene foams in compression and tension at different loading rates and temperatures. *Mater. Today Commun.* **2020**, *23*, 100917. [CrossRef]
18. AS 2498.3-1993. *Methods of Testing Rigid Cellular Plastics: Determination of Compressive Stress*; Standards Australia: Sydney, Australia, 1993.
19. AS 2498.4-1993. *Methods of Testing Rigid Cellular Plastics: Determination of Cross-Breaking Strength*; Standards Australia: Sydney, Australia, 1993.
20. Mukherjee, M.; Ramamurty, U.; Garcia-Moreno, F.; Banhart, J. The effect of cooling rate on the structure and properties of closed-cell aluminium foams. *Acta Mater.* **2010**, *58*, 5031–5042. [CrossRef]
21. Kaczyński, P.; Ptak, M.; Gawdzińska, K. Energy absorption of cast metal and composite foams tested in extremely low and high-temperatures. *Mater. Des.* **2020**, *196*, 109114. [CrossRef]
22. Yu, Y.; Cao, Z.; Tu, G.; Mu, Y. Energy Absorption of Different Cell Structures for Closed-Cell Foam-Filled Tubes Subject to Uniaxial Compression. *Metals* **2020**, *10*, 1579. [CrossRef]
23. Novak, N.; Vesenjak, M.; Duarte, I.; Tanaka, S.; Hokamoto, K.; Krstulović-Opara, L.; Guo, B.; Chen, P.; Ren, Z. Compressive Behaviour of Closed-Cell Aluminium Foam at Different Strain Rates. *Materials* **2019**, *12*, 4108. [CrossRef]

24. Cao, Z.; Li, M.; Yu, Y.; Luo, H. Fabrication of Aluminum Foams with Fine Cell Structure under Increased Pressure. *Adv. Eng. Mater.* **2016**, *18*, 1022–1026. [CrossRef]
25. Osiński, J.; Rumianek, P. The identification properties of porous foam. *Appl. Comput. Sci.* **2014**, *10*, 58–65.
26. Osiński, J.; Rumianek, P. Researches determining the properties of composites with gas phase in the car—pedestrian system: Methods and application. *Dyn. Syst. Appl.* **2013**, 385–394.
27. Frederick, G.; Kaepp, G.A.; Kudelko, C.M.; Schuster, P.J.; Domas, F.; Haardt, U.G.; Lenz, W. Optimization of Expanded Polypropylene Foam Coring to Improve Bumper Foam Core Energy Absorbing Capability. *SAE Trans.* **1995**, *104*, 394–400. [CrossRef]
28. Tang, N.; Lei, D.; Huang, D.; Xiao, R. Mechanical performance of polystyrene foam (EPS): Experimental and numerical analysis. *Polym. Test.* **2019**, *73*, 359–365. [CrossRef]
29. Abaqus Version 6.12–Theory Manual, Plasticity for Non-Metals, Large Strain Elasticity, Models; Dassault Systemes: Vélizy-Villacoublay, France. Available online: http://www.abaqus.com (accessed on 30 November 2020).
30. Machado, G.C.; Alves, M.K.; Rossi, R.; Silva, C.R.A. Numerical modeling of large strain behavior of polymeric crushable foams. *Appl. Math. Model.* **2011**, *35*, 1271–1281. [CrossRef]
31. Rivlin, R.S. Large elastic deformations of isotropic materials. I. Fundamental concepts. *Philos. Trans. R. Soc. London. Ser. A Math. Phys. Sci.* **1948**, *240*, 459–490. [CrossRef]
32. Rivlin, R.S. A note on the torsion of an incompressible highly-elastic cylinder. *Math. Proc. Camb. Philos. Soc.* **1949**, *45*, 485–487. [CrossRef]
33. Treloar, L.R. *The Physics of Rubber Elasticity*, 3rd ed.; Oxford University Press: Oxford, UK, 1975; ISBN 9780191523304.
34. Zhang, X.F.; Andrieux, F.; Sun, D.Z. Pseudo-elastic description of polymeric foams at finite deformation with stress softening and residual strain effects. *Mater. Des.* **2011**, *32*, 877–884. [CrossRef]

Article

Evaluation of the Zero Shear Viscosity, the D-Content and Processing Conditions as Foam Relevant Parameters for Autoclave Foaming of Standard Polylactide (PLA)

Tobias Standau [1], Huan Long [2], Svenja Murillo Castellón [3], Christian Brütting [1], Christian Bonten [3] and Volker Altstädt [1,*]

[1] Department of Polymer Engineering, University Bayreuth, Universitätsstraße 30, 95447 Bayreuth, Germany; tobias.standau@uni-bayreuth.de (T.S.); christian.bruetting@uni-bayreuth.de (C.B.)

[2] Bavarian Polymer Institute and Bayreuth Institute of Macromolecular Research, University of Bayreuth, Universitätsstraße 30, 95447 Bayreuth, Germany; huan.long@uni-bayreuth.de

[3] Institut für Kunststofftechnik, University of Stuttgart, Pfaffenwaldring 32, 70569 Stuttgart, Germany; svenja.murillo.castellon@ikt.uni-stuttgart.de (S.M.C.); christian.bonten@ikt.uni-stuttgart.de (C.B.)

* Correspondence: altstaedt@uni-bayreuth.de; Tel.: +49-(0)921/55-7471

Received: 3 March 2020; Accepted: 16 March 2020; Published: 18 March 2020

Abstract: In this comprehensive study, the influence of (i) material specific properties (e.g., molecular weight, zero shear viscosity, D-content) and (ii) process parameters (e.g., saturation temperature, -time, -pressure, and pressure drop rate) on the expansion behavior during the autoclave foaming process were investigated on linear Polylactide (PLA) grades, to identify and evaluate the foam relevant parameters. Its poor rheological behavior is often stated as a drawback of PLA, that limits its foamability. Therefore, nine PLA grades with different melt strength and zero shear viscosity were systematically chosen to identify whether these are the main factors governing the foam expansion and whether there is a critical value for these rheological parameters to be exceeded, to achieve low density foams with fine cells. With pressure drop induced batch foaming experiments, it could be shown that all of the investigated PLA grades could be foamed without the often used chemical modifications, although with different degrees of expansion. Interestingly, PLAs foaming behavior is rather complex and can be influenced by many other factors due to its special nature. A low molecular weight combined with a high ability to crystallize only lead to intermediate density reduction. In contrast, a higher molecular weight (i.e., increased zero shear viscosity) leads to significant increased expandability independent from the D-content. However, the D-content plays a crucial role in terms of foaming temperature and crystallization. Furthermore, the applied process parameters govern foam expansion, cell size and crystallization.

Keywords: polylactide; biofoams; zero shear viscosity; melt strength; batch foaming; pressure drop rate; low density; small cell size

---

1. Introduction

Polylactide (PLA) has drawn enormous attention in the past two decades as one of the most promising alternatives for replacing foamed products made from conventional plastics, such as polystyrene (PS), because of its similar mechanical properties [1] and because of the lower carbon foot print [2]. Though, it is well known that it is very challenging to foam PLA due to its poor melt strength and complex crystallization behavior. Different ways to improve the foamability of PLA were described in the literature and concluded in a work of Nofar et al. [3], as using chain extenders to increase the molecular weight and/or induce branched chains, varying the ratio of the so called L/D enantiomers (i.e., the composition of PLA from chiral L-lactide and D-lactide), using additives and controlling the crystallization behavior.

In the current literature, mainly the use of a wide variety of chemical modifications during processing and foaming is mainly described. Mostly, these modifiers are added to increase the molecular weight and prevent degradation. This can lead to chain extension, chain branching and/or cross-linking. Typical modifiers that are used in combination with PLA can be distinguished by the occurring reaction into two groups: (i) functional group reactions and (ii) radical reactions and have been summarized in our previous work [4]. Additionally, the rheological as well as thermal behavior is influenced by the use of these modifiers [5]. The most common used modifier is the commercial additive Joncryl® from BASF SE, which is a multi-functional epoxy based chain extender that reacts with the functional groups of PLA (i.e., carboxyl- and hydroxyl groups) during melt processing. Peroxides, such as dicumyl peroxides, lead to free radical reaction. In one of our previous studies, we could show that the melt strength could be significantly increased by the use of these modifiers, leading to foams with lower density, finer cell sizes and improved compression behavior [6,7].

Even though it is often stated in literature that the melt strength plays a vital role in the foaming of PLA, only a few publications actually quantify it at all [7–10]. An increased melt strength is beneficial for foaming, as it prevents coalescence and rupture during the cell growth step because the cell walls can withstand higher forces during stretching while expanding and can consequently lead to an improved expansion behavior (i.e., lower density and finer cells) [11].

The zero shear viscosity (ZSV) can be obtained from simple shear experiments with less experimental effort, when compared to the determination of the melt strength. The ZSV is dependent on a lot of different factors, such as molecular weight, modifications [12], chain architecture [13,14], and the amount of plasticizing agents, additives [15], or other blend partners [16]. Therefore, the zero shear viscosity can be quite easily varied. For example, Najafi et al. [17] investigated the rheological properties and foaming behavior of modified polylactide. Within this study, a semi-crystalline linear polylactide was modified with an epoxy-based chain extender. Here, the zero shear viscosity was increased and strain hardening was induced. It was shown that a higher shear viscosity in combination with the strain hardening lead to a finer cell morphology and lower foam densities. A recent work by Nofar et al. [18] dealt with rheological, thermal, and foaming behavior of unmodified and modified PLA. In both of the studies, the modification of PLA led to an increase in molecular weight and introduced a strain hardening effect at the same time. For this reason, it was not possible to separate the effect of the molecular weight from the strain hardening on the foaming behavior. Up to now, no systematic work on the influence of the zero shear viscosity on the expansion behavior of PLA was carried out. Yet, it is not clear from the literature whether a certain zero shear viscosity needs to be exceeded to obtain low density foams.

Besides the rheological behavior, the crystallinity is a key factor for processing PLA. The monomer lactic acid is chiral, thus there are two lactide isomers: L-Lactide and D-lactide. Here, the first is commonly the major component of commercial PLA grades usually with a content that is above 90%. The ratio between L-Lactide and D-lactide (L/D ratio) governs the chain architecture, hence the crystallization behavior of PLA. It should be mentioned that PLA with D-content above 10 to 12% are hindered to crystallize [19]. It is already known that PLA (with a D-content below 10%) usually crystallizes during foaming [20]. If PLA crystallize, multiple modifications are known, which occur under varied conditions; the detailed description was summarized in our previous work [4]. Besides the chain architecture, the crystallization behavior of PLA is also strongly dependent on the molecular weight itself, as the elevated molecular weight would restrict the chain mobility. During the pressure-induced batch foaming process, the crystallization behavior of PLA can be assumed as being quite complex and the following processing parameters strongly impact the final foam properties:

(1) In general, temperature most directly affects the crystallization behavior by affecting chain mobility.

(2)   The pressure dominates the solubility of the blowing agent and, thereby, the amount of it that is solved in the polymer matrix. Hence, the plasticization effect as well as the gas-induced crystallization is directly influenced by the applied pressure.

(3)   During an isothermal saturation phase (time), the crystallization kinetics of PLA is promoted by elevated temperature and the plasticization effect from dissolved blowing agent and is generally governed by the duration.

(4)   As a consequence of a sudden pressure drop, which actually induces a thermo-dynamic instability and the prompt oversaturation of the sample the foaming takes place. Hence, the pressure drop rate guides the nucleation rate and the extent of the strain-induced crystallization.

This study focuses on the foamability of the neat (i.e., unmodified) PLA, carving out which requirements for high expansion and fine morphology in an autoclave foaming process are necessary. By the comparison of a wide variety of commercial grades, different aspects, such as D-content, molecular weight, and rheological behavior, can be evaluated regarding their influence on the foaming behavior. Here, besides the often referred melt strength, the zero shear viscosity is also about to be considered as foam relevant property. Additionally, the processing parameters themselves were studied and correlated with the material-specific parameters. Yet, novel cross-relationships could be revealed, as material- and process-specific factors were systematically varied over a wide range.

2. Materials and Methods

2.1. Materials

In this study, different PLA grades from NatureWorks Ltd. (Minnetonka, MN, USA) were used. Table 1 lists the selected grades and their supposed usage, as well as the melt flow rate from the data sheet. Additionally, the D-content, the molecular weight, and the zero shear viscosity are given.

Table 1. Overview of Polylactide (PLA) grades used in this study.

Grade & Supposed Usage * (Data Sheet)	Notation	MFR * (g/10 min.) (@ 210 °C, 2.16kg)	D-Content * (%)	GPC Data ** M_w (10^3 g/mol); M_n (10^3 g/mol); PDI (-)	ZSV ** T = 180 °C, γ = 5%, 0.1 rad/s (Pas)
2003D Packaging	P_4.3D	6	4.3 [4]	232; 134; 1.73	8089
3100HP Injection Molding	IM1_2D	24	<2 [21]	162: 107; 1.51	1869
3251D Injection Molding	IM2_1.4D	80	1.4 [4]	116; 78; 1.49	243
3260HP Injection Molding	IM3_2D	65	<2 [21]	111; 75, 5; 1.47	483
4032D Film	Fi_2D	–	1.4–2.0 [4]	232; 149; 1.55	5716
4044D Extrusion	X_4D	–	~4 [21]	230; 117; 1.97	7794
4060D Hot Sealing	HS_12D	–	12–12.3 [4]	217; 118; 1.84	4200
7001D Injection Stretch Blow Molding	BM_4.4D	6	4.4 ± 0.5 [4]	242; 142; 1.71	8472
8052D Foaming	Fo_4.7D	14	4.7 [4]	178; 109; 1.64	3457

* taken from data sheet or literature. ** measured within this work, please check 2.2 Methods and Figure 4.

2.2. Methods

2.2.1. Size Exclusion Chromatography (SEC)

The SEC measurements were performed on a 1260 SECcurity by PSS (Mainz, Germany) that was equipped with a refractive index detector. A 50 mm precolumn and three 300 mm PSS SDV 10 μm columns with 103, 105, and 106 Å were used as stationary phase. Chloroform was used as a mobile phase. The column temperature was set to 30 °C and the flow rate was 1.0 mL/min. The sample concentration was 3.0 g/L. Chromatograms were interpreted while using conventional calibration against polystyrene polymer standards. Therefore, please mind that the noted molecular weights are not the absolute values and can only be compared within the test series.

2.2.2. Differential Scanning Calorimetry

The thermal properties of the samples were analyzed by differential scanning calorimetry (DSC) on a DSC 204 Phoenix from Netzsch (Selb, Germany). Samples of about 10 mg were weighed and then sealed in DSC crucibles and placed in the DSC cell. The measurements were performed under nitrogen atmosphere. First, they were heated from 20 to 200 °C at 10 K/min. Subsequently, they were cooled down from 200 to 20 °C with 10 K/min. and then heated up again from 20 to 200 °C at 10 K/min. Additional experiments with elevated heating rates of 20 K/min. were also carried out. It is well known that the crystallization of PLA is very slow. Hence, measurements were also performed at a lower cooling rate of 2 K/min to further investigate the crystallization process. The crystallinity of the materials was calculated by the use of heat of fusion of 93.1 J/g for 100% crystalline PLA homopolymers [22].

2.2.3. Rheology

(1) Complex Viscosity: The rheological characterization in shear flow was performed with a plate-plate rheometer Discovery HR-2 hybrid from TA Instruments Waters LLC (New Castle, DE, USA) at 180 °C with a diameter of 25 mm and a gap of 1 mm under nitrogen atmosphere. Dynamic mechanical experiments were carried out in a frequency range from 500 to 0.01 rad/s. The deformation amplitude was set to 5% for all of the measurements. The zero shear viscosity was determined in the frequency region of the Newtonian plateau at a frequency of 0.1 rad/s.

(2) Melt Strength: The Rheotens measurements were carried out to prove the ability of the polymer melt to withstand uniaxial strain. The melt strength was measured while using a Göttfert Rheotens device (Buchen, Germany). The melt strength test samples were prepared using a single screw extruder (L/D ratio of 26) from Göttfert (Buchen, Germany), equipped with a round die of 6 mm in diameter. For each test, a molten polymer strand was drawn down from the die by the two counter-rotating measurement wheels that were mounted on a sensitive force transducer connected to the Rheotens Göttfert 71.97 unit. The tensile force on the polymer melt strand was measured as a function of time or velocity of the measurement wheels. The melt strength is represented by the force at which the strand breaks or in the plateau phase of the rheotens curve. The measurements were carried out at 180 °C and a constant acceleration of 2.4 mm/s^2.

2.2.4. Autoclave Foaming

Foaming was carried out with the pressure drop method on a custom-made autoclave, which Figure 1 schematically illustrates. The melt pressed samples ($10 \times 20 \times 1$ mm^3) were put in the electrical heated pressure vessel, which was then purged with CO_2.

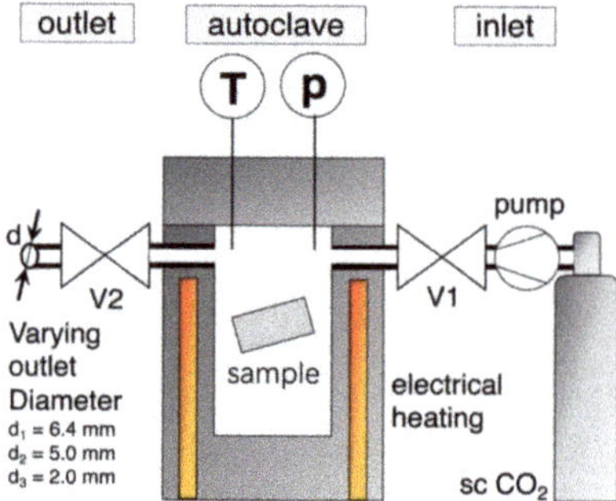

Figure 1. Sketch of foaming set up with the custom-made autoclave.

By the use of a twin set-up of Teledyne Isco 260D syringe pumps (Thousand Oaks, CA, USA), supercritical CO_2 was applied and the samples were impregnated with a constant saturation pressure of 180, 150, or 120 bar, respectively. The pump was set to constant pressure delivery mode, which ensured a constant pressure during the whole saturation period (even at small leakages). Temperature is measured in the vessel near the sample and it was varied over a wide range for each grade, to find optimum foaming condition (i.e., minimum in foam density). The saturation time was set as 30 min. (but also was varied for selected grades between 5 min. and 8 h). After saturation, the electrical controlled pneumatic valve V2 was promptly opened to start the pressure drop, which initiates foaming. Reproducibility is given for the out carried experiments, as the threefold repetition of a foaming trial usually resulted in standard deviations that were below 5%.

By changing the inner diameter of the outlet pipe (6.4, 5.0, or 2.0 mm), the pressure drop rate can be varied, as shown in Figure 2.

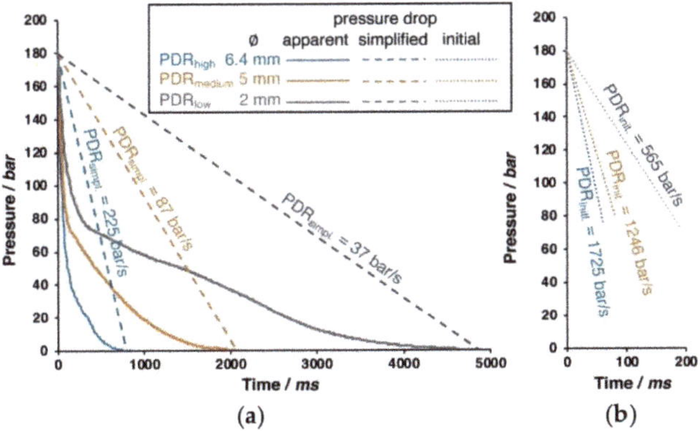

Figure 2. Different pressure drops that can be realized with different pipe diameters. (**a**) Apparent pressure drop (solid line) and simplified linear pressure drop (dashed line). (**b**) Close up of the initial pressure drop (dotted line) during the first milliseconds.

Only the initial phase of the pressure drop is nearly linear, as can be seen in the right close up in Figure 2. The apparent pressure follows a logarithmic function. The pressure drop rate (PDR) that results from different pipe diameters is also noted within the diagram. For the simplified pressure drop rate, a linear drop is assumed over the entire time. However, this is only a very rough description; and, from visualization experiments in foam extrusion process carried out by the group of C.B Park, it is known that the cell formation (i.e., cell nucleation) takes place in the first milliseconds of the pressure drop [23]. Therefore, the initial pressure drop rate is calculated from the linear slope during the depressurization to 90 bar happening in 50 ms (6.4 mm), 90 ms (5.0 mm), and 190 ms (2.0 mm), respectively. If not given otherwise, the experiments were carried out with 180 bar saturation pressure and the largest outlet of 6 mm diameter (i.e., with the highest pressure drop rate). The pressure drop rate is suggested to have impact on the foaming behavior by affecting the nucleation and cell growth [24,25]. At high pressure drop rates, the nucleation is faster, because the thermodynamic instability, which is responsible for the cell formation, happens earlier. Besides, the cell growth is also faster, while the diffusive path of dissolved gas to reach a cell is shorter with the increasing amount of nuclei. In the literature, pressure drop rates rarely can be found. Liu et al. reported 1.3 MPa/s (converts to 13 bar/s) as the pressure drop rate during autoclave foaming trials [26]. In their work, Xu et al. [27] mention that the pressure drop is not constant over the time, but quantify it by an average, as the ratio of pressure difference and depressurization time (i.e., simplified, as dashed line in Figure 2). By varying the PDR from 1.4 to 25 MPa/s (14 up to 250 bar/s), they showed that a higher PDR leads to increased cell density and a higher volume expansion ratio.

2.2.5. Foam Properties

(1)   Scanning Electron Microscopy (SEM): Cryogenic fractured foam samples were investigated by SEM JEOL JSM-6510 (Akishima, Japan). The cell sizes were determined by the use of image analysis software (ImageJ, v1.48, University of Wisconsin, Madison, WI, USA).

(2)   Density: The density was determined according to the Archimedes principle with a balance from Mettler Toledo AG245 (Columbus, OH, USA).

3. Results and Discussion

3.1. Basic Properties of the Neat Materials

First, the basic properties of the neat materials were determined in order to be able to correlate the molecular constitution (molecular weight and D-content) with the rheological and thermal properties, as well as the later foaming behavior.

The Rheotens test was performed to evaluate the influence of the melt strength on the foaming behavior. It was expected that the melt strength increases with higher molecular weight, due to an increase of polymer chain entanglements. Entanglements can lead to forces that are similar to secondary or hydrogen bonding forces. Therefore, the resistance to an applied pulling load (here draw down force) is higher with a higher molecular weight. In regards of foaming, it is expected that a certain melt strength is necessary to build a stable foam structure [28].

Figure 3a shows the Rheotens curves. Even though both grades have a molecular weight in a similar range (see Table 1), the melt strength of IM1_2D was not measurable, while it could be determined for PLA Fo_4.7D, which shows a melt strength of 0.03 N. It has to be pointed out that the injection molding grades (IM1_2D, IM2_1.4D, and IM3_2D) could not be successfully measured due to the occurring strand rupture already at very low strain. This is caused by the comparable low viscosity at the measuring temperature. As a consequence, a pronounced melt strand thinning is caused by gravity forces. This effect can be seen in Figure 3b, while comparing IM1_2D and the higher molecular weight grade X_4D.

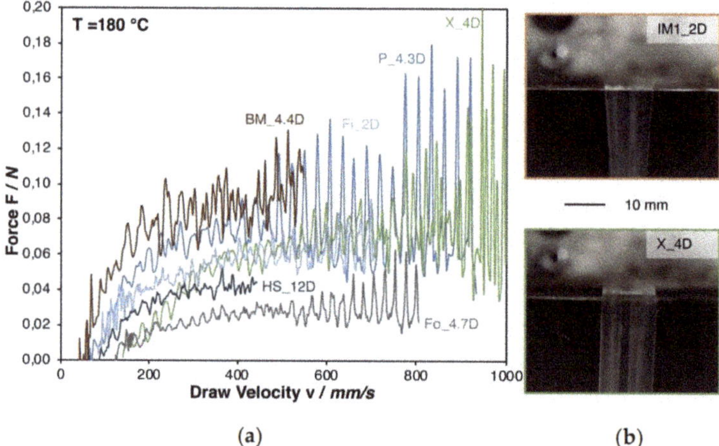

Figure 3. (**a**) Rheotens curves of all (measurable) PLA grades at 180 °C. (**b**) Die swelling behavior of a low and a high molecular PLA.

Yet, the melt strength can be a meaningful value to be correlated with the foamability, it can only be determined for selected grades. The shear behavior was also measured to generate more reliable information on all grades.

Figure 4a shows the complex viscosity as a function of the frequency for the investigated PLA grades. All of the materials show a typical shear thinning behavior, which is more pronounced with increasing complex viscosity. A direct correlation between the zero shear viscosity (complex viscosity at a frequency $\approx$ 0.1 rad/s) and the molecular weight was found, as can also be seen in Figure 4b. Besides a slight deviation of IM2_1.4D, for all grades the zero shear viscosity is proportional to the 3.7th power of molecular mass, thus confirming the linear character of the polymer chains [29,30]. A higher molecular weight and/or the existence of long chain branches are beneficial to achieve a high expansion during foaming, as it was shown for PP extrusion foams in the work of Stange et al. [31]. It can be seen that the two injection molding grades IM3_2D and IM2_1.4D possess a rather low complex viscosity, which is clearly due to their very low molecular weight. Similar to the melt strength, the complex viscosity, in general, increases with increasing molecular weight. Though, a direct correlation between the molecular weight and the melt strength was not possible due to the influence of various factors (external: e.g., temperature shifts over the down drawn melt strand and internal: e.g., different crystallization rates).

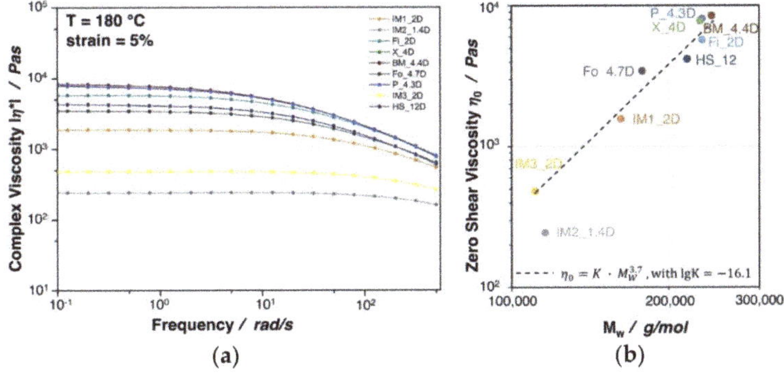

Figure 4. (**a**) Complex viscosity of neat PLA grades at 5% strain and 180 °C. (**b**) Zero shear viscosity plotted against the molecular weight.

The molecular weight and D-content [32] strongly affects the thermal properties (i.e., crystallization behavior). The DSC thermograms of the investigated PLA are shown in Figure 5. Judging from the second heating curves, all of the grades only revealing a neglectable amount of crystallinity (0.3–3%) and they can be considered as amorphous. However, the injection molding grades show the tendency of cold crystallization.

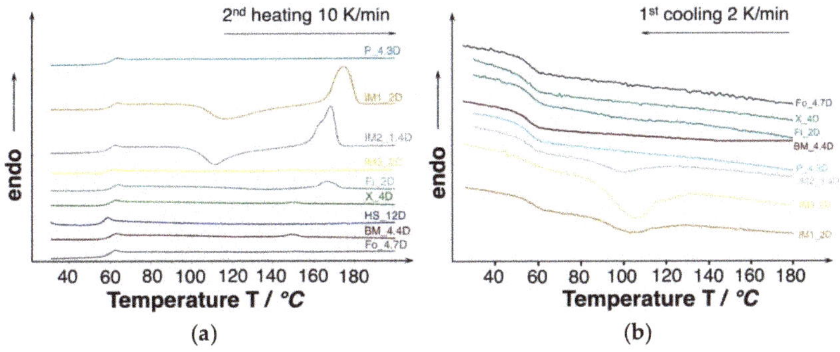

Figure 5. (**a**) Differential scanning calorimetry (DSC) curves of unprocessed PLA (10 K/min., 2nd heating). (**b**) Cooling curves of the unprocessed PLA with very low cooling rate to favor crystallization (2 K/min.).

In general, PLA show a very slow crystallization rate and most of the grades are not able to crystallize while the cooling phase during the DSC experiment, even at very slow cooling rates of 2 K/min. (see Figure 5b). However, the injection molding grades were able to crystallize partly already during the cooling. The ability to crystallize already during the cooling can be affiliated to the lower molecular weight, resulting in higher chain mobility. Consequently, higher crystallinities could be expected under certain conditions, allowing for PLA chains to come close enough to build ordered structures (i.e., crystals), such as sufficient time and/or a high strain. Additionally, it would be likely that the manufacturer already incorporated nucleating agent.

The different thermal behavior will also lead to varied crystallization during the foaming. Here, a fast crystallization rate can be beneficial, as formed crystals could increase the melt strength and they can act as nucleating points for cell growth resulting in a stable and fine celled foam structure. However, on the other side, a too high crystallization rate and degree of crystallinity can hinder the cell expansion.

3.2. Expansion of the Neat Grades

All of the grades were foamed with a saturation time of 30 min. at 180 bar CO_2 and varying temperatures. In Figure 6, the obtained foam densities, depending on the saturation temperature, are shown. Additionally, a SEM image of the morphology for each grade with the lowest achieved density is given. Most of the grades show an optimum (i.e., minimum of the foam density) at a certain temperature. The lowest densities range between 150 and 50 kg/m^3. However, the grades IM1_2D, IM2_1.4D, and IM3_2D show much more erratic behavior and increased standard deviation above 5%. Here, no clear trend is visible and foam densities are much higher (above 400 kg/m^3). The mentioned grades are injection molding grades with low molecular weight, low zero shear viscosity, and a strong ability to crystallize. The melt strength is assumed to be very low, as these grades could not be analyzed by Rheotens experiments. Furthermore, all these types possess a low number average molecular weight M_n, which indicates short polymer chains, resulting in higher crystallization rates due to a higher chain mobility. Hence, the degree of crystallization of 58% up to 81% (see Figure 7) is found to be much higher after the depressurization, as compared to the other grades. Consequently, the cell growth is hindered during expansion. Two effects are likely: (i) the sample starts to crystallize in the presence of the CO_2 during the saturation, which could lead to a reduced gas uptake by the developed crystalline phases [33,34] and/or (ii) shorter chains can be oriented in a much higher extent during the stretching, while the expansion (so called strain induced crystallization) [20]. Additionally, as these grades are commercialized, they are most probably additivated with nucleating agents by the manufacturer. Also other studies have only revealed a moderate density reduction for batch foamed PLA injection molding grades [35,36].

Apart from the injection molding grades, for P_4.3D, Fi_2D, X_4D, BM_4.4D, and Fo_4.7D foams with densities below 100 kg/m^3 were achieved at significant lower foaming temperatures, as is clearly visible from Figure 6. These five grades mainly possess a higher molecular weight and, consequently, higher melt strength, as well as higher zero shear viscosity, which allows for a more pronounced expansion. Thus, the cell growth could also be ensured. However, IM1_2D and Fo_4.7D owe molecular weights in a similar range (see Figure 4b) but they show much different foaming behavior, which allows for the conclusion that the expansion is affected by other parameters beside the molecular weight, such as the crystallization rate and amount of crystals, which appears to be much higher for the injection molding grade IM1_2D. The lowest achieved densities are 40 kg/m^3 (BM_4.4D), 52 kg/m^3 (P_4.3D), 87 kg/m^3 (X_4D), and 71 kg/m^3 (Fo_4.7D). Even though the first three mentioned grades possess a higher molecular weight and zero shear viscosity than Fo_4.7D, they all show similar expansion. Consequently, it can be stated that the molecular weight and the therefrom resulting rheological behavior seems to be sufficient enough to allow for the formation of low density foams, as the expansion is not hindered by a too strong crystallization, as it would be the case for IM1_2D. After foaming, the grades P_4.3D, Fi_2D, X_4D, BM_4.4D, and Fo_4.7D also developed significant elevated crystallinities. The degree of crystallinity ranges from 35% up to 42%, which is still lower than for the injection molding grades (> 58%). The chain mobility is expected to be significantly lower than that for the injection molded grades hindering the development of higher orientation during crystal formation, as these grades have longer chains. It should be mentioned that the crystallization is also assumed to happen during the saturation with the aid of the elevated temperature and chain mobility. Thus, it could be assumed that, underneath a critical extent, the so-formed crystals would increase the melt strength and serve as nucleation points and, thereby, be beneficial during the expansion.

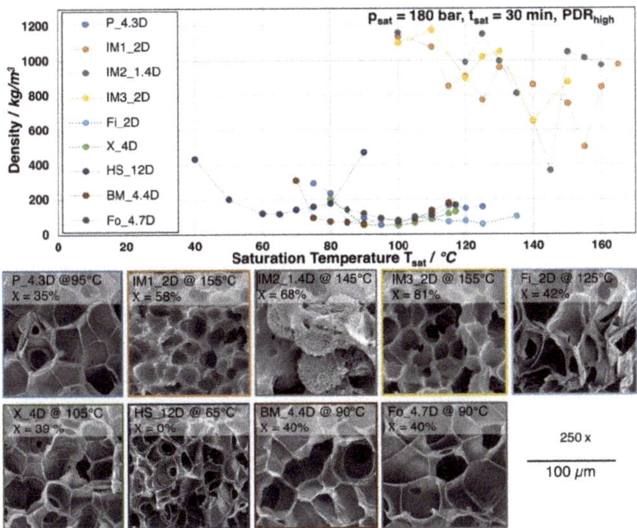

Figure 6. Top: Foam densities of the neat PLA grades foamed after a 30 min. saturation with CO_2 at 180 bar at varying temperatures. Bottom: SEM images of the foam with lowest density for each grade with crystallinity values (1st heating).

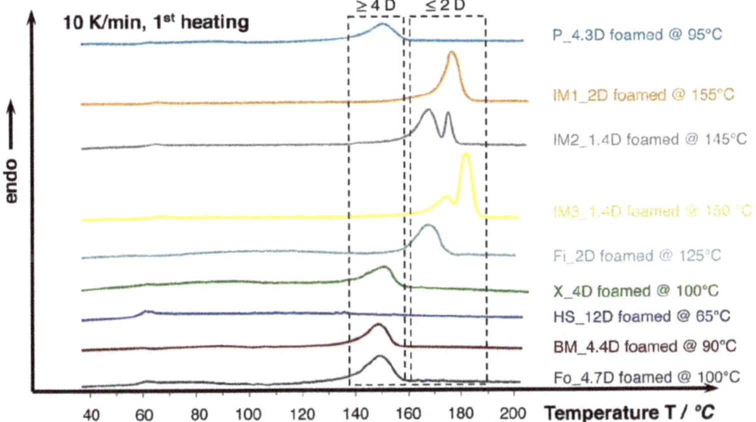

Figure 7. DSC curves (10 K/min., 1st heating) of the lowest density foams achieved in batch foam experiments with 30 min. saturation at 180 bar. Note the increased melting temperatures marked in the dashed boxes of PLA foams with lower D-content of 2 and 4, respectively.

Furthermore, the DSC curves that are shown in Figure 7 reveal that the melting temperature of the crystallized foams seems to strongly depend on the D-content. The grades with D-content of approximately 2% or lower (IM1_2D, IM2_1.4D, IM3_2D, and Fi_2D) show elevated melting temperatures of around 170 °C and higher while the grades with D-content around 4% (P_4.3D, X_4D, BM4.4D, and Fo_4.7D) clearly exhibit a lower melting point of around 150 °C. At higher D- contents, even lower melting points could be expected, as it was shown in the work of Bigg [37] for the case of unfoamed PLA. The formation of double melting peaks was also observed for the grades IM1_2D and IM2_1.4D. This effect is known for PLA and it could also happen to other grades, as it is mainly sensitively influenced by the saturation conditions (temperature, time and pressure), which was well described by Nofar et al. [38].

HS_12D is another outstanding grade, which foams at much lower temperatures. In fact, the minimum of foam density (115 kg/m^3) is reached at 65 °C. This can be traced back to the much higher D-content of approximately 12. Hence, no crystallization is expected to happen under this conditions [19,22,39] and the foaming temperature depends mainly on the T_g. The T_g of HS_12D could be expected to shift to a lower range [40], as well as the melt strength, due to the plasticization effect of the blowing agent. The molecular weight of HS_12D is between Fo_4.7D and X_4D, which both foamed to lower densities. This allows for the conclusions that the absence of crystallization due to the high D-content results in less favorable rheological behavior for foaming in the autoclave.

Most of the foams possess very thin cell walls. Here, it is evident that some of these cell walls are opened due to cell coalescence and cell rupture. A phenomenon that has been often reported for PLA foams [39,41,42] and can be traced back to the low melt strength.

3.3. Foaming Conditions vs. Expansion

In Figure 8, the influence of the saturation pressure on the achievable foam density is shown for X_4D and HS_12D (similar molecular weight but different D-content). Surprisingly, at lower saturation pressures, foams with lower densities but larger cell sizes were obtained.

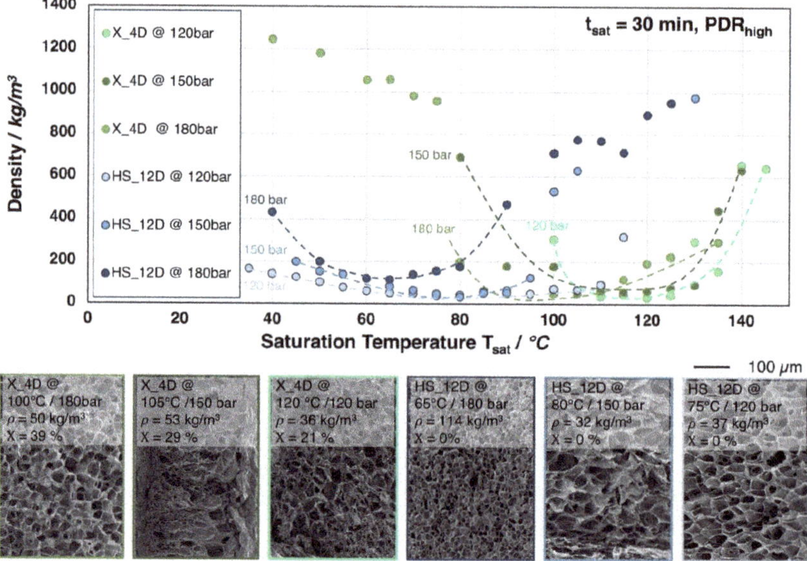

Figure 8. Top: Foam densities of the grades X_4D and HS_12D (with similar molecular weight but varying D-contents) foamed after 30 min. saturation with CO_2 at 120, 150, and 180 bar and varying temperatures. Bottom: SEM images of the foam with lowest density for both grades with crystallinity values (1st heating).

For both grades, it was observed that the saturation temperatures required for achieving the lowest possible foam densities are generally shifted to higher values with lower saturation pressure. This is a result of a less pronounced plasticization effect with a lower amount of dissolved CO_2 at lower saturation pressure. Moreover, HS_12D showed more pronounced foam density drop with lower saturation pressure and no crystallinity was observed after foaming. This could only be explained, as (i) a high expansion was ensured by a higher melt strength and (ii) that the pressure drop (rate) itself reduces with decreasing saturation pressure, thus leading to a lower cell density, as it is also described in literature [24,25]. Hereafter, the cells have the chance to grow larger and, consequently, result in lower foam density. Moreover, the gas dissolving content is considered also to have an effect on the cell nucleation. Guo et al. [43] described that an elevated saturation pressure promotes the nucleation rate; hence, a larger cell size can normally be expected from a lower saturation pressure. This is confirmed by the SEM images in Figure 8.

Similar tendency was observed for X_4D—yet much weaker—which is assumed to be mainly influenced by its thermal properties. Since X_4D possesses much lower D-content than HS_12D and showed crystallinity after foaming, the formation of crystals could be expected during CO_2 saturation. However, with less gas dissolving under lower pressure and at elevated saturation temperature, the crystallization was less promoted, leading to lower crystallinity values of 29% (at 150 bar) and 21% (at 120 bar), respectively.

In Figure 9, the foam density in dependence of the saturation time is shown for the grades Fi_2D, X_4D, and HS_12D. The foaming experiments were carried out at the optimum foaming temperatures for each grade (based on the trials summarized in Figure 6). Unusual high saturation times of up to eight hours were chosen here to identify the cause for the relenting density reduction with increasing saturation time. One could expect that a longer saturation time would lead to a higher expansion. Hence, for all grades, the lowest densities were achieved between 10 and 30 min. of saturation, respectively. For longer saturation, the achieved foam density increases almost exponential, as it can be seen from the trendlines. For the short saturation, a monomodal cell size distribution can be seen from the SEM images of X_4D. With increasing saturation times, a bimodal cell size distribution with much smaller cells is visible. After 8 h saturation, the morphology is comparable to IM2_1.4D (145 °C, 180 bar, 30 min.; cf. Figure 6), showing scattered spherical structures in between the cells.

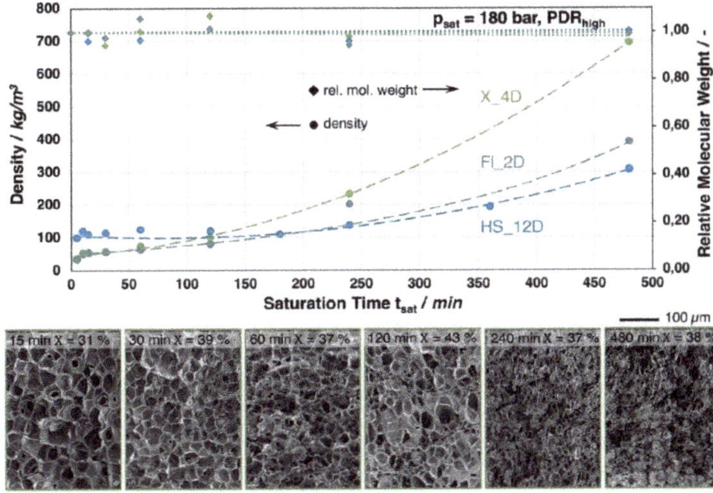

Figure 9. Top: Foam densities of the grades Fi_2D, X_4D, and HS_12D (with similar molecular weight but varying D-contents) foamed with CO_2 at 180 bar and varied saturation times. Bottom: SEM images of the X_4D foam with different saturation times and crystallinity values (1st heating, 10 K/min.).

The explanation for the decreased expansion with longer saturation time is not as trivial as expected. Of course, a decrease in molecular weight could be assumed with longer saturation to explain the decreasing expansion from the consequently anticipated less favorable rheological behavior during the expansion. However, the samples do not undergo any significant thermal degradation during saturation at different temperatures, as it can be clearly judged by the GPC results. Here, the relative molecular weight, as well as the polydispersity, did not significantly change. Another possible explanation would be that the absorption is ongoing over the time, meaning that the samples are not fully saturated with CO_2. The consequence would be that the T_g (and also the viscosity) would be decreased by a more pronounced plasticizing effect with increased saturation time, meaning that the optimum temperature for foaming would shift to ever lower values. Repeating the experiments at a lower temperature could impose this, which did not lead to lower densities. This allows for the conclusion that the samples can be assumed as being fully saturated, even after short saturation times and that there must be another reason for the less pronounced density reduction with an increasing saturation time. It is striking that HS_12D shows slightly divergent behavior when compared to the other grades (with lower D-content), as the density does not change significantly up to a saturation time of 240 min. When considering this and the crystallinity values (cf. Figure 9) of the foams from Fi_2D and X_4D, we develop the hypothesis that crystallization seems to play a crucial role for the expansion at different isothermal saturation under high pressure CO_2 loading (even for PLA with high D-content, which is strongly hindered to form crystals). There is strain-induced crystallization during the foaming, as was mentioned earlier. This leads to relatively high crystallinity values with high expansion rates. Henceforth, it would be assumed that, with less expansion, less (strain induced) crystallinity would be seen. Nevertheless, the values for the long saturated but less expanded samples stay rather high. Yet, this can be explained by the CO_2 induced crystallization, as was described by Xu et al. [44], showing elevated crystallinity for PLA saturated with super critical CO_2 for more than ten minutes. This allows for the assumption for Fi_2D and X_4D, that the CO_2-induced and strain induced crystallization are overlapping effects and with ongoing saturation the CO_2-induced crystallization will prevail. Hence, the crystals that formed during the saturation can hinder the expansion during depressurization, as the rheological behavior can be expected as being less favorable for expansion.

For HS_12D, this effect does not seem to play a role, which is plausible, as it is considered as total amorphous. However, this only seems to be valid for the saturation approximately up to 240 min. The DSC curves from the foams that were saturated longer reveal an additional phase transition that is close to the T_g and the formation of slight melting peaks at 153 and 166 °C (see Figure 10), which could be enabled from the harsh conditions of the applied high pressure and the long saturation phase at elevated temperature. After 16 h, this effect levels out, as judged from the associated density change. Hence, it is not unlikely that the autoclave treatment of usually amorphous polymers results in crystallization, as it was also shown before for polycarbonate [45]. To further describe the responsible mechanisms, more advanced techniques (e.g., high pressure DSC, modulated DSC, rheology of gas loaded melt, wide-angle X-ray scattering WAXS) would be helpful, but they could not be applied within this study. Still, it can be concluded that a long saturation phase under isothermal conditions leads to crystallization that is favored by the presence of CO_2, which is not beneficial for the expansion; and, even PLA with high D-content, which can normally be considered as totally amorphous is affected, as it shows additional phase transitions.

The foaming behavior (i.e., density) of Fi_2D, X_4D, and HS_12D, depending on the pressure drop rate, is shown in Figure 11. The foaming experiments were carried out under the optimal conditions of each grade (i.e., saturation time, -temperature, and -pressure based on the findings of Figure 6), yet different gas outlet pipe diameters were used to vary the PDR.

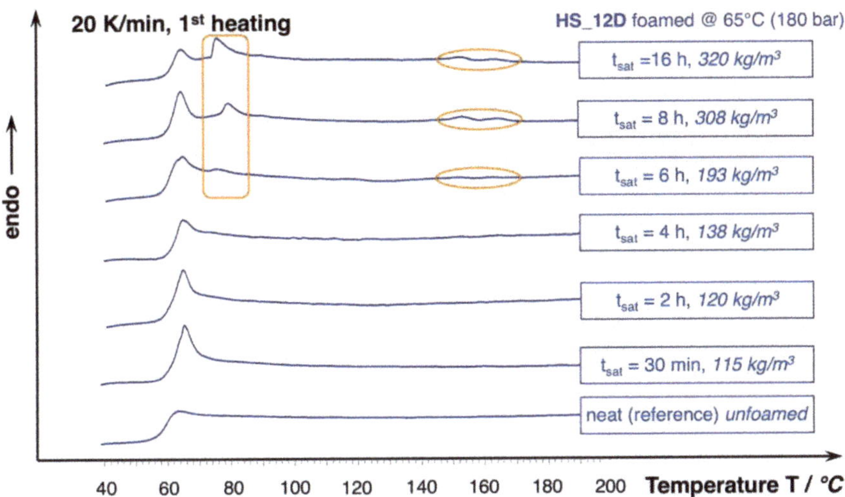

Figure 10. DSC curves of HS_12D foams and unfoamed reference (1st heating, 20 K/min.).

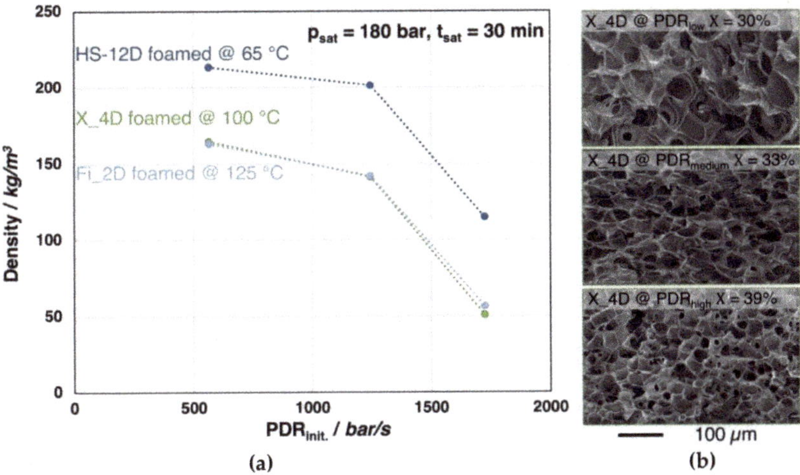

Figure 11. (a) Foam densities of the grades Fi_2D, X_4D and HS_12D (with similar molecular weight but varying D-contents) foamed after 30 min. saturation with CO_2 at 180 bar and varied pressure drop rates. (b) SEM images of the X_4D foams with different pressure drop rates and crystallinity values (1st heating).

Clearly, density reduction is more pronounced with an increasing pressure drop rate for all the three investigated grades. It should be noted that a similar tendency could be found in all three PLA samples with different D-content, which indicates that PDR plays a similar role in the foaming of PLA, regardless of D-content. However, for the low D-content grades (i.e., Fi_2D and X_4D), the crystallinity is also increased, most likely as the strain induced crystallization becomes more obtruded with the higher expansion. Furthermore, as it can be seen from the SEM images of the foam structures of X_4D, clearly the cell size becomes decreased with increased pressure drop rate, as the nucleation is promoted. As a conclusion, it can be said that the result shows a clear trend, where a higher PDR generally leads to a lower foam density and smaller cell size, as cell nucleation is promoted [43,46].

4. Conclusions

A wide variety of commercial PLA grades was investigated in this work. The grades can firstly be distinguished by their molecular weight, which leads secondly to different rheological behavior (i.e., melt strength and zero shear viscosity) and varied thermal properties (i.e., crystallization kinetics). Additionally, different D-contents were considered, which also determine the thermal properties (i.e., crystallinity). Besides these material-specific properties, different process parameters were changed during the pressure induced batch foaming (i.e., saturation temperature, -pressure, -time, and pressure drop rate). Consequently, a wide variety of factors that could influence the expansion behavior of PLA were conclusively studied.

Low density foams (volume expansion ratio VER $\geq$ 10) could be achieved for most of the PLA grades. Exceptions were the injection molding grades, whose low molecular weights lead to lower zero shear viscosities, vanishing low melt strengths, and high crystallization kinetics, which, in total, limit the expansion. It also could be shown that the rheological properties (melt strength and zero shear viscosity) are not the only factors governing the expandability as the crystallization behavior, which is known to be very complex for PLA can tremendously hinder the expansion, as it is the case for the injection molding grades. The D-content governs the foaming temperature and the melt temperatures of the crystallized foams. PLA with lower D-content needs to be foamed at higher temperatures, as it was also observed that a D-content of approx. 4% leads to melting temperature of around 150 °C, while at 2% and lower it is around 170 °C. A D-content of 12% commonly leads to no crystallization and consequently to much lower foaming temperatures that were way below 100 °C. The saturation pressure strongly governs the foaming temperature on which the lowest density can be achieved, as well as the nucleation, and thereby the fineness of the cells. Surprisingly, a shorter saturation time leads to lower densities, because the CO_2-induced crystallization intensifies with ongoing saturation, allowing for less expansion, as we hypothesized. Even for PLA with a high D-content of 12%, which can usually be considered as totally amorphous, a divergent foaming behavior ($t_{sat} > 4$ h) was shown, which is presumably due to an additional phase transition, induced by the long saturation. Furthermore, a high pressure drop rate is favorable for achieving low density foams with fine morphology, as the nucleation rate is increased independently from the D-content.

Author Contributions: T.S.: conceptualization, writing—review and editing sections "1. Introduction", "2. Materials and Methods" and Sections 3.2 and 3.3, abstract and conclusions, review, editing and preparation of original draft. H.L.: investigation and data curation of foam experiments, writing Sections 3.2 and 3.3, S.M.C.: investigation and data curation of rheological experiments and writing Section 3.1. C.B. (Christian Brütting) writing—review and editing section "1. Introduction" and "3 Results and Discussion". C.B. (Christian Bonten) and V.A. supervision, reviewing and funding acquisition. All authors have read and agreed to the published version of the manuscript.

Funding: This research was funded by German Research Foundation (DFG), grant number AL474/33-1, and BO1600/40-1. Open access charges were funded by the German Research Foundation (DFG) and the University of Bayreuth in the funding program Open Access Publishing.

Acknowledgments: The authors would like to acknowledge the Bavarian Polymer Institute (BPI) for providing access to different analysis methods. Furthermore, we thank mechanical workshop at the University of Bayreuth for construction and installation of the high pressure autoclave used in this work. Also, we would like to acknowledge Rika Schneider (MC II, University of Bayreuth) for additional GPC measurements.

Conflicts of Interest: The authors declare no conflict of interest.

References

1. Dorgan, J.R.; Lehermeier, H.; Mang, M. Thermal and Rheological Properties of Commercial-Grade Poly (Lactic Acid)s. *J. Polym. Environ.* **2000**, *8*, 1–9. [CrossRef]
2. Vink, E.T.H.; Davies, S. Life Cycle Inventory and Impact Assessment Data for 2014 Ingeo [TM] Polylactide Production. *Ind. Biotechnol.* **2015**, *11*, 167–180. [CrossRef]
3. Nofar, M.; Park, C.B. Poly (lactic acid) foaming. *Prog. Polym. Sci.* **2014**, *39*, 1–21. [CrossRef]

4. Standau, T.; Zhao, C.; Murillo Castellón, S.; Bonten, C.; Altstädt, V. Chemical Modification and Foam Processing of Polylactide (PLA). *Polymers* **2019**, *11*, 306.

5. Göttermann, S.; Standau, T.; Weinmann, S.; Altstädt, V.; Bonten, C. Effect of chemical modification on the thermal and rheological properties of polylactide. *Polym. Eng. Sci.* **2017**, *57*, 1242–1251. [CrossRef]

6. Standau, T.; Goettermann, S.; Weinmann, S.; Bonten, C.; Altstädt, V.; Di Maio, E. Autoclave foaming of chemically modified polylactide. *J. Cell. Plast.* **2017**, *53*, 481–489. [CrossRef]

7. Standau, T.; Murillo Castellón, S.; Delavoie, A.; Bonten, C.; Altstädt, V. Effects of Chemical Modifications on the Rheological and the Expansion Behavior of Polylactide (PLA) in Foam Extrusion. *e-Polymers* **2019**, *19*, 297–304. [CrossRef]

8. Yu, L.; Toikka, G.; Dean, K.; Bateman, S.; Yuan, Q.; Filippou, C.; Nguyen, T. Foaming behaviour and cell structure of poly(lactic acid) after various modifications. *Polym. Int.* **2013**, *62*, 759–765. [CrossRef]

9. Dean, K.M.; Petinakis, E.; Meure, S.; Yu, L.; Chryss, A. Melt Strength and Rheological Properties of Biodegradable Poly(Lactic Aacid) Modified via Alkyl Radical-Based Reactive Extrusion Processes. *J. Polym. Environ.* **2012**, *20*, 741–747. [CrossRef]

10. Liu, W.; Wang, X.; Li, H.; Du, Z.; Zhang, C. Study on rheological and extrusion foaming behaviors of chain-extended poly (lactic acid)/clay nanocomposites. *J. Cell. Plast.* **2013**, *49*, 535–554. [CrossRef]

11. Standau, T.; Hädelt, B.; Schreier, P.; Altstädt, V. Development of a Bead Foam from an Engineering Polymer with Addition of Chain Extender: Expanded Polybutylene Terephthalate. *Ind. Eng. Chem. Res.* **2018**, *57*, 17170–17176. [CrossRef]

12. Mihai, M.; Huneault, M.A.; Favis, B.D. Rheology and extrusion foaming of chain-branched poly(lactic acid). *Polym. Eng. Sci.* **2010**, *50*, 629–642. [CrossRef]

13. Dorgan, J.R.; Williams, J.S.; Lewis, D.N. Melt rheology of poly(lactic acid): Entanglement and chain architecture effects. *J. Rheol.* **1999**, *43*, 1141–1155. [CrossRef]

14. Locati, G.; Pegoraro, M.; Iti, D.N.; Milan, V.M.B. A Model for the Zero Shear Viscosity. *Polym. Eng. Sci.* **1999**, *39*, 741–748. [CrossRef]

15. Ding, W.; Kuboki, T.; Wong, A.; Park, C.B.; Sain, M. Rheology, thermal properties, and foaming behavior of high d-content polylactic acid/cellulose nanofiber composites. *RSC Adv.* **2015**, *5*, 91544–91557. [CrossRef]

16. Bagheriasl, D.; Carreau, P.J.; Riedl, B.; Dubois, C.; Hamad, W.Y. Shear rheology of polylactide (PLA)–cellulose nanocrystal (CNC) nanocomposites. *Cellulose* **2016**, *23*, 1885–1897. [CrossRef]

17. Najafi, N.; Heuzey, M.C.; Carreau, P.J.; Therriault, D.; Park, C.B. Rheological and foaming behavior of linear and branched polylactides. *Rheol. Acta* **2014**, *53*, 779–790. [CrossRef]

18. Nofar, M. Rheological, thermal, and foaming behaviors of different polylactide grades. *Int. J. Mater. Sci. Res.* **2018**, *1*, 16–22. [CrossRef]

19. Saeidlou, S.; Huneault, M.A.; Li, H.; Park, C.B. Poly(lactic acid) crystallization. *Prog. Polym. Sci.* **2012**, *37*, 1657–1677. [CrossRef]

20. Mihai, M.; Huneault, M.A.; Favis, B.D. Crystallinity development in cellular poly(lactic acid) in the presence of supercritical carbon dioxide. *J. Appl. Polym. Sci.* **2009**, *113*, 2920–2932. [CrossRef]

21. Gendron, R.; Mihai, M. *Polymeric Foams*; Lee, S.-T., Ed.; Taylor & Francis Group, 6000 Broken Sound Parkway NW, Suite 300; CRC Press: Boca Raton, FL, USA, 2016; ISBN 978-1-4987-3887-3.

22. Lim, L.-T.; Auras, R.; Rubino, M. Processing technologies for poly(lactic acid). *Prog. Polym. Sci.* **2008**, *33*, 820–852. [CrossRef]

23. Wong, A.; Guo, Y.; Parka, C.B. Fundamental mechanisms of cell nucleation in polypropylene foaming with supercritical carbon dioxide—Effects of extensional stresses and crystals. *J. Supercrit. Fluids* **2013**, *79*, 142–151. [CrossRef]

24. Tammaro, D.; Iannace, S.; Di Maio, E. Insight into bubble nucleation at high-pressure drop rate. *J. Cell. Plast.* **2017**, *53*, 551–560. [CrossRef]

25. Antunes, M.; Realinho, V.; Velasco, J.I. Study of the influence of the pressure drop rate on the foaming behavior and dynamic-mechanical properties of CO_2 dissolution microcellular polypropylene foams. *J. Cell. Plast.* **2010**, *46*, 551–571. [CrossRef]

26. Liu, J.; Lou, L.; Yu, W.; Liao, R.; Li, R.; Zhou, C. Long chain branching polylactide: Structures and properties. *Polymer* **2010**, *51*, 5186–5197. [CrossRef]

27. Xu, Z.M.; Jiang, X.L.; Liu, T.; Hu, G.H.; Zhao, L.; Zhu, Z.N.; Yuan, W.K. Foaming of polypropylene with supercritical carbon dioxide. *J. Supercrit. Fluids* **2007**, *41*, 299–310. [CrossRef]

28. Bonten, C. *Plastics Technology*; Carl Hanser Verlag GmbH & Co. KG: München, Germany, 2019; ISBN 978-1-56990-767-2.

29. Fleissner, M. Characterization of polymer molecular mass distribution from rheological measurements. *Makromol. Chemie. Macromol. Symp.* **1992**, *61*, 324–341. [CrossRef]

30. Auhl, D.; Stange, J.; Münstedt, H.; Krause, B.; Voigt, D.; Lederer, A.; Lappan, U.; Lunkwitz, K. Long-chain branched polypropylenes by electron beam irradiation and their rheological properties. *Macromolecules* **2004**, *37*, 9465–9472. [CrossRef]

31. Stange, J.; Münstedt, H. Rheological properties and foaming behavior of polypropylenes with different molecular structures. *J. Rheol.* **2006**, *50*, 907–923. [CrossRef]

32. Siebert-Raths, A. Modifizierung von Polylactid (PLA) für Technische Anwendungen Verfahrenstechnische Optimierung der Verarbeitungs- und Gebrauchseigenschaften. Dissertation, Universität Rostock, Rostock, Germany, 2012.

33. Nofar, M.; Zhu, W.; Park, C.B. Effect of dissolved CO_2 on the crystallization behavior of linear and branched PLA. *Polymer* **2012**, *53*, 3341–3353. [CrossRef]

34. Zhai, W.; Ko, Y.; Zhu, W.; Wong, A.; Park, C.B. A study of the crystallization, melting, and foaming behaviors of polylactic acid in compressed CO_2. *Int. J. Mol. Sci.* **2009**, *10*, 5381–5397. [CrossRef] [PubMed]

35. Zhang, X.; Ding, W.; Zhao, N.; Chen, J.; Park, C.B. Effects of Compressed CO_2 and Cotton Fibers on the Crystallization and Foaming Behaviors of Polylactide. *Ind. Eng. Chem. Res.* **2018**, *57*, 2094–2104. [CrossRef]

36. Zafar, M.T.; Kumar, S.; Singla, R.K.; Maiti, S.N.; Ghosh, A.K. Surface Treated Jute Fiber Induced Foam Microstructure Development in Poly(lactic acid)/Jute Fiber Biocomposites and their Biodegradation Behavior. *Fibers Polym.* **2018**, *19*, 648–659. [CrossRef]

37. Bigg, D.M. Polylactide copolymers: Effect of copolymer ratio and end capping on their properties. *Adv. Polym. Technol.* **2005**, *24*, 69–82. [CrossRef]

38. Nofar, M.; Ameli, A.; Park, C.B. Development of polylactide bead foams with double crystal melting peaks. *Polymer* **2015**, *69*, 83–94. [CrossRef]

39. Reignier, J.; Gendron, R.; Champagne, M.F. Extrusion foaming of poly(lactic acid) blown with CO_2: Toward 100% green material. *Cell. Polym.* **2007**, *26*, 83–115. [CrossRef]

40. Li, R.; Zeng, D.; Liu, Q.; Jiang, Z.; Fang, T. Glass Transition Temperature in Microcellular Foaming Process with Supercritical Carbon Dioxide: A Review. *Polym. Plast. Technol. Eng.* **2015**, *54*, 119–127. [CrossRef]

41. Larsen, A.; Neldin, C. Physical Extruder Foaming of Poly (lactic acid)—Processing and Foam Properties. *Polym. Eng. Sci.* **2013**, *53*, 941–949. [CrossRef]

42. Mihai, M.; Huneault, M.A.; Favis, B.D.; Li, H. Extrusion foaming of semi-crystalline PLA and PLA/thermoplastic starch blends. *Macromol. Biosci.* **2007**, *7*, 907–920. [CrossRef]

43. Guo, Q.; Wang, J.; Park, C.B.; Ohshima, M. A microcellular foaming simulation system with a high pressure-drop rate. *Ind. Eng. Chem. Res.* **2006**, *45*, 6153–6161. [CrossRef]

44. Xu, L.; Huang, H. Foaming of Poly(lactic acid) Using Supercritical Carbon Dioxide as Foaming Agent: In fl uence of Crystallinity and Spherulite Size on Cell Structure and Expansion Ratio. *Ind. Eng. Chem. Res.* **2014**, *53*, 2277–2286. [CrossRef]

45. Mascia, L.; Re, G.D.; Ponti, P.P.; Bologna, S.; Giacomo, G.D.; Haworth, B. Crystallization effects on autoclave foaming of polycarbonate using supercritical carbon dioxide. *Adv. Polym. Technol.* **2006**, *25*, 225–235. [CrossRef]

46. Lee, J.W.S.; Park, C.B. Use of nitrogen as a blowing agent for the production of fine-celled high-density polyethylene foams. *Macromol. Mater. Eng.* **2006**, *291*, 1233–1244. [CrossRef]

 *materials*

Article

Enhancement of the Antibacterial Activity of Natural Rubber Latex Foam by Blending It with Chitin

Nanxi Zhang and Hui Cao *

Beijing Key Laboratory of Bioprocess, College of Life Science and Technology, Beijing University of Chemical Technology, Beijing 100029, China; nancyzhang512@outlook.com
* Correspondence: caohui@mail.buct.edu.cn; Tel.: +86-1369-115-4072

Received: 21 January 2020; Accepted: 17 February 2020; Published: 26 February 2020

Abstract: To enhance the antibacterial activity of natural rubber latex foam (NRLF), chitin was added during the foaming process in amounts of 1–5 phr (per hundred rubber) to prepare an environmentally friendly antibacterial NRLF composite. In this research, NRLF was synthesized by the Dunlop method. The swelling, density, hardness, tensile strength, elongation at break, compressive strength and antibacterial activity of the NRLFs were characterized. FTIR and microscopy were used to evaluate the chemical composition and microstructure of the NRLFs. The mechanical properties and antibacterial activity of the NRLF composites were tested and compared with those of pure NRLF. The antibacterial activity was observed by the inhibition zone against *E. coli*. NRLF composite samples were embedded in a medium before solidification. The experimental results of the inhibition zone indicated that with increasing chitin content, the antibacterial activity of the NRLF composites increased. When the chitin content reached 5 phr, the NRLF composite formed a large and clear inhibition zone in the culture dish. Moreover, the NRLF–5 phr chitin composite improved the antibacterial activity to 281.3% of that of pure NRLF against *E. coli*.

Keywords: natural rubber latex foam; chitin; antibacterial activity

---

1. Introduction

Natural rubber latex (NRL), the viscous liquid from *Hevea brasilensis*, is the main source of commercial natural rubber. Fresh NRL comprises an aqueous colloid with 44–70 wt% water. Generally, latex is processed into high ammonia natural rubber latex (HA-NRL) for convenience of preservation and transportation [1]. The main component of natural rubber is poly(cis-1,4-isoprene). In addition to water and rubber hydrocarbons, natural rubber latex also includes proteins, lipids and other substances [2,3].

Natural rubber latex products can be divided into four categories: impregnated products, molded products, extruded products and foam products. Compared with the former three, natural rubber latex foam (NRLF) is a porous material with a low density [4,5]. NRLF has the characteristics of elastic resilience, absorbency, providing sound insulation, being shockproof and providing ventilation [6–12]. At present, the commercial products made from NRLF mainly include mattresses and pillows.

NRLF, widely used as bedding, is one of the economic mainstays of Southeast Asian countries [13]. However, NRLF is easily attacked by ultraviolet light due to its unsaturated carbon–carbon bonds [14]. NRLF bedding is not suitable for long-term exposure to sunlight, and slow bacterial growth might occur on the cover [15–17]. Therefore, it is necessary to improve the antibacterial activity of NRLF and to choose appropriate antibacterial agents to reduce the cost.

Chitin [($C_8H_{13}O_5N$)n], which widely exists in the shells of crustaceans such as shrimps and crabs, is a renewable, low-cost, antibacterial and well-sourced resource [18–22]. The chemical structure of chitin is shown in Figure 1. Chitin is a cationic natural polymer due to its amide groups. The positively charged polymer neutralizes the negatively charged functional groups on the surface of bacteria. Once the cell wall is damaged, the cell osmotic pressure would be destroyed and finally the bacteria would die [23,24]. Therefore, chitin could be considered an environmentally friendly antibacterial agent [25–29].

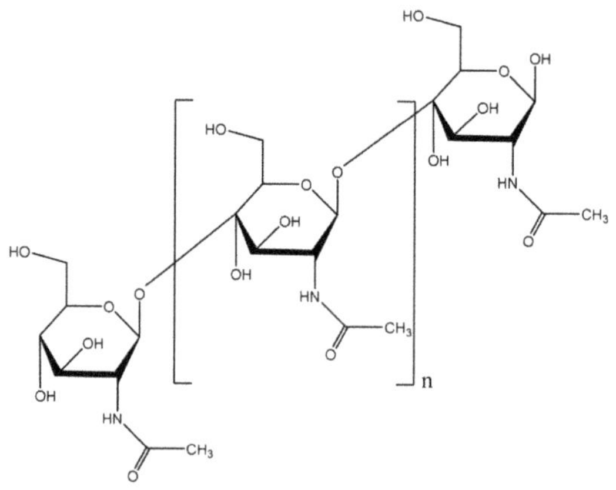

Figure 1. Chemical structural formula of chitin.

Most of the research on NRLF composites involves filler loading [30–33]. The focus is on exploring the mechanical properties of composite foams and emphasizing the reuse of natural fibers. Currently, the research on antibacterial NRLF mainly focuses on zinc oxide [34] and silver particles [35–39]. Khemara Mama [40] demonstrated that NRLF treated with silver nanoparticles of only 0.2 per hundred rubber (phr) had improved antibacterial ability by 43.8% against *E. coli* and 25% against *S. aureus* compared to NRLF that was not treated with silver nanoparticles. Few reports have been published on the utilization of natural renewable materials to prepare antibacterial NRLF composites.

Compared with zinc oxide and silver particles, chitin has the advantages of renewability, biodegradability and biocompatibility. Furthermore, the preparation of silver antimicrobials is quite complex due to the synthesis of nanoparticles [39] while blending with chitin is more simple and convenient. The poor solubility of chitin provides the NRLF composite with more durability within its service life. Therefore, chitin–NRLF composites can not only enhance the antibacterial activity of foam, but are also inexpensive and environmentally friendly [19].

In this research, 0, 1, 2, 3, 4 and 5 phr chitin-loaded NRLFs were prepared using the Dunlop method. Morphology, swelling, density, chemical composition, hardness, tensile strength, elongation at break and compressive strength are characterized to verify the antibacterial activity vs. mechanical properties of chitin–NRLF composites. This natural antibacterial composite foam has considerable prospects in commercial applications of natural rubber, which is consistent with the concept of sustainability and the goal of a green economy.

2. Materials and Methods

2.1. Materials

The Dunlop method was used to prepare the NRLFs. High ammonia natural rubber latex (HA-NRL) was supplied by Yunnan Natural Rubber Industry Group Jingyang Co., Ltd (Yunnan, China). The chemicals (potassium hydroxide, 2-mercaptobenzimidazole, 2-mercaptobenzothiazole, sulfur, potassium oleate, ammonium sulfate, zinc oxide, sodium fluorosilicone), Luria-Bertani solid medium (deionized water, tryptone, sodium chloride, yeast extract, agar) and chitin were supplied by Sinopharm Chemical Reagent Co., Ltd. (China). *E. coli* BL21 were supplied by Qincheng Biotechnology

Co., Ltd. (Shanghai, China). The names, concentrations and proportions of the chemicals used in the synthesis experiment are shown in Table 1.

Table 1. Formulation parameters for the synthesis of the chitin–natural rubber latex foam (NRLF) composites.

Ingredient	TSC	phr
natural rubber latex	0.6	100
potassium hydroxide	1	0.5
2-mercaptobenzimidazole	0.5	1.5
2-mercaptobenzothiazole	0.5	2
sulfur	0.5	2.5
potassium oleate	0.33	1
ammonium sulfate	0.4	2.5
zinc oxide	0.5	3
sodium fluorosilicone	0.5	1
chitin	1	0, 1, 2, 3, 4, 5

TSC: total solid content; phr: per hundred rubber.

2.2. Sample Preparation

The Dunlop method [41,42] involves prevulcanization, foaming, gelation, vulcanization and demolding. First, potassium hydroxide, 2-mercaptobenzimidazole, 2-mercaptobenzothiazole and sulfur were added into the HA-NRL. Prevulcanization of the NRL was carried out by stirring at 30 °C for 48 h at 50 rpm. Second, potassium oleate, ammonium sulfate and chitin were blended with the prevulcanized latex (PV-NRL). The PV-NRL was stirred for 8 min at 200 rpm to form a foam, and this process continued until the volume remained the same. Third, zinc oxide was added while the whisking machine was adjusted to 100 rpm. After 10 min, the sodium silicofluonate was slowly added, and the foam was stirred to the gelling point at 70 rpm. Once the gelling point was reached, the foam was injected into the mold and solidified at 90 °C for 30 min. Finally, the NRLF was vulcanized at 100 °C for 2 h. After demolding, washing and drying, the NRLF was obtained. The NRLFs loaded with chitin at 0, 1, 2, 3, 4 and 5 phr were prepared using the above procedure. Five samples were prepared for each condition.

2.3. Methods

2.3.1. Morphology

The microstructure and morphology of the NRLFs were observed with optical microscopy (Leica DM500, Heidelberg, German). The magnification (objective × eyepiece) was 40 ×. The samples were sliced for light transmission and observation. From the resulting graphs, the size of the micropores and the dispersion of the chitin particles were evaluated.

2.3.2. Swelling

The NRLFs were made into 25 mm × 25 mm × 25 mm samples. The dried samples were weighed and then hermetically immersed in deionized water for 72 h. The surfaces of the samples were wiped with kitchen paper before the swollen samples were weighed. The testing temperature was 25 °C. The swelling is indicated as S (%).

$$S = (\text{mass of swollen sample} - \text{mass of initial sample})/\text{mass of initial sample} \tag{1}$$

2.3.3. Density

The dry NRLFs were made into 50 mm × 50 mm × 25 mm (length × width × height) samples for measurement. The testing temperature was 25 °C. Five samples were tested under each loading condition, and the density is indicated as ρ (kg/m^3).

2.3.4. Chemical Composition

The chemical composition was characterized by FTIR (Nicolet iS50, Madison, WI, America). The transmittance of the samples to infrared light was measured. The chemical composition of the NRLFs with chitin loading (0, 1, 2, 3, 4 and 5 phr) was investigated in a mid-infrared region (500–4000 cm^{-1}).

2.3.5. Hardness

The hardness was tested according to the Shore C standard. A Shore C hardness tester (LX-C, Jiangdu, China) was used to measure the hardness of the soft foam materials. The dry NRLFs were made into 10 mm × 10 mm × 6 mm (length × width × height) samples for measurement. The testing temperature was 25 °C. Five samples were tested under each loading condition.

2.3.6. Tensile Strength and Elongation at Break

Tensile strength and elongation at break were tested according to the Chinese National Standard GB/T 6344-2008. The NRLFs were made into 13 mm × 152 mm dumbbell-shaped samples with a gauge length of 50 mm. The speed of the universal material testing machine (XWW-20A, Beijing, China) used herein was 500 mm/min. The testing temperature was 25 °C. Five samples were tested under each loading condition.

2.3.7. Compressive Strength

The NRLFs were made into 50 mm × 50 mm × 25 mm (length × width × height) samples. The compression ratio was 50%. The speed of the universal material testing machine (XWW-20A, Beijing, China) was 50 mm/min. The testing temperature was 25 °C. Five samples were tested under each loading condition.

2.3.8. Antibacterial Activity

The antibacterial activity was characterized by the inhibition zone against *E. coli*. The NRLFs were made into cylindrical samples with a diameter of 8 mm and a height of 2 mm. The LB solid medium was poured into the culture dishes at 60 °C, and the samples were embedded in the medium. After the culture medium was completely solidified, bacterial liquid was added. All the dishes were incubated at 37 °C for 12 h.

3. Results

3.1. Morphology

Figure 2 shows the micrographs of the NRLFs in different compositions. The microstructure, shape of the cells and dispersion of the chitin can be observed. Figure 2a–f shows the NRLFs loaded with 0–5 phr chitin. From Figure 2a–c, it is obvious that the shape of the cells changes little with a low chitin content. When the loading is 3 phr, the chitin begins to agglomerate on the walls. At the same time, the walls attached to the chitin would gain traction on the surrounding foam. Due to the influence of the chitin, the cells are no longer circular in shape as the loading increases.

The deformation and fracture of the cells were detected, as shown in Figure 2e. When the loading reaches 4 phr, the rubber walls fracture due to the over-expansion of the bubbles and the obstacles provided by the chitin. In Figure 2f, the number of cell cracks and the pore size increase. The broken foam wall adheres to the chitin. The pores are not stable enough, and chitin becomes aggregated, so the rubber begins to break under an uneven force. Macroscopically, small particles can be seen on the surface of the NRLF. The pores become large and weak with high filler contents.

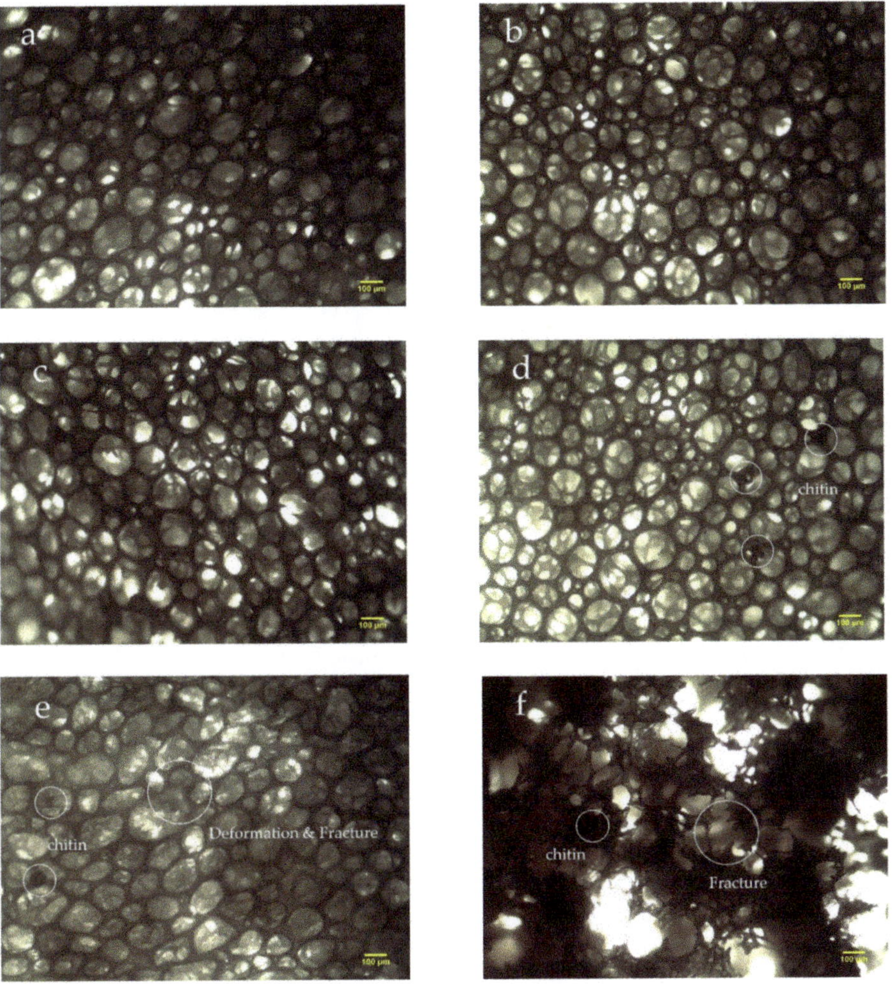

Figure 2. Micrographs of chitin–NRLF composites. (**a**) pure NRLF; (**b**) NRLF–1 phr chitin; (**c**) NRLF–2 phr chitin; (**d**) NRLF–3 phr chitin; (**e**) NRLF–4 phr chitin; (**f**) NRLF–5 phr chitin.

The cell diameters of the NRLFs are shown in Figure 3. The relationship between pore size and filler loading can be seen in the diagram. First, as the loading of chitin increases, the pore size of the NRLF increases gradually. The large surface tension of the foam around the chitin results in the bursting and merging of the cells. However, the cell diameter in the NRLF–5 phr chitin case decreases. Under this loading, many bubbles become large enough to burst, and the remaining cells are weak and easily deformed. Therefore, only small cells could be measured completely. Second, with the increase in chitin loading, the standard deviation of the cell diameters increases as well. This indicates that the random dispersion of the chitin in NRLF leads to an uneven pore size. Compared with the cells in the

NRLFs with a low chitin load, the difference among the cells in the NRLFs with a high chitin load is too large to be accurately estimated.

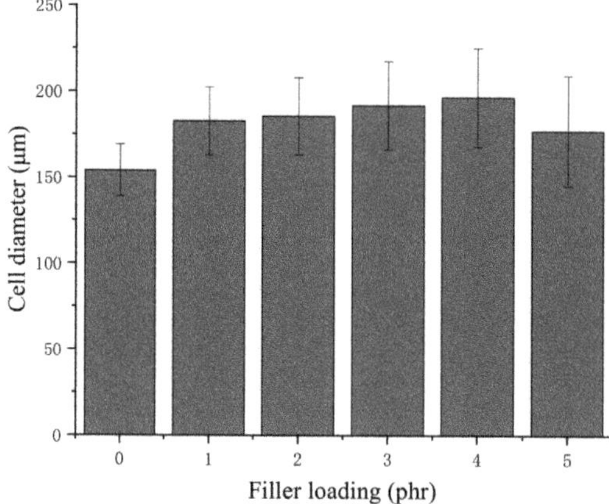

Figure 3. Cell diameters of chitin–NRLF composites.

3.2. Swelling

The swelling ability could directly explain the porous structure of the NRLFs according to Equation (1). The NRLF composites were immersed in deionized water for 72 h, and the mass of the NRLF composites was measured. As shown in Figure 4, with increasing chitin content, the swelling percentage decreases. The reason might be the collapse and adhesion of the rubber foam when the loading increases. For NRLFs with a higher loading, the number of pores decreases sharply even though the cells merge and expand. Therefore, the low porosity of the NRLF composites diminishes the absorption capacity.

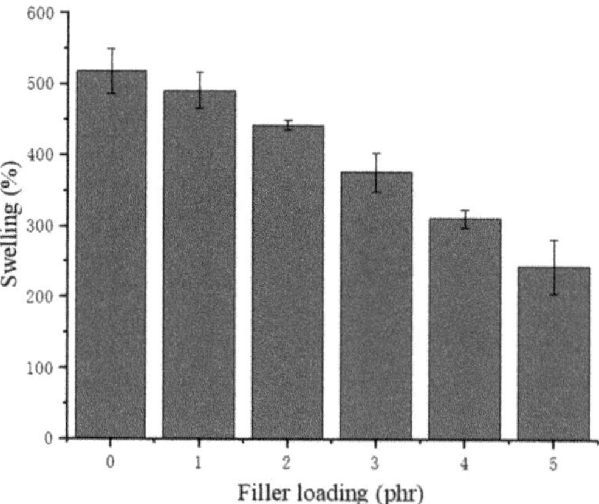

Figure 4. Swelling of chitin–NRLF.

3.3. Density

As shown in Figure 5, the density of the NRLF increases with increasing chitin loading. The density of chitin particles is 1370 kg/m^3, which is much higher than that of the pure NRLF. In addition, agglomerates destroy the original bubble structure in the NRLF and cause collapse of the cells.

A decrease in bubble volume leads to a decreased proportion of air, so the NRLFs with a high loading contain additional rubber and chitin. Although the density of the NRLF composite increases as the filler content increases, it is still lower than that of the other materials. This is attributed to the unique porous structure of the NRLF. The density of the NRLF composite is still less than 0.2 kg/m^3.

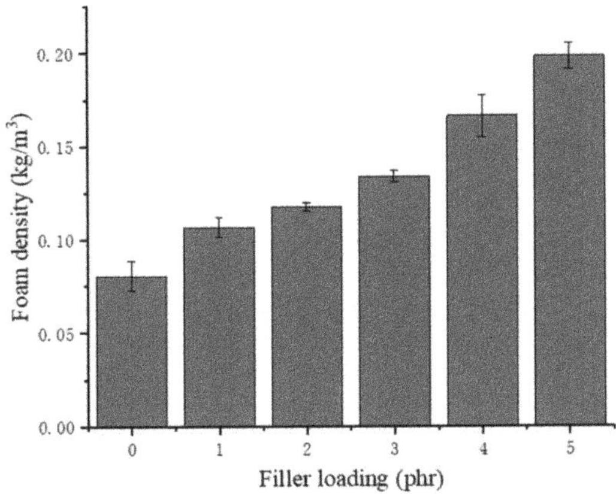

Figure 5. Density of chitin–NRLF composites.

3.4. Chemical Composition

The FTIR transmittance spectra of the NRLFs are shown in Figure 6. FTIR spectra are used to analyze the chemical compositions and to identify the structural transformation that occurs during blending. The peak at 2900 cm^{-1} is due to the stretching vibration of the C-H bonds in methyl, the peak at 1660 cm^{-1} is due to the stretching vibration of C = C bonds, the peak at 2850 cm^{-1} is due to the stretching vibration of C-H bonds in methylene, the peak at 1460 cm^{-1} is due to the bending vibration of C-H bonds in methylene, and the peak at 842 cm^{-1} is due to the out-of-plane bending of C-H bonds [43–45]. These peaks are the key features in the spectra for the rubber hydrocarbons. Peaks at 3280 cm^{-1} are probably connected to moisture in the samples.

Natural rubber latex originally contains proteins, carbohydrates, lipids and other substances, hence pure NRL possesses hydroxyl and amide peaks as well. From 1200 cm^{-1} to 1000 cm^{-1}, there are ether bond peaks [46,47]. This characteristic peak differentiates the chitin from the polyisoprene. As shown in the second local graph, the transmittance peaks of ether bonds are enhanced compared with pure NRLF. This demonstrates that chitin blends with the NRLF and does not significantly change the chemical structure of the natural rubber. It can be seen that with an increase in the chitin content, the natural rubber peaks in the transmittance spectra do not change. This indicates that there is no reaction between the chitin and natural rubber latex during blending and only physical loading occurs. In other words, the chitin has no significant impact on the spatial structure of the natural rubber.

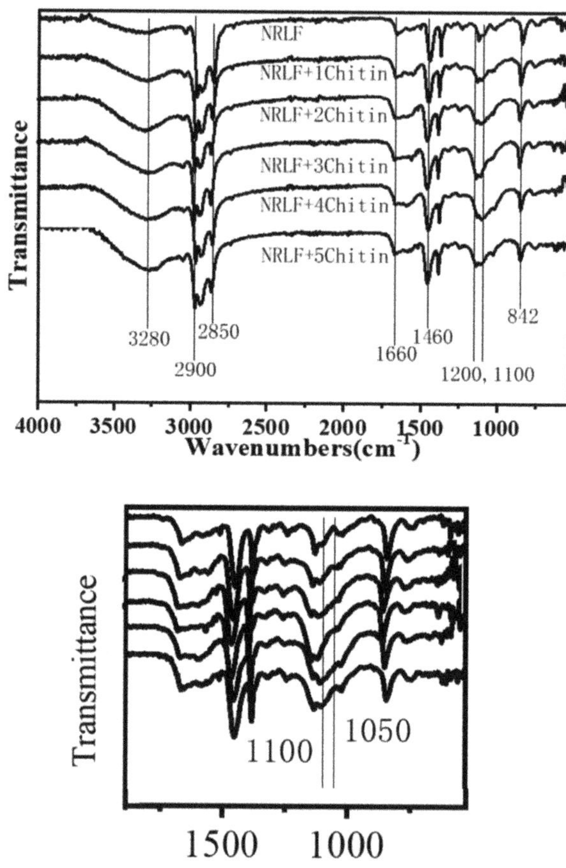

Figure 6. FTIR spectra of chitin–NRLF.

3.5. Hardness

The effects of the chitin on the hardness are presented in Figure 7. The hardness (Shore C) mainly represents the ability of rubber to resist the pressing or intrusion of hard objects. The NRLF samples with an increased chitin loading have an increased hardness, according to the trend in Figure 7. The increase in chitin content leads to the collapse and partial adhesion of the NR, so the rubber clots provide additional resistance to the external intrusion. Chitin is a linear polymer composed of (1–4) linked 2-acetamido-2-deoxy-D-glucosamine. The hardness of chitin is 7–7.5 (Mohs) due to its regularly arranged structure. The Mohs hardness is used to describe the hardness of minerals, which is absolutely higher than that of rubber. Therefore, chitin could obviously enhance the hardness of NRLF composites.

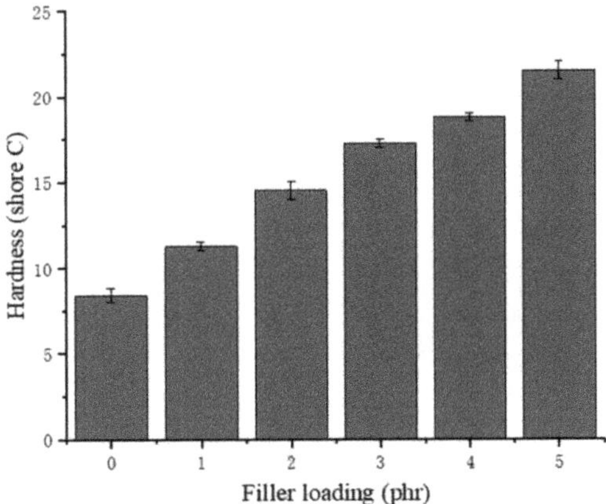

Figure 7. Hardness of chitin–NRLF composites.

3.6. Tensile Strength and Elongation at Break

As shown in Figure 8, the tensile strength of the chitin–NRLF composite decreases with increasing chitin content. For NRLF–5 phr chitin, the tensile strength decreases to approximately half that of the pure NRLF. This decline proves that the structure of the NRLF is destroyed during the foaming process and that additional cells appear to burst before solidification. The low compatibility between the chitin and NR leads to a large surface tension on the foam, resulting in an easy bursting of the bubbles. With increasing pore size, the cells gradually become more fragile. The increase in the filler loading could cause poor dispersion and agglomeration of the chitin. The precipitation of the chitin on the surface would form stress concentrations during tensile testing.

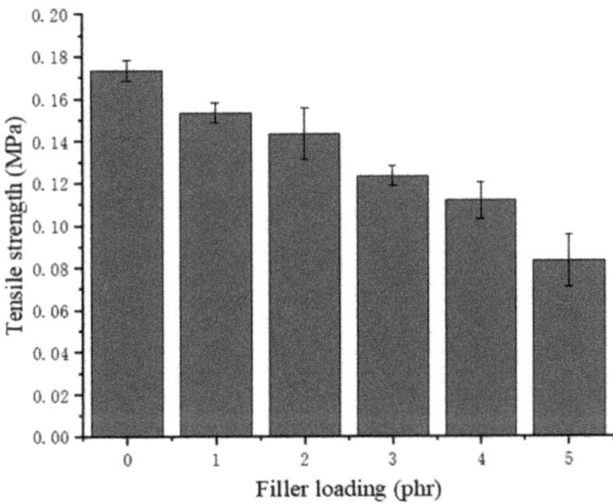

Figure 8. Tensile strength of chitin–NRLF.

The elongation at break of NRLFs normally depends on the crosslinking density and the foam structure. Given that the NRLF composites herein were vulcanized under the same conditions [48–50], the difference in elongation at break would mostly be attributed to the size and quantity of the pores. As shown in Figure 9, the elongation at break of the NRLF decreases with increasing chitin loading. Similar to the trend for the tensile strength, the elongation at break of NRLF decreases remarkably when the chitin reaches 3 phr or more. The trend of these results is analogous to other literature [30].

Stress concentrations would suddenly develop at the agglomerations of the chitin, and the thinned foam walls could accelerate the breaking. Moreover, the increased hardness of the NRLF composite might contribute to the brittleness that appears during the tensile testing, which means a decreased amount of deformation before a break occurs.

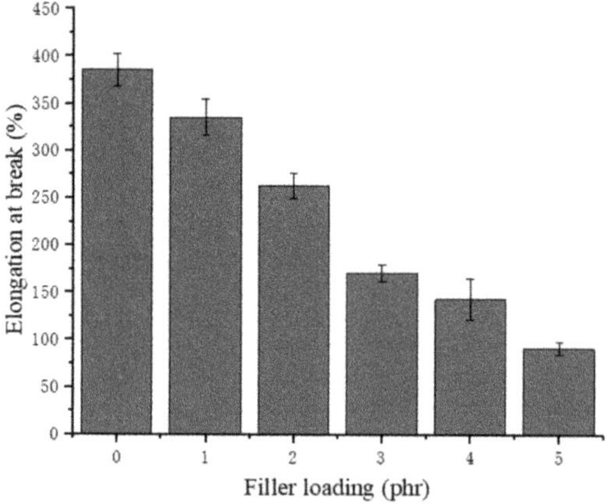

Figure 9. Elongation at break of chitin–NRLF.

3.7. Compression Strength

As shown in Figure 10, the compressive strength of the NRLF increases substantially with increasing chitin loading. Chitin has a much higher compression strength than natural rubber, and a small amount of chitin could dramatically enhance the compressive strength of NRLF composites. In addition, when the chitin content continues to increase, the fracture of the cell might lead to adhesion of the natural rubber. For the same volume, the NRLFs with a high loading allocate an increased proportion to rubber but a decreased proportion to air. During compression, NRLFs with a high loading would have a decreased amount of inner space available to shrink. Therefore, when the samples are compressed at 50% during the compression testing, the additional rubber and chitin particles in the NRLFs with a high loading provide an increased resistance to the pressure and increased compressive strength as well.

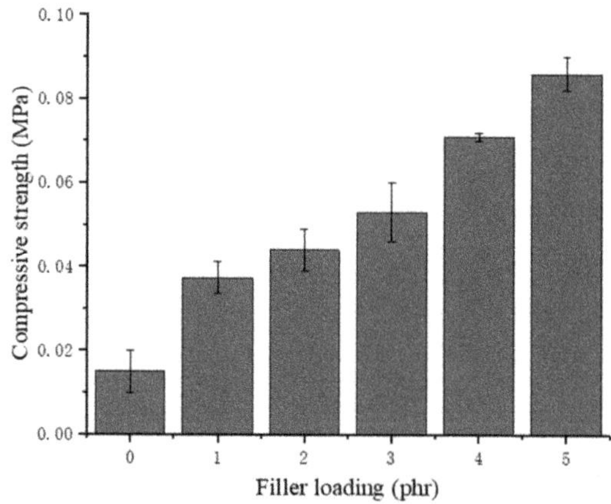

Figure 10. Compression strength of chitin–NRLF.

3.8. Antibacterial Activity

As shown in Figure 11, the antibacterial activity of the NRLFs on *E. coli* was characterized by inhibition zone testing against *E. coli*. Figure 11a–f represents the NRLF composites filled with chitin from 0 to 5 phr.

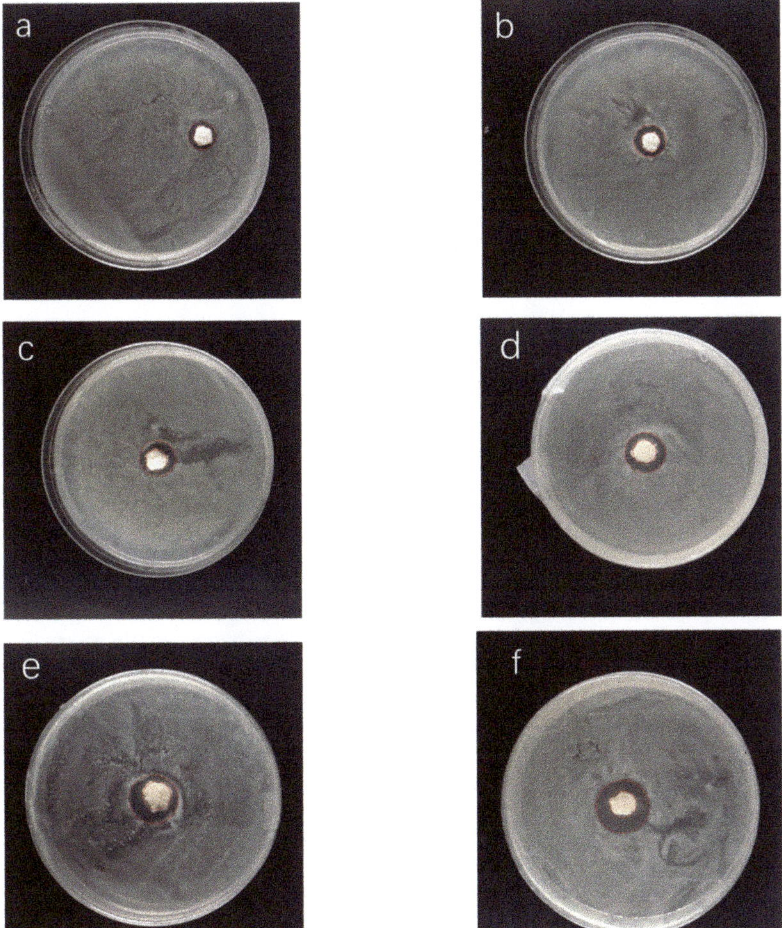

Figure 11. Inhibition zone against *E. coli* of chitin–NRLF composites. (**a**) pure NRLF; (**b**) NRLF–1 phr chitin; (**c**) NRLF–2 phr chitin; (**d**) NRLF–3 phr chitin; (**e**) NRLF–4 phr chitin; (**f**) NRLF–5 phr chitin.

From Figure 11a, pure NRLF without chitin already has antibacterial activity against *E. coli*. This result is consistent with the previous study [40]. There are two possible explanations for this phenomenon. First, the proteins in natural rubber latex might be bacteriostatic [51]. Second, 3 phr of zinc oxide was added, as listed in Table 1, and zinc oxide is a proven antibacterial [52,53].

The increase in the inhibition zone of each of the NRLF composites can be seen in Figure 12. When the chitin loading increases, the antibacterial activity of the NRLF gradually increases. The samples were tightly embedded in the solid medium so they contacted the *E. coli* completely. The antibacterial activity of NRLF–3 phr chitin appears to be twice that of pure NRLF. When loaded with 5 phr, the inhibition zone is obviously enlarged and is quite clear compared with that of the former samples. Furthermore, the NRLF–5 phr chitin composite improved the antibacterial activity by 181.3% against *E. coli* compared to that of the pure NRLF.

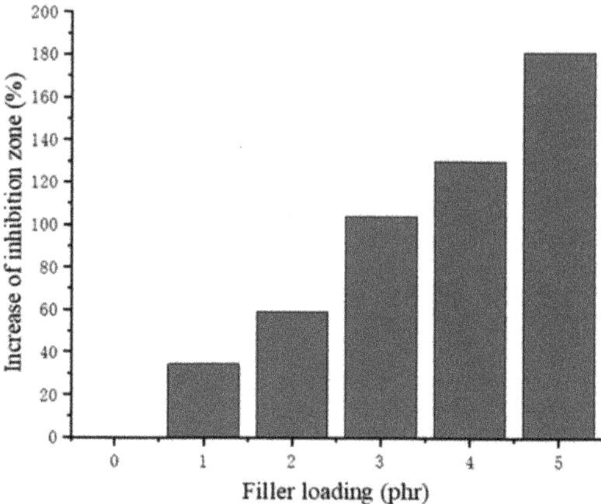

Figure 12. Increase in the inhibition zone of chitin–NRLF.

4. Conclusions

In this research, chitin was chosen as an antibacterial agent to enhance the antibacterial activity of NRLF. NRLF composites were prepared by the Dunlop method, and chitin was blended with foam during the foaming process. In addition to the antibacterial activity, morphology, swelling, density, chemical composition, hardness, tensile strength, elongation at break and compression strength were characterized as well.

Compared with that of pure NRLF, the performance of the NRLF composite is related to the loading of chitin. When the chitin content increases, the cells expand and deform with the chitin. Overexpansion could lead to bursting of the bubbles and fracture of the rubber walls. The tensile strength, elongation at break and water swelling decrease gradually, resulting in uneven force and stress concentrations.

The chitin is inclined to aggregate due to the poor interaction between the chitin and natural rubber latex. The cracked cells and broken walls stick to the central chitin agglomerations. A decreased number of bubbles contribute to the collapse and contraction of the foam materials. With a decreased proportion of air and additional space for chitin, when the chitin loading increases, the compressive strength, density, hardness and antibacterial activity increase.

Using the natural antibacterial agent chitin as a loading filler, environmentally friendly and antibacterial NRLFs were prepared herein. The antibacterial activity of NRLF composites could be increased by up to 181.3% compared to pure NRLF. When loaded with 3 phr chitin, the microstructure and mechanical properties of the chitin–NRLF composite change moderately and the antibacterial activity reaches more than double that of pure NRLF. Therefore, chitin–NRLF composites could be widely applied to household products, such as pillows, mattresses or cushions, and possess good practical value for future applications.

Author Contributions: Conceptualization, N.Z.; Data curation, N.Z.; Formal analysis, N.Z.; Methodology, N.Z.; Supervision, H.C.; Writing—Original draft, N.Z.; Writing—Review and editing, H.C. All authors have read and agreed to the published version of the manuscript.

Funding: This research was funded by the National Natural Science Foundation of China (21865026); Xinjiang Production and Construction Corps Project (2018BC005); Hebei key research and development project (18273401D); Hebei Province's Innovation Capacity Improvement Project (18952814D).

Conflicts of Interest: The authors declare no conflict of interest.

References

1. Singh, M. The Colloidal Properties of Commercial Natural Rubber Latex Concentrates. *J. Rubber Res.* **2018**, *21*, 119–134. [CrossRef]
2. Sakdapipanich, J. Current Study on Structural Characterization and Unique Film Formation of Hevea brasiliensis Natural Rubber Latex. *Adv. Mater. Res.* **2014**, *844*, 498–501. [CrossRef]
3. Nawamawat, K.; Sakdapipanich, J.T.; Ho, C.C.; Ma, Y.; Song, J.; Vancso, J.G. Surface nanostructure of Hevea brasiliensis natural rubber latex particles. *Colloids Surf. A Physicochem. Eng. Asp.* **2011**, *390*, 157–166. [CrossRef]
4. Suksup, R.; Sun, Y.; Sukatta, U.; Smitthipong, W. Foam rubber from centrifuged and creamed latex. *J. Polym. Eng.* **2019**, *39*, 336. [CrossRef]
5. Oliveira-Salmazo, L.; Lopez-Gil, A.; Silva-Bellucci, F.; Job, A.E.; Rodriguez-Perez, M. A Natural rubber foams with anisotropic cellular structures: Mechanical properties and modeling. *Ind. Ind. Crop. Prod.* **2016**, *80*, 26–35. [CrossRef]
6. Sandhu, I.; Kala, M.; Thangadurai, M.; Singh, M.; Alegaonkar, P.; Saroha, D.R. Experimental Study of Blast Wave Mitigation in Open Cell Foams. *Mater. Today Proc.* **2018**, *5*, 28170–28179. [CrossRef]
7. Ratcha, A.; Samart, C.; Yoosuk, B.; Sawada, H.; Reubroycharoen, P.; Kongparakul, S. Polyisoprene modified poly(alkyl acrylate) foam as oil sorbent material. *J. Appl. Polym. Sci.* **2015**, *132*. [CrossRef]
8. Najib, N.N.; Ariff, Z.M.; Bakar, A.A.; Sipaut, C.S. Correlation between the acoustic and dynamic mechanical properties of natural rubber foam: Effect of foaming temperature. *Mater. Des.* **2011**, *32*, 505–511. [CrossRef]
9. Riyajan, S.-A.; Keawittarit, P. A novel natural rubber-graft-cassava starch foam for oil/gasohol absorption. *Polym. Int.* **2016**, *65*, 491–502. [CrossRef]
10. Ratcha, A.; Yoosuk, B.; Kongparakul, S. Grafted Methyl Methacrylate and Butyl Methacrylate onto Natural Rubber Foam for Oil Sorbent. *Adv. Mater. Res.* **2014**, *844*, 385–390. [CrossRef]
11. Huang, D.; Zhang, M.; Guo, C.; Shi, L.; Lin, P. Experimental investigations on the effects of bottom ventilation on the fire behavior of natural rubber latex foam. *Appl. Therm. Eng.* **2018**, *133*, 201–210. [CrossRef]
12. Ratcha, A.; Samart, C.; Yoosuk, B.; Kongparakul, S. Reusable Modified Natural Rubber Foam for Petroleum-Based Liquid Removal. *Macromol. Symp.* **2015**, *354*, 177–183. [CrossRef]
13. Fong, Y.; Khin, A.; Seong, L. Conceptual Review and the Production, C.onsumption and Price Models of the Natural Rubber Industry in Selected ASEAN Countries and World Market. *Econ. Model.* **2018**, *6*, 403–418.
14. Gu, H.S.; Itoh, Y. Aging Behaviors of Natural Rubber in Isolation Bearings. *Adv. Mater. Res.* **2011**, *163–167*, 3343–3347. [CrossRef]
15. Bode, H.B.; Kerkhoff, K.; Jendrossek, D. Bacterial Degradation of Natural and Synthetic Rubber. *Biomacromolecules* **2001**, *2*, 295–303. [CrossRef]
16. Ali Shah, A.; Hasan, F.; Shah, Z.; Kanwal, N.; Zeb, S. Biodegradation of natural and synthetic rubbers: A review. *Int. Biodeterior. Biodegrad.* **2013**, *83*, 145–157. [CrossRef]
17. Rose, K.; Steinbüchel, A. Biodegradation of Natural Rubber and Related Compounds: Recent Insights into a Hardly Understood Catabolic Capability of Microorganisms. *Appl. Environ. Microbiol.* **2005**, *71*, 2803. [CrossRef]
18. Benhabiles, M.S.; Salah, R.; Lounici, H.; Drouiche, N.; Goosen, M.F.A.; Mameri, N. Antibacterial activity of chitin, chitosan and its oligomers prepared from shrimp shell waste. *Food Hydrocoll.* **2012**, *29*, 48–56. [CrossRef]
19. Khattak, S.; Wahid, F.; Liu, L.-P.; Jia, S.-R.; Chu, L.-Q.; Xie, Y.-Y.; Li, Z.-X.; Zhong, C. Applications of cellulose and chitin/chitosan derivatives and composites as antibacterial materials: Current state and perspectives. *Appl. Microbiol. Biotechnol.* **2019**, *103*, 1989–2006. [CrossRef]
20. Khoushab, F.; Yamabhai, M. Chitin Research Revisited. *Mar. Drugs* **2010**, *8*, 1988–2012. [CrossRef]
21. Muzzarelli, R.A. *Chitin*; Elsevier: Amsterdam, The Netherlands, 2013.
22. Roberts, G.A. *Chitin Chemistry*; Macmillan International Higher Education: Nottingham, UK, 1992.
23. Hale, J.D.; Hancock, R.E. Alternative mechanisms of action of cationic antimicrobial peptides on bacteria. *Expert Rev. Anti Infect. Ther.* **2007**, *5*, 951–959. [CrossRef] [PubMed]
24. Hoque, J.; Akkapeddi, P.; Yarlagadda, V.; Uppu, D.S.; Kumar, P.; Haldar, J. Cleavable cationic antibacterial amphiphiles: Synthesis, mechanism of action, and cytotoxicities. *Langmuir* **2012**, *28*, 12225–12234. [CrossRef] [PubMed]

25. Qin, Y.; Zhang, S.; Yu, J.; Yang, J.; Xiong, L.; Sun, Q. Effects of chitin nano-whiskers on the antibacterial and physicochemical properties of maize starch films. *Carbohydr. Polym.* **2016**, *147*, 372–378. [CrossRef] [PubMed]

26. Wei, J.; Liu, J.; Qiang, J.; Yang, L.; Wan, Y.; Wang, H.; Gao, W.; Ko, F. Antibacterial Performance of Chitin Nanowhisker Reinforced Poly(lactic acid) Composite Nanofiber Membrane. *Adv. Sci. Lett.* **2012**, *10*, 649–651. [CrossRef]

27. Tamura, H.; Furuike, T.; Nair, S.V.; Jayakumar, R. Biomedical applications of chitin hydrogel membranes and scaffolds. *Carbohydr. Polym.* **2011**, *84*, 820–824. [CrossRef]

28. Shankar, S.; Reddy, J.P.; Rhim, J.-W.; Kim, H.-Y. Preparation, characterization, and antimicrobial activity of chitin nanofibrils reinforced carrageenan nanocomposite films. *Carbohydr. Polym.* **2015**, *117*, 468–475. [CrossRef]

29. Bashir, A.S.M.; Manusamy, Y.; Chew, T.L.; Ismail, H.; Ramasamy, S. Mechanical, thermal, and morphological properties of (eggshell powder)-filled natural rubber latex foam. *J. Vinyl Addit. Technol.* **2017**, *23*, 3–12. [CrossRef]

30. Kudori, S.N.I.; Ismail, H.; Shuib, R.K. Kenaf Core and Bast Loading vs. Properties of Natural Rubber Latex Foam (NRLF). *BioResources* **2019**, *14*, 1765–1780.

31. Surya, I.; Kudori, S.N.I.; Ismail, H. Effect of Partial Replacement of Kenaf by Empty Fruit Bunch (EFB) on the Properties of Natural Rubber Latex Foam (NRLF). *BioResources* **2019**, *14*, 9375–9391.

32. Panploo, K.; Chalermsinsuwan, B.; Poompradub, S. Natural rubber latex foam with particulate fillers for carbon dioxide adsorption and regeneration. *RSC Adv.* **2019**, *9*, 28916–28923. [CrossRef]

33. Sun, Y.; Du, Z. A Flexible and Highly Sensitive Pressure Sensor Based on AgNWs/NRLF for Hand Motion Monitoring. *Nanomaterials* **2019**, *9*, 945. [CrossRef] [PubMed]

34. Rathnayake, W.G.I.U.; Ismail, H.; Baharin, A.; Bandara, I.M.C.C.D.; Rajapakse, S. Enhancement of the antibacterial activity of natural rubber latex foam by the incorporation of zinc oxide nanoparticles. *J. Appl. Polym. Sci.* **2014**, *131*. [CrossRef]

35. Rathnayake, I.U.; Ismail, H.; De Silva, C.R.; Darsanasiri, N.D.; Bose, I. Antibacterial effect of Ag-doped TiO2 nanoparticles incorporated natural rubber latex foam under visible light conditions. *Iran. Polym. J.* **2015**, *24*, 1057–1068. [CrossRef]

36. Rathnayake, W.G.I.U.; Ismail, H.; Baharin, A.; Darsanasiri, A.G.N.D.; Rajapakse, S. Synthesis and characterization of nano silver based natural rubber latex foam for imparting antibacterial and anti-fungal properties. *Polym. Test.* **2012**, *31*, 586–592. [CrossRef]

37. Rathnayake, I.; Ismail, H.; Azahari, B.; Bandara, C.; Rajapakse, S. Novel Method of Incorporating Silver Nanoparticles into Natural Rubber Latex Foam. *Polym. Plast. Technol. Eng.* **2013**, *52*, 885–891. [CrossRef]

38. Rathnayake, I.; Ismail, H.; Azahari, B.; Darsanasiri, N.D.; Rajapakse, S. Synthesis and Characterization of Nano-Silver Incorporated Natural Rubber Latex Foam. *Polym. Plast. Technol. Eng.* **2012**, *51*, 605–611. [CrossRef]

39. Rathnayake, I.; Ismail, H.; Azahari, B.; De Silva, C.; Darsanasiri, N. Imparting antimicrobial properties to natural rubber latex foam via green synthesized silver nanoparticles. *J. Appl. Polym. Sci.* **2014**, *131*. [CrossRef]

40. Mam, K.; Dangtungee, R. Effects of silver nanoparticles on physical and antibacterial properties of natural rubber latex foam. *Mater. Today Proc.* **2019**, *17*, 1914–1920. [CrossRef]

41. Osborne, C.K. Production of Foam Rubber. U.S. Patent No. 2,845,659, 5 August 1958.

42. Klempner, D.; Frisch, K.C. *Handbook of Polymeric Foams and Foam Technology*; Hanser Munich etc.: Birmingham, UK, 1991; Volume 404.

43. Rolere, S.; Liengprayoon, S.; Vaysse, L.; Sainte-Beuve, J.; Bonfils, F. Investigating natural rubber composition with Fourier Transform Infrared (FT-IR) spectroscopy: A rapid and non-destructive method to determine both protein and lipid contents simultaneously. *Polym. Test.* **2015**, *43*, 83–93. [CrossRef]

44. Chaikumpollert, O.; Yamamoto, Y.; Suchiva, K.; Kawahara, S. Protein-free natural rubber. *Colloid Polym. Sci.* **2012**, *290*, 331–338. [CrossRef]

45. Narathichat, M.; Sahakaro, K.; Nakason, C. Assessment degradation of natural rubber by moving die processability test and FTIR spectroscopy. *J. Appl. Polym. Sci.* **2010**, *115*, 1702–1709. [CrossRef]

46. Yamaguchi, Y.; Nge, T.T.; Takemura, A.; Hori, N.; Ono, H. Characterization of uniaxially aligned chitin film by 2D FT-IR spectroscopy. *Biomacromolecules* **2005**, *6*, 1941–1947. [CrossRef] [PubMed]

47. Min, B.-M.; Lee, S.W.; Lim, J.N.; You, Y.; Lee, T.S.; Kang, P.H.; Park, W.H. Chitin and chitosan nanofibers: Electrospinning of chitin and deacetylation of chitin nanofibers. *Polymer* **2004**, *45*, 7137–7142. [CrossRef]

48. Milani, G.; Hanel, T.; Donetti, R.; Milani, F. A closed form solution for the vulcanization prediction of NR cured with sulphur and different accelerators. *J. Math. Chem.* **2015**, *53*, 975–997. [CrossRef]

49. Milani, G.; Milani, F. Comprehensive numerical model for the interpretation of cross-linking with peroxides and sulfur: Chemical mechanisms and optimal vulcanization of real items. *Rubber Chem. Technol.* **2012**, *85*, 590–628. [CrossRef]

50. Milani, G.; Hanel, T.; Donetti, R.; Milani, F. Combined experimental and numerical kinetic characterization of NR vulcanized with sulfur, N.-terbutyl, 2-benzothiazylsulfenamide, and N, N.-diphenylguanidine. *J. Appl. Polym. Sci.* **2016**, *133*. [CrossRef]

51. Kanokwiroon, K.; Teanpaisan, R.; Wititsuwannakul, D.; Hooper, A.B.; Wititsuwannakul, R. Antimicrobial activity of a protein purified from the latex of Hevea brasiliensis on oral microorganisms. *Mycoses* **2008**, *51*, 301–307. [CrossRef]

52. Raghupathi, K.R.; Koodali, R.T.; Manna, A.C. Size-dependent bacterial growth inhibition and mechanism of antibacterial activity of zinc oxide nanoparticles. *Langmuir* **2011**, *27*, 4020–4028. [CrossRef]

53. Sirelkhatim, A.; Mahmud, S.; Seeni, A.; Kaus, N.H.M.; Ann, L.C.; Bakhori, S.K.M.; Hasan, H.; Mohamad, D. Review on zinc oxide nanoparticles: Antibacterial activity and toxicity mechanism. *Nano Micro Lett.* **2015**, *7*, 219–242. [CrossRef]

MDPI

St. Alban-Anlage 66

4052 Basel

Switzerland

Tel. +41 61 683 77 34

Fax +41 61 302 89 18

www.mdpi.com

Materials Editorial Office

E-mail: materials@mdpi.com

www.mdpi.com/journal/materials